Neighboring Group Participation

Volume 1

Neighboring Group Participation

Volume 1

Brian Capon
*The University
Glasgow, Scotland*

and

Samuel P. McManus
*The University of Alabama in Huntsville
Huntsville, Alabama*

Plenum Press · New York and London

CHEMISTRY

Library of Congress Cataloging in Publication Data

Capon, Brian.
  Neighboring group participation.

  Includes bibliographical references and index.
  1. Reactivity (Chemistry) 2. Chemistry, Physical organic. I. McManus, Samuel P., joint author. II. Title.
  QD505.5.C36   547'.1'39   76-17812
  ISBN 0-306-35027-0 (v. 1)

© 1976 Plenum Press, New York
A Division of Plenum Publishing Corporation
227 West 17th Street, New York, N.Y. 10011

All rights reserved

No part of this book may be reproduced, stored in a retrieval system, or transmitted, in any form or by any means, electronic, mechanical, photocopying, microfilming, recording, or otherwise, without written permission from the Publisher

Printed in the United States of America

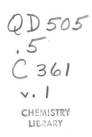

Preface

Neighboring group participation is a term which encompasses all intramolecular reactions and all reactions which involve nonelectrostatic through-space interactions between groups within the same molecule. The term was invented in 1942 by Saul Winstein, whose many contributions to the growth and maturing of the subject are unequaled. Although the inventor of the term, Winstein was not the first worker to study neighboring group participation. An examination of Beilstein will show that many intramolecular reactions were known to the synthetic organic chemist well before the turn of the century, and as early as 1891 W. P. Evans, working at Giessen, described a kinetic investigation of the base-promoted cyclization of ethylene chlorohydrins to ethylene oxides—an important intramolecular reaction. He was followed some twenty years later by Freundlich, whose valuable studies on participation by the amino group began to appear in 1911. Freundlich was later joined by Salomon, who by the mid-thirties had developed a reasonable understanding of the efficiency of the neighboring amino group in acyclic systems. In the late twenties to mid-thirties the subject began to expand with the work of Bennett on participation by thioether groups, Nilsson and Smith on neighboring hydroxyl, and Caldin and Wolfenden on neighboring carboxylate, and with discussions of the dependence of cyclization rates on ring size by Ruzicka, Salomon, and Bennett. This background, along with the elegant studies of Hughes and Ingold on aliphatic nucleophilic substitution, set the stage for the entrance of Winstein. Also about this time Isbell proposed a process similar to that proposed by Winstein for the neighboring acetoxy group, viz., "the intramolecular orthoester reaction," but it was the influence of Winstein that caused the subject to become recognized as all-pervasive in organic chemistry.

Examples of neighboring group participation constantly occur with all classes of organic compounds, including organometallic compounds and natural products. The analogy between neighboring group participation and the intracomplex interactions of the functional groups of an enzyme and those of the substrate is well known. In addition, some steps in enzyme reactions are true examples of neighboring group participation, such as the deacylation of an acyl enzyme or the deprotonation of the Schiff base formed by a class I aldolase and its substrate.

It is obviously impossible to write a comprehensive monograph on a subject as broad as this, since too many examples are hidden away in papers whose titles (and, in many cases, abstracts) give no mention of a relationship to neighboring group participation. We have not, therefore, set out to write a book which is comprehensive, but one which treats the subject in a systematic and critical way, while trying not to omit any significant examples. We realize of course that some readers will not agree with our interpretation of what is significant and we welcome comments on this score. Perhaps because of a different perspective our interpretation of the experimental results differs in some instances from that of the original investigators; we hope that from the way in which the discussion is presented the reader will be able to tell this.

To make this undertaking manageable we have chosen to divide the material into two volumes. The first part of the present volume is a general introduction and it includes a discussion of some factors which may account for the fast rates observed for reactions occurring with neighboring group participation and some ways in which neighboring group participation is experimentally evaluated. The second part of this volume contains a systematic treatment of simple oxygen, sulfur, and nitrogen groups. The second volume will include a treatment of other simple groups, e.g., the halogens, carbon, hydrogen, etc., and of complex groups, e.g., amides, esters, organometallic compounds, etc. In addition, a survey of neighboring group participation in key reactions, e.g., reduction, hydrolysis, etc., will be included. In the present volume we have tried to cover the literature through the middle of 1975, although a few papers appearing later than this were inserted during the production process.

We are indebted to the individuals whom we acknowledge in the text for supplying unpublished results or for discussion of results. One of us (S.P.M.) would particularly like to acknowledge the Chemistry faculty of the University of South Carolina for the warm hospitality afforded him during his tenure as a visiting faculty member during 1974–75. It was during this time that much of his material for this volume and some for the next was drafted or revised. That task was helped along by discussions with R. S. Bly, R. L. Cargill, and especially P. E. Peterson, and by the provision of excellent facilities. S.P.M. also expresses his appreciation to his colleague J. M. Harris for critically reviewing the entire volume. Finally, S.P.M. acknowledges with special gratitude his wife Nancy for typing and proofreading and for checking many of the references.

September, 1976　　　　　　　　　　　　　　　　　BRIAN CAPON
　　　　　　　　　　　　　　　　　　　　　　　　SAMUEL P. MCMANUS

Contents

Part 1. **General Considerations** 1

Chapter 1. Introduction 3
    1.1. Historical and General Description of
         Neighboring Group Effects 3
    1.2. Terms and Definitions 11

Chapter 2. Some Factors that Influence Anchimeric Assistance ... 19
    2.1. The Stabilities of Bridged Carbocations—Some
         Molecular Orbital Calculations 20
    2.2 Hyperconjugation—An Alternative Explanation?. 31
    2.3. The Effect of Intramolecularity—Ring Size
         and *gem*-Dialkyl Effects 43
        2.3.1. Ring Size 49
        2.3.2. The *gem*-Dialkyl Effect 58

Chapter 3. Some Experimental Methods Used in the Study of
         Neighboring Group Participation 77
    3.1. Kinetic Methods 79
        3.1.1. Estimation of Reaction Rates 79
        3.1.2. Participation as a Function of Electron
             Demand 89
        3.1.3. Kinetic Isotope Effects 95
    3.2. Investigation of Solvent Effects 101
    3.3. Isolation or Trapping of Intermediates 107
    3.4. Spectroscopic Observation of Intermediates 110
    3.5. Product Analysis 113

Part 2. Participation by Simple Oxygen, Sulfur, and Nitrogen Groups ... 123

Chapter 4. Participation by Oxygen Groups ... 125
    4.1. Ether Groups ... 125
        4.1.1. In Solvolytic Displacement Reactions ... 125
        4.1.2. In Reactions at Carbonyl Carbon ... 158
        4.1.3. In Electrophilic Addition Reactions ... 160
    4.2. Hydroxyl Groups ... 162
        4.2.1. In Solvolytic Displacement Reactions ... 163
        4.2.2. In Addition Reactions ... 177
        4.2.3. In Elimination Reactions ... 181
        4.2.4. In Hydrolysis and Related Reactions ... 182
        4.2.5. In Reduction Reactions ... 187
    4.3. Hydroperoxide Groups ... 187
    4.4. Oxime Groups ... 188

Chapter 5. Participation by Sulfur Groups ... 195
    5.1. Thioether Groups ... 195
        5.1.1. In Solvolytic Displacement Reactions ... 195
        5.1.2. In Electrophilic Addition Reactions ... 210
        5.1.3. In Elimination Reactions ... 216
        5.1.4. In Free-radical, Carbene, and Photochemical Reactions ... 218
    5.2. Thiol Groups ... 220

Chapter 6. Participation by Nitrogen Groups ... 227
    6.1. Amino Groups ... 227
        6.1.1. Anchimeric Assistance in Ring Closure Reactions ... 227
        6.1.2. R_2N-3 Participation ... 231
        6.1.3. R_2N-4 Participation ... 243
        6.1.4. R_2N-5 and R_2N-6 Participation ... 249
        6.1.5. Miscellaneous Modes of Participation ... 252
    6.2. Nitrile Groups ... 258
    6.3. Hydrazone Groups ... 259

Author Index ... 263

Subject Index ... 273

Part 1
General Considerations

In this part we shall illustrate the common types of neighboring group participation and shall introduce terms and definitions that are identified with this area of research and some more general terms that are used in discussing neighboring group effects. We shall also treat here some factors that may be important in understanding neighboring group participation, and we shall discuss some experimental and theoretical methods that have been used in the pursuit of an understanding of neighboring-group-assisted reactions. In our development of the subject we hope the reader is provided with sufficient detail that he may sense the reasons why some neighboring-group-assisted reactions are reasonably well understood and accepted and why the claim of participation in others has excited controversy.

1
Introduction

1.1. Historical and General Description of Neighboring Group Effects

The most widely investigated substituent effects on organic reactions are electronic effects transmitted through the carbon skeleton and steric effects. Substituents, however, may influence reactivity in other ways. When a substituent stabilizes a transition state or intermediate by becoming bonded to the reaction center this effect is called *neighboring group participation*,[1]* and if such participation leads to an enhanced reaction rate, the group is said to provide *anchimeric assistance* (derived from the Greek: *anchi*, "adjacent"; *meros*, "part").[2]

The term *intramolecular catalysis* introduced by Bender[3] is also widely used to describe neighboring group effects, especially when analogous *intermolecular catalysis* is observed. Thus this term is commonly used when referring to reactions that are subject to acid, base, and/or nucleophilic catalysis such as hydrolysis of esters, amides, and acetals; the mutarotation of aldoses; or the enolization of ketones. It is rarely used when referring to nucleophilic substitution reactions at saturated carbon.

Some general types of neighboring group participation are illustrated below:

---

* In this book the term *neighboring group participation* is reserved for "through-space" interactions between the substituents on the reaction center. Hyperconjugation is regarded as a separate phenomenon. It is recognized that both effects may operate at once, but we think it preferable to define them as separate effects. This usage is not universally followed but was the one adopted by J. M. Harris, *Prog. Phys. Org. Chem.*, **11**, 92 (1974).

1. Nucleophilic participation:

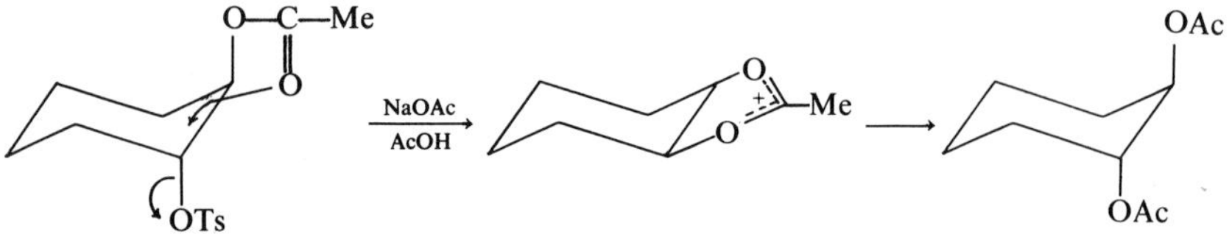

(Ref. 4)

2. Nucleophilic participation without anchimeric assistance:

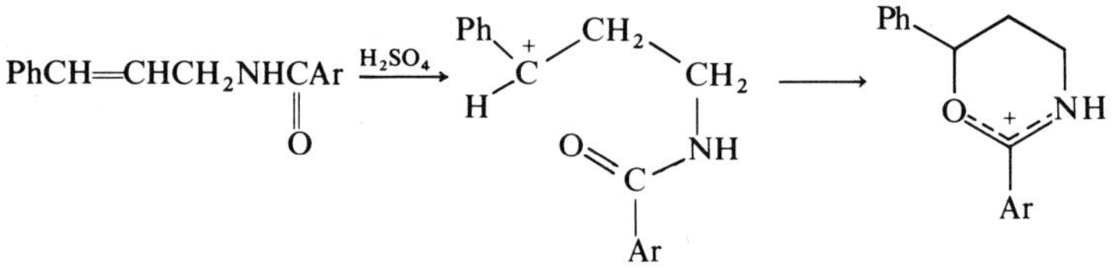

$$PhCH=CHCH_2NHCAr \xrightarrow{H_2SO_4}$$

(Ref. 5)

3. Electrophilic participation:

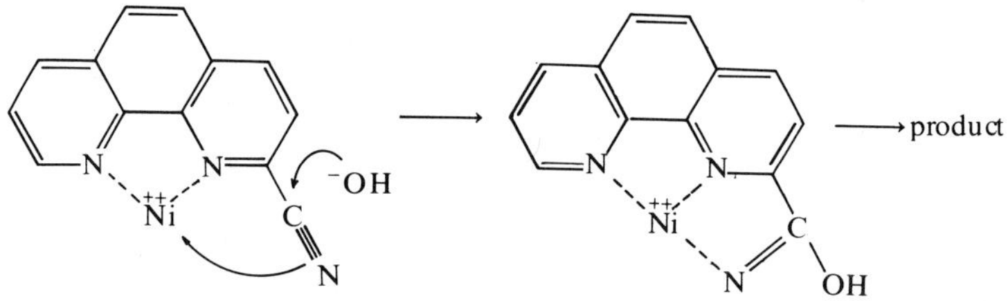

⟶ product   (Ref. 6)

4. Acidic catalysis:

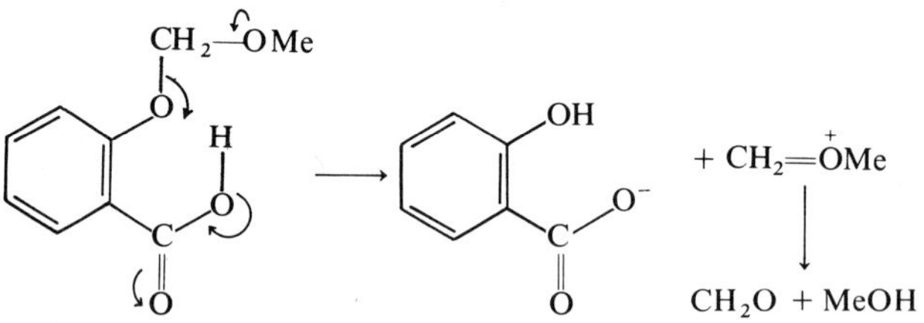

(Ref. 7)

5. Basic catalysis:

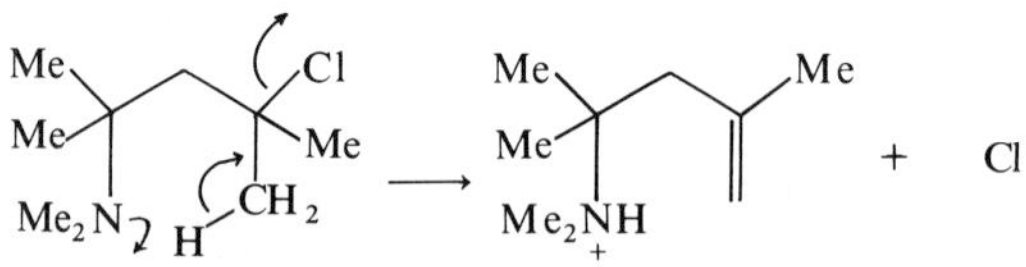

(Ref. 8)

Introduction

Presumably, most of the early investigators were attracted to studies of neighboring group effects in order to improve our understanding of basic chemical reactivity as well as to unravel some anomalous results of long standing.[9,10] As the field has developed, however, neighboring group participation has been studied in order to throw light on other fields such as enzyme reaction mechanisms[11] and biogenesis.[12] Because neighboring group participation is so widespread, the fullest understanding of it should be sought.

The studies of Hughes and Ingold and their co-workers on nucleophilic displacement reactions in the 1930s laid the foundation for studies of neighboring groups. Culminating with six[13] papers on "reactions kinetics and the Walden inversion," these studies suggested, for the first time, that a neighboring carboxylate group "may stabilize a pyramidal configuration... and lead to eventual substitution with retention" [cf. Eq. (1)].

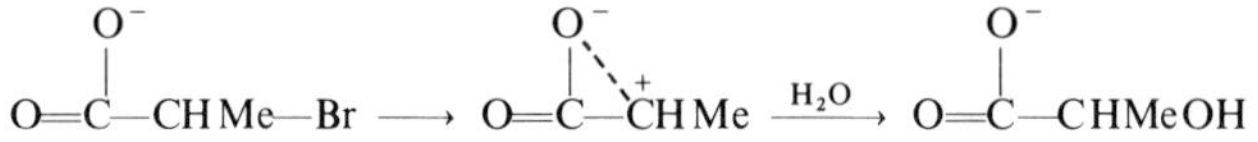

$$O=C(O^-)-CHMe-Br \longrightarrow O=C(O^-)\cdots \overset{+}{C}HMe \xrightarrow{H_2O} O=C(O^-)-CHMeOH \qquad (1)$$

At about the same time Roberts and Kimball[14] proposed, in analogy to ethylene oxides, that a positively charged bromine could bridge two carbons to yield structure (1), which was expected to be more stable than carbocation (2). The original proposal was advanced to account for the

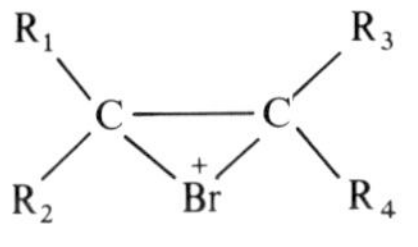

(1) 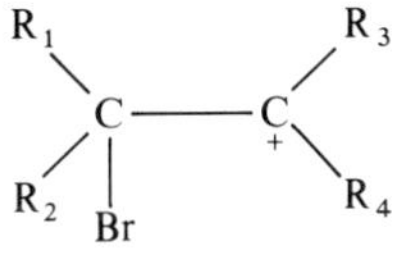   (2)

stereochemistry of bromine additions to olefins, but Winstein and Lucas[15] quickly applied the concept of bromine bridging to the reactions of the diastereoisomeric 3-bromo-2-butanols with hydrogen bromide to yield dibromocompounds with retention of configuration [e.g., Eq. (2)]. This

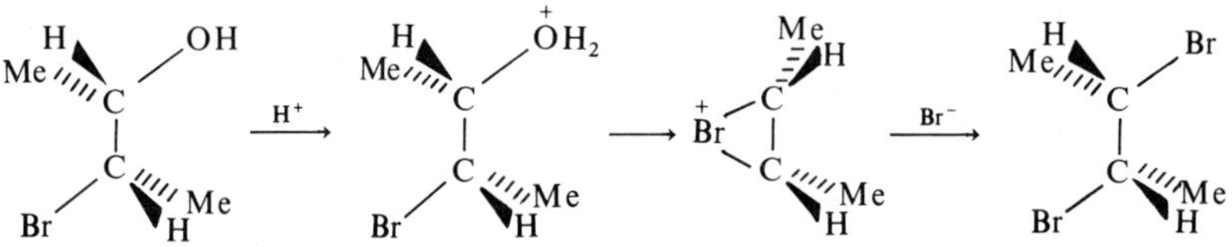

(2)

work marked Winstein's entry into the field of neighboring group participation, to which he was a major contributor for 30 years.[9]

At about this time also, Nevell et al.[16] made their controversial proposal that the intermediate in the rearrangement of camphene hydrochloride (3) into isobornyl chloride (5) is a mesomeric σ-delocalized cation (5). This

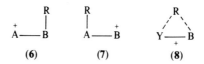

concept was elaborated by Ingold in his theory of *synartesis*.[17] As defined by Ingold, synartesis (fastening together) is the name for resonance between

$$\overset{+}{A} \text{—} B \qquad A \text{—} \overset{+}{B} \qquad Y \underset{+}{\text{—}} B$$
$$\text{(6)} \qquad \text{(7)} \qquad \text{(8)}$$

(with R substituent on B in (6) and (7), and R bridging to the Y—B bond in (8))

two structures (6) and (7) that leads to a σ-delocalized structure (8), a *synartetic* ion, more stable than either (6) or (7). Thus the solvolysis of isobornyl chloride (4), which yields "camphene products" and occurs 10^5 times faster than solvolysis of its *endo* isomer, bornyl chloride, was formulated as proceeding *via* the synartetic ion (10) rather than via ions (9) and (11). The free-energy reaction coordinate diagram was formulated as shown in Fig. 1.

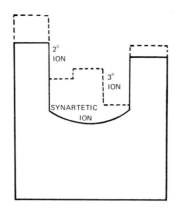

Fig. 1.

Introduction

Here the dotted line shows the postulated solvolysis course in the absence of synartesis, involving first the formation of the secondary 1,7,7-trimethyl-2-norbornyl cation, which rearranges after the rate-determining transition

![structures (9), (10), (11)]

(9)　　(10)　　(11)

state to the tertiary 2,3,3-trimethyl-2-norbornyl cation (in Winstein's symbolism, this is a k_c process). The full line shows the postulated preferred reaction course via a resonance-stabilized synartetic ion (a k_Δ process). The transition state is stabilized because it partakes of the character of the synartetic ion.

It was considered, quite reasonably, in view of stereoelectronic considerations, that concerted migration of the σ bond on departure of the Cl^- would not be possible with bornyl chloride, the leaving group being *endo*. Hence there would be no synartetic stabilization of the transition state in this case. The trouble with this interpretation is that it is possible to explain the greater rate of solvolysis of isobornyl chloride (**4**) compared to bornyl chloride without involving synartesis at all, as illustrated by the free-energy reaction coordinate diagram (Fig. 2). Again the dotted line shows the postulated solvolysis course when the rate-limiting step involves formation of the secondary ion (**9**), which rearranges to the tertiary ion (**10**). The full line

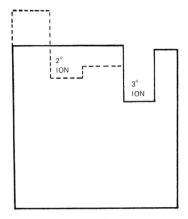

Fig. 2.

represents a postulated solvolysis course involving σ-bond migration concerted with departure of the leaving group, leading directly to the tertiary cation (11), the secondary cation (9) being bypassed (a k_R process). The rate enhancement results not because the transition state partakes of the character of a synartetic ion but because it partakes of the character of a tertiary ion. Thus if this formulation were correct, we would have neighboring group participation (by a carbon–carbon σ bond), anchimeric assistance, but no synartetic acceleration. There would be σ bridging in the transition state but not in the intermediate cation. Thus synartetic acceleration is a subdivision of anchimeric assistance.[18]

An important difference between the theories behind Figs. 1 and 2 is that one expects from the former that the solvolysis of camphene hydrochloride (12) would be accelerated because it should lead to a synartetic ion (enter and leave from the right-hand side of Fig. 1), whereas it would not be accelerated if the latter theory is correct. Ethanolysis of camphene hydrochloride (12) occurs about 6000 times faster than that of t-butyl chloride[17] but only 2.5 times faster than that of 1,2,2,5,5-pentamethyl cyclopentyl chloride (13).[19] Hence the decision as to whether the ethanolysis of (4) is accelerated depends on which model reaction is chosen for the nonassisted

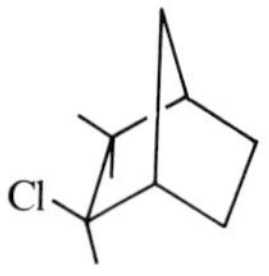

(12)   (13)
$10^3 k$(EtOH) 25°C, s^{-1}: 1.16   0.458

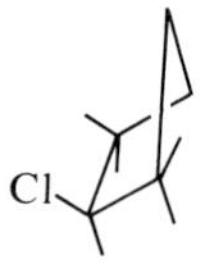

solvolysis of (12). We consider that the solvolysis of (13) is a better model reaction than solvolysis of t-butyl chloride and hence prefer the second formulation (Fig. 2) for the solvolysis of isobornyl chloride (4). Other evidence also suggests that tertiary 2-norbornyl cations have an unbridged structure.[20]

In the late 1940s and early 1950s Winstein and Trifan published the results of their investigation on the solvolysis of the 2-norbornyl p-bromo-

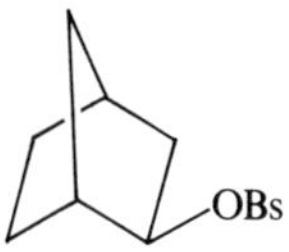

(14) 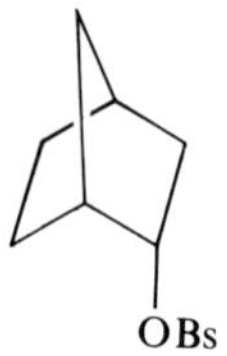   (15) 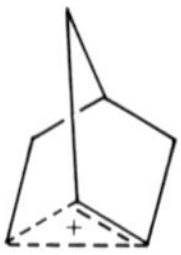   (16)

Introduction

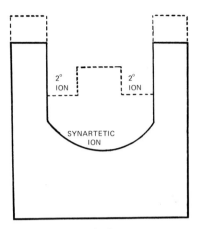

Fig. 3.

benzenesulfonates.[21] They showed that the *exo* isomer (**14**) undergoes acetolysis 350 times faster than the *endo* isomer (**15**) and yields, as the product of substitution, almost exclusively *exo*-norbornyl acetate. These results led Winstein and Trifan to propose that the reaction involved a σ-bridged ion (**16**) that has a structure similar to Wilson and Ingold's synartetic ion (**5**). Therefore, the free-energy reaction coordinate diagram would be as shown in Fig. 3. The explanation of the rate enhancement is again that the transition

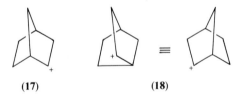

(**17**)   (**18**)

state partakes of the stabilization of the σ-bridged ion. With this system the alternative explanation (analogous to that based on Fig. 2) is not possible since the second unbridged ion (**18**) that would be formed on rearrangement of the first unbridged ion (**17**) has the same structure and energy content. The only difference between them is that they are enantiomers. Therefore a transition state that leads directly to the second ion should not be specially stabilized, and in the norbornyl system synartesis or some other stabilizing process[22] is apparently needed to explain the results. It is this proposal that has led to a long and complex controversy[23] among organic chemists, with some workers, notably H. C. Brown, preferring to explain these results as arising from steric factors.

In the late 1940s and early 1950s neighboring group participation by carbon groups in the solvolyses of other compounds was discovered, viz., cholesteryl chloride (Shoppee[24]), 3-phenyl-2-butyl toluene-*p*-sulfonate (Cram[25]), *exo*-2-norbornenyl halides (Roberts et al.[26]), and cyclopropylcarbinyl chloride* (Roberts and Mazur[27]) and benzenesulfonate* (Bergstrom and Siegel[28]). Frequently these reactions have been formulated as proceeding via bridged carbocations, and the observed rate enhancements has been ascribed to a special stabilization of the cations and the transition states leading to them. Such bridged ions are often described as *nonclassical*, a term first introduced by Roberts and Mazur.[29] A little later Roberts and Lee[30] used the term *bridged nonclassical*, but the term *nonclassical* unqualified by *bridged* has become common usage, and the more important of the two adjectives is usually dropped. The main disadvantage of the term *nonclassical* is that its meaning is imprecise, and, as pointed out by Ingold,[29] any nonclassical idea, if correct, becomes classical with the passage of time. Throughout this book we shall therefore avoid the term *nonclassical* and use instead terms such as *bridged* (σ- and π-) and *synartetic*, which provide a more "positive description."[31]

By the early 1950s a firm experimental basis for neighboring group participation by groups with n, π, and σ electrons in substitution reactions at saturated carbon had been provided by the above-mentioned work on carbon participation together with the work of Winstein and Lucas on neighboring methoxy,[32] acetoxy,[1,33] and halogen[15,34] groups; of Tipson[35] and Isbell[36] on neighboring acetoxy groups; of Freundlich,[37] Salomon,[38] Bartlett,[39] Ross,[40] Rydon,[41] Cohen,[42] and Winkler[43] on neighboring nitrogen groups; of Evans,[44] Smith,[45] Warner,[46] Porret,[47] Winstein and Lucas,[15,48] Twigg,[49] and Heine[50] on neighboring oxide ion and hydroxyl groups; of Bennett,[51] Mohler,[52] Böhme,[53] Ogston,[54] and Bartlett and Swain[55] on neighboring thioether groups; of Caldin and Wolfenden,[56] Hughes and Ingold,[13] Chadwick and Pacsu,[57] Grunwald and Winstein,[58] and Heine[59] on neighboring carboxylate groups; and of Winstein[60] on neighboring amido groups.

Also at this time examples of neighboring group participation in addition reactions were known,[61] and work was commencing on neighboring group participation in reactions of derivatives of carboxylic acids[62] and of phosphate esters.[63]

---

* As discussed on pp. 39–43 the high rates of solvolyses of these compounds are probably best explained in terms of hyperconjugation rather than neighboring group participation.

1.2. Terms and Definitions

The most widely studied type of neighboring group participation is where the neighboring group acts as a nucleophile. One of the steps of such reactions always involves an intramolecular nucleophilic attack, and if the carbon atom at which the attack occurs is sp^3 hybridized, the product is usually different from what would be expected in the absence of participation. In the course of such a process a new ring is formed, and it may suffer three different fates: (1) ring opening at the same point where the ring closure took place, leading to an unrearranged product with the same configuration as the starting material [e.g., Eq. (3)][64]; (2) ring opening at a different point from which the ring closure took place, leading to a rearranged product [e.g., Eq. (4)][65]; (3) no ring opening so that the overall reaction is a cyclization [e.g., Eq. (5)].[37]

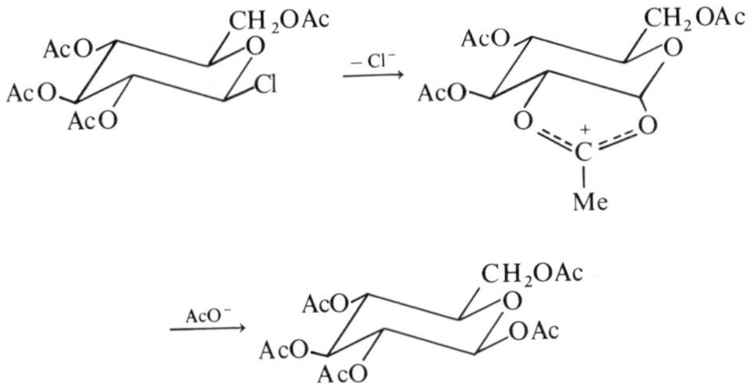

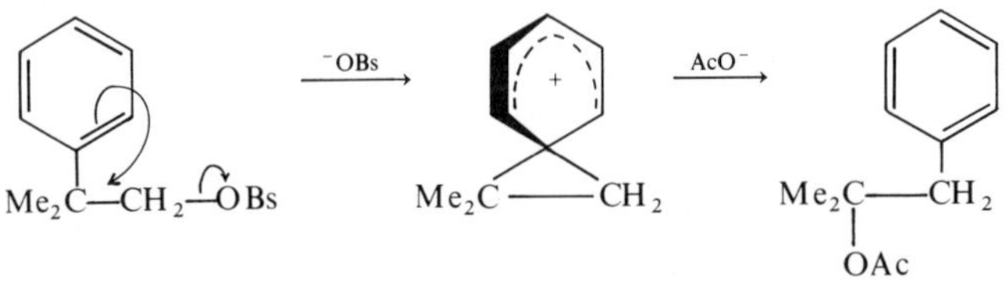

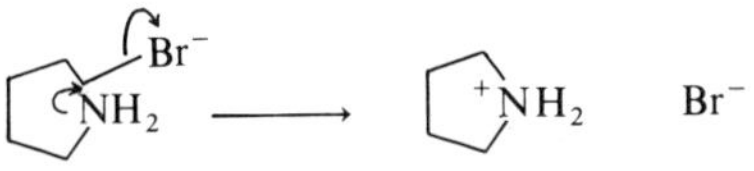

Sometimes the cyclic intermediate undergoes ring opening at two points competitively so that some of the product is rearranged and some unrearranged [see Eq. (6)].[66] A special situation arises when the starting material is chiral and the cyclic intermediate achiral since then the product is also achiral. Thus optically active *trans*-2-acetoxycyclohexyl toluene-*p*-

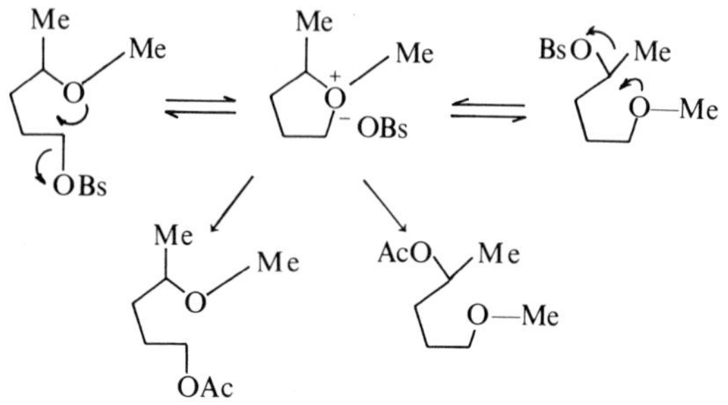

$$\text{(6)}$$

sulfonate yields racemic *trans*-1,2-diacetoxycyclohexane.[33] Here the cyclic intermediate undergoes fission at two points at equal rates [Eq. (7)].

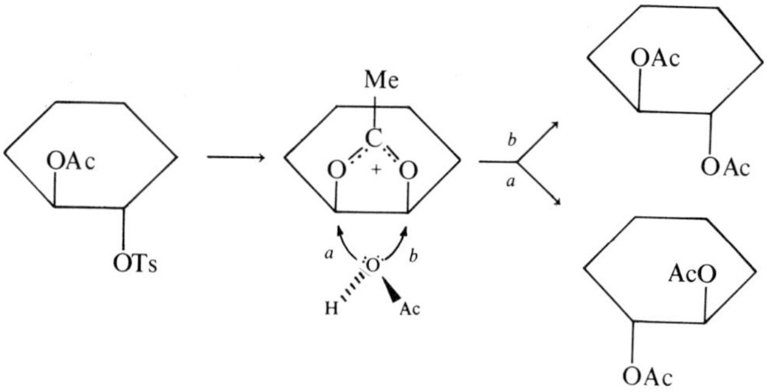

$$\text{(7)}$$

When describing nucleophilic participation it is frequently convenient to use the symbol G-*n*, where G is the participating group and *n* the size of the ring that is formed in the transition state. Equations (3)–(5) therefore depict examples of AcO-5, Ar-3, and N-5 participation, respectively.[66,67] The mode of participation shown in Eq. (4) may be more precisely symbolized as Ar_1-3 participation; the subscript 1 symbolizes that the nucleophilic attack is carried out by carbon-1 of the phenyl group. An example of Ar_2-6 participation leading to a cyclic product is shown in Eq. (8).[68]

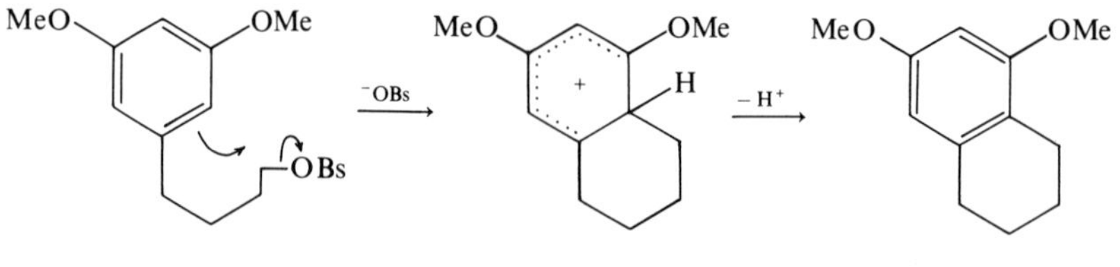

$$\text{(8)}$$

Introduction

Table 1. *Classification of Neighboring Groups*

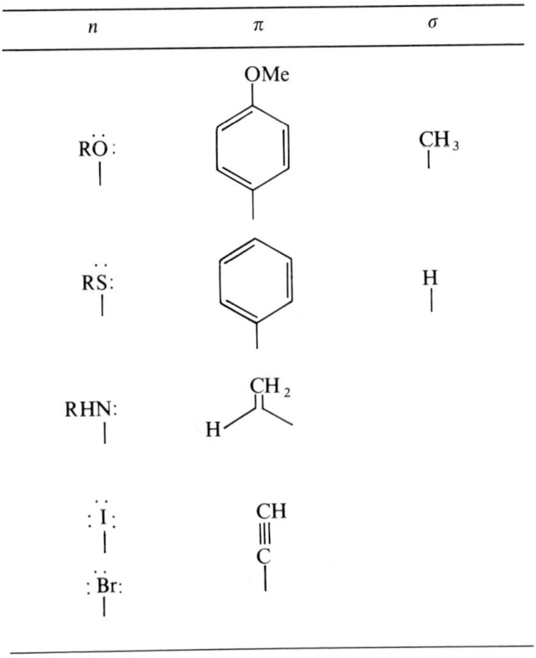

n	π	σ
RÖ:—	4-MeO-C₆H₄—	CH₃—
RS:—	C₆H₅—	H—
RHN:—	H₂C=CH—CH₂— (allyl)	
:Ï:—	HC≡C—	
:Br:—		

In nucleophilic participation the electrons of the neighboring group that become bonded to the reaction center may be nonbonding electrons [as in Eq. (5)], π electrons [as in Eq. (4) and (8)], or σ electrons (as in the solvolysis of isobornyl chloride). These three modes are referred to as n, π, and σ participation, respectively (see Table 1).

Winstein introduced several symbols that are often used when comparing the rates of reactions that involve nucleophilic neighboring group

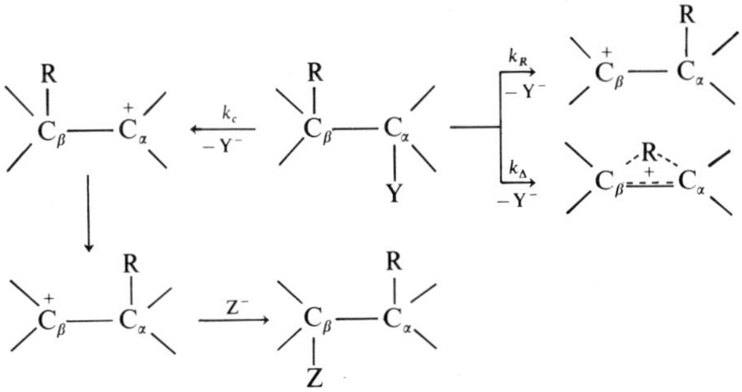

Scheme 1

participation. The original formulation is as shown in Scheme 1.[19] Here k_c, k_R, and k_Δ are all rate constants for processes that involve neighboring group participation but only k_R and k_Δ are for anchimerically assisted processes. The difference between k_Δ and k_R is that k_Δ is the rate constant for a process that leads to a bridged ion (via a bridged transition state), whereas k_R is the rate constant for a process that leads to a rearranged ion (also via a bridged transition state). These constants are therefore the rate constants for the processes symbolized by the full lines in Figs. 1 and 2, respectively. As written in Scheme 1, k_c is the rate constant for a reaction that involves neighboring group participation after the rate-determining formation of an unrearranged ion corresponding to the dotted lines in Figs. 1 and 2. Later it was realized that a process that involves nucleophilic assistance by the solvent is frequently an important competing pathway, and the symbol k_s was given to this, with k_c being used for ionization without any form of nucleophilic assistance, irrespective of whether it was followed by a group migration (see Ref. 66 and Lancelot *et al.* in Ref. 23). We shall follow the latter usage in this book.

Winstein and his co-workers wrote various equations that interrelate these constants. Thus for primary and secondary substrates k_c is thought to be unimportant so that the overall rate constant for solvolysis k_t is given by

$$k_t = k_\Delta + k_s$$
$$\therefore k_t/k_s - 1 = k_\Delta/k_s$$
$$= (k_\Delta/k_c)/(k_s k_c)$$

The quantity $k_t/k_s - 1$ therefore represents the ratio of the rate enhancement provided by neighboring group participation compared to a hypothetical nonnucleophilically assisted process divided by the rate enhancement provided by nucleophilic solvent assistance.[66] In the most general situation:

$$k_t = k_\Delta + k_R + k_s + k_c \tag{9}$$

but rarely, if ever, will more than two of the constants on the right-hand side of Eq. (9) be significant. This treatment ignores ion-pair return, and it has been shown that for primary and secondary substrates the quantity F, the fraction of ion pairs that proceeds to products, is sometimes necessary, so that (see Ref. 70 and Lancelot *et al.* in Ref. 23)

$$k_t = Fk_\Delta + k_s$$

Introduction

In attempting to quantify the propensity of various neighboring groups to participate nucleophilically, Winstein termed the quantity $L = RT\ln(k_\Delta/k_c)$ the driving force for participation.[71] It is the free-energy difference between the transition states for formation of the bridged and unbridged ions. This terminology has been used infrequently in recent years.

According to current theory, nucleophilic assistance is usually not important in solvolyses of tertiary substrates,[72] so that the relative importance of k_s, k_Δ, and k_c may generally be written:

Type of substrate	k_c	k_s	k_Δ
Primary	Rare	Strong	Strong
Secondary	Unusual	Strong (more steric problems than primary)	Strong
Tertiary	Common	Rare	Reduced

From this discussion it is seen that in order to make a quantitative measure of anchimeric assistance it is necessary to know k_c or k_s, the (often hypothetical) rate constants for reaction of the substrates without any form of nucleophilic assistance or with nucleophilic assistance from the solvent. The problems in estimating these constants will be discussed in Chapter 3.

The above constants are usually only used in discussions of nucleophilic substitution reactions at saturated carbon. In other reactions a quantity known as the *effective molarity* is frequently used as a measure of anchimeric assistance.[73] This can be determined only if the rate constant for an analogous intermolecular reaction has been measured. Consider two analogous reactions, one intramolecular [Eq. (10)] and one intermolecular [Eq. (11)]. These are, respectively, first- and second-order pro-

$$G\text{—}D \longrightarrow \text{product}, \quad \text{rate} = k_1[GD] \qquad (10)$$

$$G + D \longrightarrow \text{product}, \quad \text{rate} = k_2[G][D] \qquad (11)$$

cesses, and the ratio k_1/k_2, which has the dimensions of concentration, is defined as the effective molarity. The greater the rate enhancement on going from the inter- to the intramolecular reaction, the larger is the effective molarity. If the concentration of one of the reactants in the intermolecular reaction is made much larger than that of the other, the effective molarity is the hypothetical molarity of the former, which will give the intermolecular reaction a pseudo-first-order constant equal to the measured first-order constant of the intramolecular reaction. The effective molarity is of course frequently unattainable in practice.

The difference between these two methods of expressing the magnitude of anchimeric assistance is that the ratios k_Δ/k_s or k_Δ/k_c involve a comparison of two different types of processes: an intramolecular displacement, which involves a specific neighboring group as nucleophile (k_Δ) and an intermolecular process, which involves the solvent as nucleophile (k_s) or one that occurs without nucleophilic assistance (k_c). On the other hand, the effective molarity involves a comparison of two analogous processes that differ only in that one is intramolecular and the other intermolecular.[74]

References

1. S. Winstein and R. E. Buckles, *J. Am. Chem. Soc.*, **64**, 2780 (1942).
2. S. Winstein, C. R. Lindegren, H. Marshall, and L. L. Ingraham, *J. Am. Chem. Soc.*, **75**, 148 (1953).
3. M. L. Bender, *J. Am. Chem. Soc.*, **79**, 1258 (1957).
4. S. Winstein, C. Hansen, and E. Grunwald, *J. Am. Chem. Soc.*, **70**, 812 (1948).
5. S. P. McManus, C. U. Pittman, and P. E. Fanta, *J. Org. Chem.*, **37**, 2353 (1972).
6. R. Breslow, R. Fairweather, and J. Keana, *J. Am. Chem. Soc.*, **89**, 2135 (1967).
7. B. Capon and M. C. Smith, *Chem. Commun.*, 523 (1965); B. Capon, M. C. Smith, E. Anderson, R. H. Dahm, and G. H. Sankey, *J. Chem. Soc. B*, 1038 (1969).
8. A. M. Braun, C. E. Ebner, C. A. Grob, and F. A. Jenny, *Tetrahedron Lett.*, 4733 (1965).
9. P. D. Bartlett, *J. Am. Chem. Soc.*, **94**, 2161 (1972); A. Streitwieser, Jr., *Prog. Phys. Org. Chem.* **9**, 1 (1972).
10. C. K. Ingold, *Structure and Mechanism in Organic Chemistry*, 2nd ed., Cornell University Press, Ithaca, N.Y. (1969), pp. 509–544, 762–787.
11. M. L. Bender, *Mechanisms of Homogeneous Catalysis from Protons to Proteins*, John Wiley & Sons, Inc (Interscience Division), New York (1971); W. P. Jencks, *Catalysis in Chemistry and Enzymology*, McGraw-Hill Book Company, New York, (1969); T. C. Bruice and S. J. Benkovic, *Bioorganic Mechanisms*, Vols. I and II, W. A. Benjamin, Inc., Reading, Mass. (1966).
12. W. S. Johnson, *Acc. Chem. Res.*, **1**, 1 (1968); E. E. van Tamelen, *Acc. Chem. Res.*, **1**, 111 (1968).
13. E. D. Hughes, C. K. Ingold *et al.*, *J. Chem. Soc.*, 1196–1271 (1937).
14. I. Roberts and G. E. Kimball, *J. Am. Chem. Soc.*, **59**, 947 (1937).
15. S. Winstein and H. J. Lucas, *J. Am. Chem. Soc.*, **61**, 1576, 2845 (1939).
16. T. P. Nevell, E. de Salas, and C. L. Wilson, *J. Chem. Soc.*, 1188 (1939); see also H. B. Watson, *Annu. Rep. (Chem. Soc. London)*, **36**, 197 (1939); H. B. Watson, *Modern Theories of Organic Chemistry*, 2nd ed., Oxford University Press, London (1941), pp. 208–209.
17. F. Brown, E. D. Hughes, C. K. Ingold, and J. F. Smith, *Nature*, **168**, 65 (1951); C. K. Ingold, Ref. 10, p. 768.
18. J. Berson, in: *Molecular Rearrangements* (P. De Mayo, ed.), Vol. 1, p. 176, Ref. 126a, John Wiley & Sons, Inc. (Interscience Division), New York (1963).
19. H. C. Brown and F. J. Chloupek, *J. Am. Chem. Soc.*, **85**, 2322 (1963).

20. H. L. Goering and J. V. Clevenger, *J. Am. Chem. Soc.*, **94**, 1010, (1972); R. Haseltine, E. Huang, K. Ranganayakulu, T. S. Sorensen, and N. Wong, *Can. J. Chem.*, **53**, 1876 (1975).
21. S. Winstein and D. Trifan, *J. Am. Chem. Soc.*, **71**, 2953 (1949); **74**, 1147, 1154 (1952).
22. F. R. Jensen and B. E. Smart, *J. Am. Chem. Soc.*, **91**, 5688 (1969); N. A. Clinton, R. S. Brown, and T. G. Traylor, *ibid.*, **92**, 5228 (1970); T. G. Traylor, W. Hanstein, H. J. Berwin, N. A. Clinton, and R. S. Brown, *ibid.*, **93**, 5715 (1971); G. D. Sargent and T. J. Mason, *ibid.*, **96**, 1063 (1974).
23. P. D. Bartlett, *Nonclassical Ions*, W. A. Benjamin, Inc., Reading Mass. (1965); H. C. Brown, *Boranes in Organic Chemistry*, Cornell University Press, Ithaca, N. Y. (1972), Chaps. IX–XI; H. C. Brown, *Chem. Br.*, **2**, 199 (1966); H. C. Brown, *Chem. Eng. News*, **45**, 87 (Feb. 13, 1967); H. C. Brown, *Chem. Soc. Spec. Publ.*, No. **16**, 140 (1962); R. Bernhard, *Sci. Res.* (McGraw-Hill, NY), 26 (Aug. 18, 1969); H. C. Brown, *ibid.*, 5 (Dec. 22, 1969); G. D. Sargent, *Quart. Rev. Chem. Soc.*, **20**, 301 (1966); G. D. Sargent, *in: Carbonium Ions* (G. A. Olah and P. v. R. Schleyer, eds.), Vol. III, John Wiley & Sons, Inc. (Interscience Division), New York (1972), Chap. 24; K. B. Wiberg, B. A. Hess, Jr., and A. J. Ashe, III, *ibid.*, Chap. 26; C. J. Lancelot, D. J. Cram, and P. v. R. Schleyer, *ibid.*, Chap. 27; H. G. Richey, Jr., *ibid.*, Chap. 25; H. C. Brown, *Acc. Chem. Res.*, **6**, 377 (1973).
24. C. W. Shoppee, *J. Chem. Soc.*, 1147 (1946).
25. D. J. Cram, *J. Am. Chem. Soc.*, **71**, 3863 (1949).
26. J. D. Roberts, W. Bennett, and R. Armstrong, *J. Am. Chem. Soc.*, **72**, 3329 (1950).
27. J. D. Roberts and R. H. Mazur, *J. Am. Chem. Soc.*, **73**, 2509 (1951).
28. C. G. Bergstrom and S. Siegel, *J. Am. Chem. Soc.*, **74**, 145 (1952).
29. J. D. Roberts and R. H. Mazur, *J. Am. Chem. Soc.*, **73**, 3542 (1951).
30. J. D. Roberts and C. C. Lee, *J. Am. Chem. Soc.*, **73**, 5009 (1951).
31. C. K. Ingold, *Structure and Mechanism in Organic Chemistry*, 2nd ed., Cornell University Press, Ithica, N. Y. (1969). p. 768, footnote 132.
32. S. Winstein and R. B. Henderson, *J. Am. Chem. Soc.*, **65**, 2196 (1943); S. Winstein and L. L. Ingraham, *ibid.*, **74**, 1160 (1952); S. Winstein, C. R. Lindegren, and L. L. Ingraham, *ibid.*, **75**, 155 (1953); S. Winstein, C. R. Lindegren, H. Marshall, and L. L. Ingraham, *ibid.*, **75**, 147 (1953).
33. S. Winstein, H. V. Hess, and R. E. Buckles, *J. Am. Chem. Soc.*, **64**, 2796 (1942); S. Winstein, E. Grunwald, R. E. Buckles, and C. Hanson, *ibid.*, **70**, 816 (1948); H. J. Lucas, F. W. Mitchell, and H. K. Garner, *ibid.*, **72**, 2138 (1950).
34. S. Winstein, *J. Am. Chem. Soc.*, **64**, 2791 (1942); S. Winstein, E. Grunwald, and L. L. Ingraham, *ibid.*, **70**, 821 (1948); H. J. Lucas and C. W. Gould, *ibid.*, **63**, 2541 (1950); H. J. Lucas and H. K. Garner, *ibid.*, **72**, 2145 (1950).
35. R. S. Tipson, *J. Biol. Chem.*, **130**, 55 (1939).
36. H. S. Isbell, *Annu. Rev. Biochem.*, **9**, 65 (1940); H. S. Isbell and H. L. Frush, *J. Res. Nat. Bur. Stand.*, **27**, 413 (1941); **43**, 161 (1949); H. S. Isbell, *Chem. Soc. Rev.*, **3**, 1 (1974).
37. H. Freundlich and H. Krestovnikoff, *Z. Phys. Chem.*, **76**, 79 (1911); H. Freundlich and M. B. Richards, *ibid.*, **79**, 681 (1912); H. Freundlich and W. Neumann, *ibid.*, **87**, 69 (1914); H. Freundlich and R. Bartels, *ibid.*, **101**, 177 (1922); H. Freundlich and H. Kroepelin, *ibid.*, **122**, 39 (1926); H. Freundlich and G. Salomon, *ibid.*, **166**, 161 (1933); *Ber.*, **66**, 355 (1933).
38. G. Salomon, *Helv. Chim. Acta*, **16**, 1361 (1933); **17**, 851 (1934); **19**, 743 (1936); G. Salomon, *Trans. Faraday Soc.*, **32**, 153 (1936).
39. P. D. Bartlett, S. D. Ross, and C. G. Swain, *J. Am. Chem. Soc.*, **69**, 2971 (1947); **71**, 1415 (1949); P. D. Bartlett, J. W. Davis, S. D. Ross, and C. G. Swain, *ibid.*, **69**, 2977 (1947).
40. S. D. Ross, *J. Am. Chem. Soc.*, **69**, 2982 (1947).
41. W. E. Hanby, G. S. Hartley, E. O. Powell, and H. N. Rydon, *J. Chem. Soc.*, 519 (1947).
42. B. Cohen, E. R. Van Artsdalen, and J. Harris, *J. Am. Chem. Soc.*, **70**, 281 (1948).

43. A. W. Hay, A. L. Thompson, and C. A. Winkler, *Can. J. Res. Sect. B.*, **26**, 175 (1948); A. L. Thompson, T. J. Hardwick, and C. A. Winkler, *ibid.*, **26**, 181 (1948).
44. W. P. Evans, *Z. Phys. Chem.*, **7**, 337 (1891).
45. H. Nilsson and L. Smith, *Z. Phys. Chem.*, **166**, 136 (1933).
46. L. O. Winstrom and J. C. Warner, *J. Am. Chem. Soc.*, **61**, 1205 (1939); J. E. Stevens, C. L. McCabe, and J. C. Warner, *ibid.*, **70**, 2449 (1948), C. L. McCabe and J. C. Warner, *ibid.*, **70**, 4031 (1948).
47. D. Porret, *Helv. Chim. Acta*, **24**, 80E (1941); **27**, 1321 (1944).
48. S. Winstein and R. E. Buckles, *J. Am. Chem. Soc.*, **64**, 2787 (1942).
49. G. H. Twigg, W. S. Wise, H. J. Lichtenstein, and A. R. Philpotts, *Trans. Faraday Soc.*, **48**, 699 (1952).
50. H. W. Heine, A. D. Miller, W. H. Barton, and R. W. Greiner, *J. Am. Chem. Soc.*, **75**, 4778 (1953).
51. G. M. Bennett and A. L. Hock, *J. Chem. Soc.*, 477 (1927); G. M. Bennett and W. A. Berry, *ibid.*, 1676 (1927); G. M. Bennett, F. Heathcoat, and A. N. Mosses, *ibid.*, 2567 (1929); G. M. Bennett, *Chem. Ind. London*, **51**, 776 (1932); G. M. Bennett and E. G. Turner, *J. Chem. Soc.*, 813 (1938); G. M. Bennett, *Trans. Faraday Soc.*, **37**, 794 (1941).
52. H. Mohler and J. Hartnagel, *Helv. Chim. Acta*, **24**, 564 (1941); **25**, 859 (1942).
53. H. Böhme and K. Sell, *Chem. Ber.* **81**, 123 (1948).
54. A. G. Ogston, E. R. Holiday, J. St. L. Philpot, and L. A. Stocken, *Trans. Faraday Soc.*, **44**, 45 (1948).
55. P. D. Bartlett and C. G. Swain, *J. Am. Chem. Soc.*, **71**, 1406 (1949).
56. E. F. Caldin and J. H. Wolfenden, *J. Chem. Soc.*, 1239 (1936).
57. A. F. Chadwick and E. Pacsu, *J. Am. Chem. Soc.*, **65**, 392 (1943).
58. E. Grunwald and S. Winstein, *J. Am. Chem. Soc.*, **70**, 841 (1948).
59. J. F. Lane and H. W. Heine, *J. Am. Chem. Soc.*, **73**, 1348 (1951); H. W. Heine, E. Becker, and J. F. Lane, *ibid.*, **75**, 4514 (1953).
60. S. Winstein, L. Goodman, and R. Boschan, *J. Am. Chem. Soc.*, **72**, 2311 (1950); S. Winstein and W. Boschan, *ibid.*, **72**, 4669 (1950).
61. D. S. Tarbell and P. D. Bartlett, *J. Am. Chem. Soc.*, **59**, 407 (1937); R. T. Arnold, M. de Moura Campos, and K. L. Lindsay, *ibid.*, **75**, 1044 (1953); R. T. Arnold and K. L. Lindsay, *ibid.*, **75**, 1048 (1953).
62. L. J. Edwards, *Trans. Faraday Soc.*, **46**, 723 (1950); **48**, 696 (1952); S. J. Leach and H. Lindley, *ibid.*, **49**, 921 (1953); V. Prelog and S. Heimbach-Juhász, *Ber.*, **74**, 1702 (1941); F. F. Blicke, W. B. Wright, and M. F. Zienty, *J. Am. Chem. Soc.*, **63**, 2488 (1941).
63. J. D. Chanley, E. M. Gindler, and H. Sobotka, *J. Am. Chem. Soc.*, **74**, 4347 (1952).
64. R. U. Lemieux and C. Brice, *Can. J. Chem.*, **33**, 109 (1955).
65. R. Heck and S. Winstein, *J. Am. Chem. Soc.*, **79**, 3432 (1957).
66. S. Winstein, E. Allred, R. Heck, and R. Glick, *Tetrahedron*, **3**, 1 (1958).
67. S. Winstein, *Chem. Ind. (London)*, 562 (1954); S. Winstein, R. Heck, S. Lapporte, and R. Baird, *Experientia*, **12**, 138 (1956); R. Heck and S. Winstein, *J. Am. Chem. Soc.*, **79**, 3105 (1957).
68. R. Heck and S. Winstein, *J. Am. Chem. Soc.*, **79**, 3114 (1957).
69. S. Winstein, B. K. Morse, E. Grunwald, K. C. Schreiber, and J. Corse, *J. Am. Chem. Soc.*, **74**, 1113 (1952).
70. E. F. Jenny and S. Winstein, *Helv. Chim. Acta*, **41**, 807 (1958).
71. S. Winstein, E. Grunwald, and L. L. Ingraham, *J. Am. Chem. Soc.*, **70**, 821 (1948).
72. J. M. Harris, *Prog. Phys. Org. Chem.*, **11**, 163 (1974); C. A. Grob, K. Seckinger, S. W. Tam, and R. Traber, *Tetrahedron Lett.*, 3051 (1973).
73. B. Capon, *Essays Chem.*, **3**, 148 (1972).
74. M. I. Page, *Chem. Soc. Rev.*, **2**, 295 (1973).

2
Some Factors that Influence Anchimeric Assistance

It is obvious that a number of requirements are necessary for anchimeric assistance to occur. First, the neighboring group must be configurationally and conformationally situated to provide proper orbital overlap in the transition state. Second, the reaction that involves neighboring group participation must not be very much slower than any competing reactions. Thus, solvolyses of tertiary halides and esters usually occur with greatly reduced anchimeric assistance (as measured by k_Δ/k_c) even when a good neighboring group is appropriately situated,[1] because the introduction of the extra α substituent lowers the free energy of the transition state for forming a carbocation while hardly affecting the free energy of the transition state of the neighboring group process. In fact the transition state for the pathway that involves neighboring group participation may be of higher energy with a tertiary than with a secondary halide owing to steric crowding.

In the reactions of secondary halides and esters that contain neighboring groups an important, frequently the most important, competing pathway is solvolysis with nucleophilic solvent assistance, the k_s process.[2] This means that compounds that react without anchimeric assistance or with only weak anchimeric assistance in strongly nucleophilic solvents may react with strong anchimeric assistance in poorly nucleophilic solvents. Obviously the quantity $k_t/k_s - 1 = (k_\Delta/k_c)/(k_s/k_c)$ (see p. 14) depends on k_s as well as on k_Δ.

We shall now consider the stability of bridged or synartetic ions, which bears on anchimeric assistance. It is obvious that if the relative energies of related bridged and unbridged carbocations could be calculated this should provide a fairly good theoretical estimate of the quantity k_Δ/k_c, and we shall now review attempts to do this.

2.1. The Stabilities of Bridged Carbocations—Some Molecular Orbital Calculations*

There have been a number of molecular orbital calculations that provide a comparison between the energies of related bridged and unbridged carbocations. All these calculations are of course for isolated molecules in the gas phase and may not provide a very accurate guide to the state of affairs in solution.[4] Also it is disconcerting for the layman to find that different methods of calculation, each strongly advocated by its originators, sometimes lead to opposing conclusions and that for some species the theoretically "preferred" structure comes and goes rapidly.

One of the simplest systems for which calculations of corresponding bridged and unbridged cations have been reported is the ethyl cation. The unbridged structure (**1**) and the hydrogen-bridge structure (**2**), sometimes considered to be better described as a protonated ethylene, have been

$$CH_3-CH_2^+ \qquad\qquad CH_2^+\overset{\overset{H}{\diagup\diagdown}}{-}CH_2$$

(**1**) (**2**)

considered. In 1964 Hoffman reported, with certain reservations, extended Hückel calculations that gave the bridged structure 2 kcal mole^{-1} higher energy than the unbridged structure.[5] In 1965 Davis and Murthy reported a similar conclusion from further EHT calculations but with the bridged structure now having 23.5 kcal mole^{-1} greater energy.[6] In 1968, Yonezawa *et al.* reported SCF calculations that suggested that the bridged structure was more stable that the unbridged structure,[7] and a similar conclusion was reached by Dannenberg and Berke in 1969 using the INDO method.[8] Also in 1969, Sustmann *et al.* reported NNDO calculations that indicated that the bridged structure was 33.24 kcal mole^{-1} more stable and at the same time reported *ab initio* calculations, using a limited basis set without full optimization, that indicated that it was 9.0 kcal mole^{-1} less stable.[9] They preferred the results of the *ab initio* calculations. Other calculations in 1969 using an SCF–MO–LC method indicated similar energies for the

---

* EHT, extended Hückel theory; CNDO, complete neglect of differential overlap; INDO, intermediate neglect of differential overlap; NNDO, neglect of diatomic differential overlap; SCF, self-consistent field; MINDO, modified INDO; *ab initio*, without the use of independently derived parameters. For an independent assessment of the different molecular orbital methods applied to carbocations, see Ref. 3.

bridged and unbridged structures.[10] In 1970, Williams et al. reported further *ab initio* calculations, still with a small basis set (STO-3G) but now with full optimization, and obtained a result similar to the preceding ones, with the bridged form calculated to be 11.4 kcal mole^{-1} less stable than the unbridged.[11] A similar result was obtained by Pfeiffer and Jewett, who calculated the energy difference to be 12.1 kcal mole^{-1}.[12] Other *ab initio* calculations in 1970 by Clark and Lilley, who also used a "contracted" basis set but who included polarization functions for hydrogen, again indicated the bridged ion to be less stable, but only by 3.39 kcal mole^{-1}, and Clark and Lilley suggested that this structure corresponded to an energy maximum.[13] In contrast, modified CNDO calculations indicated the bridged ion to be about 10 kcal mole^{-1} more stable.[14] In 1971, Latham et al. extended the basis set [4-31G(B)] of their *ab initio* calculations and again concluded that the unbridged structure was the most stable, but the energy difference was reduced to 6.76 kcal mole^{-1}.[15] A similar conclusion was reached by Williams et al., who calculated the energy difference to be 8.8 kcal mole^{-1}.[16] These workers also stated that they expected "that this energy difference would not change significantly if we were to make the refinements necessary to push our calculations to the Hartree–Fock limit" and considered that correlation energies for the two structures would not be very different. In 1972, Hariharan et al. calculated that when polarization functions (*d* functions on C and *p* functions on H) were added to the basis set their *ab initio* calculations now indicated the bridged ion to be the more stable but by only 0.9 kcal mole^{-1}.[17] They also considered (in contrast to Williams et al.[16]) that when correlation energies were included in the calculation the bridged structure would probably be favored still further. Dixon and Lipscomb made a similar point when reporting the results of their SCF calculations, which favored the unbridged form by 12 kcal mole^{-1}.[18] Calculations that took correlation energies into account were reported in 1973 by Zurauski et al., who calculated the bridged structure to be 9 kcal mole^{-1} more stable.[19] These authors also considered whether their calculations gave the "final answer." They concluded that their calculations contained two sources of error: (1) limitation of the basis set and (2) neglect of additivity corrections. It was thought that source (1) would probably not have an important effect but that source (2) leads their method to underestimate the relative stability of the bridged form. They therefore concluded that the result of their calculation is qualitatively but not quantitatively correct and that the bridged cation (**2**) is more stable than the unbridged form by more than the 9 kcal mole^{-1} they calculated.

This historical survey is summarized graphically in Fig. 1.

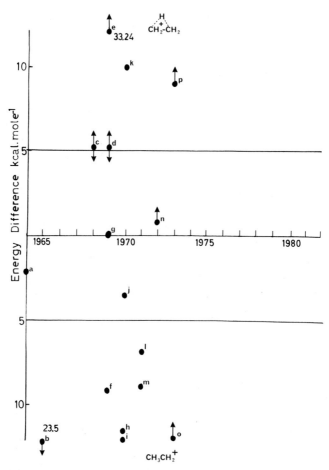

Fig. 1. Plot of the calculated relative energies of the bridged and unbridged $C_2H_5^+$ cations against the year of publication of the calculation. The horizontal lines indicate the experimental limits according to Ausloos and co-workers.[20] The years to 1982 have been left blank to enable the reader who so wishes to make additions. a, EHT, Ref. 5; b, EHT, Ref. 6; c, SCF, Ref. 7; d, INDO, Ref. 8; e, NNDO, Ref. 9; f, ab initio, limited basis set, incomplete optimization, Ref. 9; g, SCF–MO–LC, Ref. 10; h, ab initio, STO-3G basis set, complete optimization, Ref. 11; i, ab initio, limited basis set, Ref. 12; j, ab initio, (7, 3, 3/3, 1, 1) basis set and with polarization functions, Ref 13; k, modified CNDO, Ref 14; 1, ab initio, 4–31G(B) basis set, Ref. 15; m, ab initio Ref. 16; n, ab initio, with polarization functions, Ref. 17; o, SCF, Ref. 18; p, ab initio, with polarization functions and correlation energies, Ref. 19.

The most extensive investigation that provides information on the structure of the ethyl cation in the gas phase is that of Ausloos and his co-workers, who generated ion $CD_3CH_2^+$ and $CH_3CD_2^+$ from a variety of precursors that were themselves generated by photoabsorption decomposition, electron impact, or CO_2 irradiation.[20] They studied proton and deuteron transfer to bases by determining the H and D content of the resulting ethylene mass spectrometrically. Typical results indicated that $CD_3CH_2^+$ transferred D^+ 0.55 and H^+ 0.45 and that $CH_3CD_2^+$ transferred D^+ 0.33 and H^+ 0.67. It was concluded that there was overall randomization of the deuterium. The results from reactions of the ions with H^- donors also indicated extensive scrambling of deuterium. Scrambling occurred at all pressures up to 1000 torr, which was interpreted as indicating that the rate constant for scrambling is $> 10^{10}$ sec^{-1}. These results can be explained only if the bridged and unbridged ions have fairly similar energies (difference < 5 kcal mole^{-1}). If either of them were a lot more stable than the other, there would be no scrambling or only partial scrambling. Hence the most recent molecular orbital calculations do not appear to be in agreement with this experimental work.

The first attempt to detect a bridged ethyl cation in solution was that of Roberts and Yancey in 1952. These workers showed that deamination of ethylamine-1-^{14}C in aqueous solution yields ethylene and 38% ethanol that contains 1.5% ethanol-2-^{14}C.[21] Clearly the bridged ion is not an important intermediate in this reaction. It has also been shown that solvolyses of ethyl-1-^{14}C toluene-*p*-sulfonate in acetic acid, formic acid, and 75% dioxane–water[22] and of specifically deuterated ethyl toluene-*p*-sulfonate in trifluoroacetic acid[23] and 96% sulfuric acid[24] proceed without appreciable rearrangement. It seems that bimolecular solvolysis is the predominating reaction in all these solvents. However, solvolysis of deuterium-labeled ethyl toluene-*p*-sulfonate in fluorosulfuric acid yields a product with 30–40% rearrangement,[24] and it is possible that a bridged ion is an intermediate in this reaction.

It has also been found that when CD_3CH_2F is allowed to react with excess SbF_5–SO_2, FSO_3H–SbF_5–SO_2, or HF–SbF_5–SO_2 at $-78°C$ rapid deuterium scrambling occurs.[25] If such solutions are quenched with cyclohexane or cyclohexane-d_{12}, the resulting ethane has the deuterium completely randomized. It was concluded that the barrier to randomization was less than 3.5 kcal mole^{-1}, and from similar experiments in which CD_3CH_2F was added to SbF_5 and cyclohexane at 77°K it was concluded that it was less than 1.9 kcal mole^{-1}.[26] It therefore seems that if the ethyl cation is generated in solution it is in rapid equilibrium with a bridged species under these conditions as well as in the gas phase.

A similar complex situation is found with the propyl cations $C_3H_7^+$. Molecular orbital calculations have been made for nine states, the 2-propyl cation (3); three conformations of the 1-propyl cation (4), (5), and (6); a a hydrogen-bridged cation (7); two conformations of a corner-protonated cyclopropane (8) and (9); and edge- and face-protonated cyclopropanes (10) and (11). In 1964 Hoffmann reported on the basis of extended Hückel calculations that edge-protonated cyclopropane "does not want to go over to" a corner-protonated cyclopropane,[5] and in 1968 Petke and Whitten reported *ab initio* calculations that gave the edge-protonated form as 125.2 kcal mole^{-1} more stable than the face-protonated form.[27] Also in 1968, Fisher *et al.* reported INDO calculations that gave the relative energies of structures (10), (8), (9), and (11) to be 0, 22, 90, and 120 kcal mole^{-1}, respectively, and calculations using a modified CNDO method were also reported.[28,29] The results of these and subsequent calculations by other workers are given in Table 1. The most recent *ab initio* [30] and semiempirical[31] calculations both give the 2-propyl cation as the most stable species, but there is still disagreement as to which is the second most stable one.

The $C_3H_7^+$ cation has been studied in the gas phase by mass spectrometry, and the heats of formation and relative energies are given in Table 2. The 2-propyl cation is the most stable species, but it is difficult to make an exact comparison between the calculated and experimental energies because of the large experimental errors.

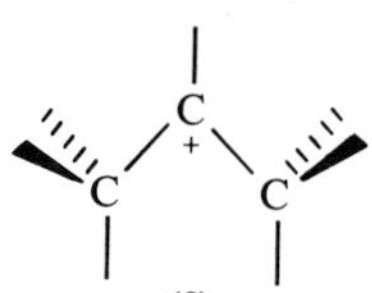

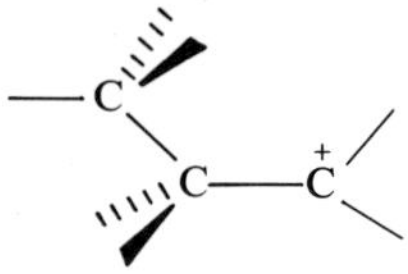

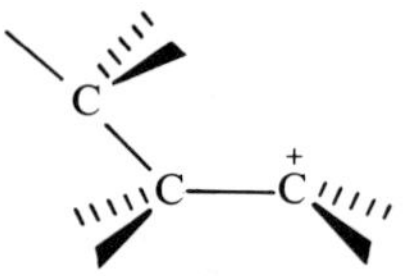

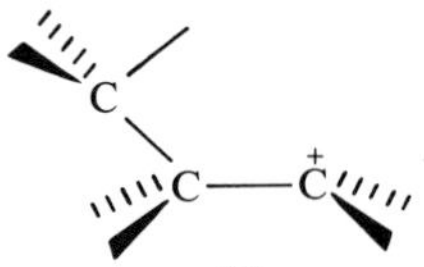

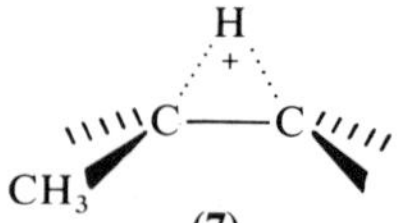

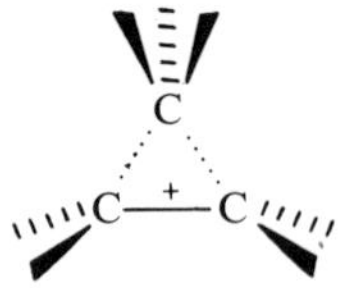

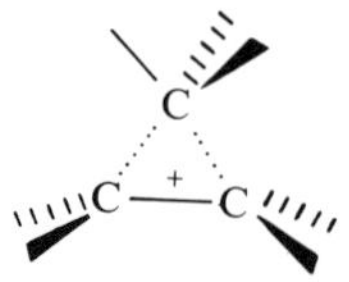

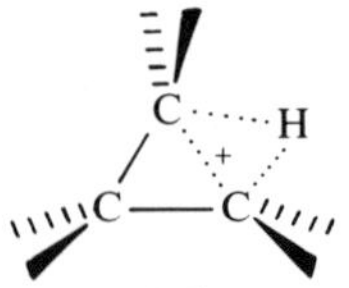

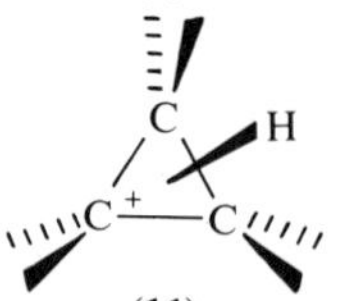

Table 1. Molecular Orbital Calculations on the Relative Energies (kcal mole^{-1}) of Different States of the $C_3H_7^+$ Cation

	CNDO[a]	NDDO[b]	Ab initio[c] STO-3G	Ab initio[c] 4-31G	MINDO/2[d]	MINDO/2[e]	Ab initio[f] 6-31G	MINDO/3[g]
2-Propyl (3)	0	0	0	0	0	0	0	0
Methyl-staggered 1-propyl (4)	25	18.47[i]	21.0	19.4	26[i]	24.5	18.3	18.6[i]
Methyl-staggered 1-propyl (5)	25	18.47[i]	19.7	17.4	26[i]		17.0	18.6[i]
Methyl-eclipsed 1-propyl (6)	25	18.47[i]	20.5	16.9	26[i]		14.1	18.6[i]
Hydrogen-bridged cation (7)	6	−17.36	23.9	18.2			11[h]	
Corner-protonated cyclopropane (8)	−3	−60.80	22.8	17.3			13.0	
Corner-protonated cyclopropane (9)	43		22.9	17.4	5	3.5	13.1	12.3
Edge-protonated cyclopropane (10)	−14	−81.01	27.1	27.1	−2.4	−3.9	19.1	7.5
Face-protonated cyclopropane (11)	62	55.48	161.0	139.6	57.5	56	130.1	88.3

[a] References 28 and 29.
[b] Reference 9.
[c] L. Radom, J. A. Pople, V. Buss, and P. von R. Schleyer, J. Am. Chem. Soc., **93**, 1813 (1971); **94**, 311 (1972).
[d] N. Bodor and M. J. S. Dewar, J. Am. Chem. Soc., **93**, 6685 (1971).
[e] N. Bodor, M. J. S. Dewar, and D. H. Lo, J. Am. Chem. Soc., **94**, 5303 (1972).
[f] Reference 30.
[g] Reference 31.
[h] Estimated.
[i] The exact conformation was not clearly stated.

Table 2. Heats of Formation and Relative Energies of $C_3H_7^+$ Species (kcal mole^{-1} at 340°K)

	ΔH_f	Relative energies
Isopropyl cation	191.7 ± 2.1[a]	0
n-Propyl cation	208 ± 3[b]	16.3 ± 5.1
Protonated cyclopropane	199.8 ± 2.2[a]	8.1 ± 4.3
	199 ± ?[c]	7.3 ± ?

[a] S.-L. Chong and J. L. Franklin, *J. Am. Chem. Soc.*, **94**, 6347 (1972).
[b] F. P. Lossing and G. P. Semeluk, *Can. J. Chem.*, **48**, 955 (1970); heat of formation of radical ±2 kcal mole^{-1}; ionization potential ±1 kcal mole^{-1}.
[c] D. J. McAdoo, F. W. McLafferty, and P. F. Bente, *J. Am. Chem. Soc.*, **94**, 2027 (1972).

The $C_3H_7^+$ cation has also been studied in SbF$_5$–SO$_2$ClF. At low temperatures ($< -5°$C) the NMR spectrum indicated that the 2-propyl cation is present with less than 1% of any other species. Between 0° and 40°C the spectrum indicates rapid exchange between the two types of protons.[32] This could be explained by interconversion with a 1-propyl cation, but observations of ^{13}C scrambling on the ion prepared from 2-chloropropane-2-^{13}C require a protonated cyclopropane as an intermediate instead of, or as well as, the 1-propyl cation.[33] From studies of scrambling of the label in the specifically deuterated CD$_3$CH$^+$CH$_3$ ion it was considered that this was an edge-protonated and not a corner-protonated species.[34]

The role of protonated cyclopropanes as reaction intermediates has been reviewed by Collins,[35] Fry and Karabatsos,[36] and Lee.[37]

There have also been many MO calculations of the energies of more complex ions usually using the less exact methods. In view of the way in which the computed relative energies of the different $C_2H_5^+$ and $C_3H_7^+$ species have varied over the years it is clear that not too much reliance should be placed on the results, and it seems to us that the methods available at present cannot be used to predict* with certainty the relative stabilities of bridged and unbridged cations.

Of historical interest are the calculations of Simonetta and Winstein on the stabilization of a carbocationic center by β-dienyl, β-vinyl, and β-phenyl substituents.[38] They used the Hückel method, to which many objections have been made, especially for charged species,[39] and concluded

---

* "To predict": to declare in advance, prophesy (*Webster's Third New International Dictionary*); to tell in advance, prophesy (*The Random House Dictionary of the English Language*); to foretell, prophesy, announce beforehand (OED).

that the stabilization energies for the dienyl, vinyl, and phenyl substituents that arise from formation of bridged cations are 10, 6, and 4 kcal mole^{-1}. These calculations were extended by Piccolini and Winstein,[40] and Howden and Roberts[41] applied a simple LCAO method to the cyclopropylcarbinyl cation. Subsequently Winstein reported many HMO calculations on *homoconjugated* and *homoaromatic* systems.[42] These calculations are probably not very accurate but were important in that they provided a stimulus for further experimental work.

Since the work of Winstein and Roberts there have been many molecular orbital calculations on various $C_4H_7^+$ ions, *cf.* **12–19** (Table 3). The majority of these give the bisected form of the cyclopropylmethyl cation as the most stable structure. It seems certain that the tricyclobutonium ion in either conformation (**15**) or (**16**) is very unstable as its energy is calculated to be 92.8–331.6 kcal mole^{-1} greater than that of the bisected form. The NMR spectrum of the dimethyl cyclopropylmethyl cation in SO_2–SbF_5–HSO_3F shows that it has the bisected conformation, and the barrier to rotation is 13.7 ± 0.4 kcal mole^{-1}.[43] Unfortunately the NMR spectrum of the cyclopropylmethyl cation itself is much more difficult to interpret since only a time-averaged spectrum can be observed. However, on the basis of the ^{13}C chemical shifts of the time-averaged spectrum it was thought that most likely the ion was present as a set of rapidly equilibrating bicyclobutonium ions.[44] It seems to us that the last word has yet to be said on the structure of the cyclopropylmethyl cation, either theoretically or experimentally.

Molecular orbital calculations on the phenonium ions **20–24** have been made using the EHT, CNDO/2, and *ab initio* methods. The EHT method gives the unbridged ion as being 20 kcal mole^{-1} more stable than the bridged form, whereas the CNDO/2 method gives the bridged form as being 119.4 kcal mole^{-1} more stable. It was thought that the CNDO/2 method overestimates the stabilities of cyclic structures, whereas the Hückel method underestimates them.[45] The *ab initio* calculations were made using the contracted STO-3G basis set, with limited geometric optimization and without taking polarizations and electron correlation into account. The results are shown in Table 4. It was thought that extension of the basis set and further geometric optimization would reduce the relative energies of the unbridged ions by 10–15 kcal mole^{-1}.[46]

Hehre and Hiberty have also made *ab initio* calculations on the $C_2H_4F^+$ and $C_2H_4Cl^+$ cations. When the 4-31G basis set was used the 1-haloethyl cations were calculated to be the most stable with both halogens (see Table 5). The bridged fluoronium ion was calculated to be much less stable than the

Table 3. Molecular Orbital Calculations on the Relative

	EHT[a]	ASMO[b]	CNDO[c]	EHT[d]	CNDO[e]	CNDO[f]
Cyclopropylmethyl bisected (12)	0	0	0	0	0	0
Cyclopropylmethyl perpendicular (13)	8.96[n]	18.7	25.1		129.7	17.1
Bicyclobutonium (14)				11		
Tricyclobutonium (15)	p			150	331.6	
Tricyclobutonium (16)						
Cyclobutyl planar (17)				46	−28.6	−64.8
Cyclobutyl puckered (18)					−15.8[g]	
Allylcarbinyl (19)						

[a] Reference 5.
[b] T. Yonezawa, H. Nakatsuji, and H. Kato, *Bull. Chem. Soc. Jpn.*, **39**, 2788 (1966).
[c] K. B. Wiberg, *Tetrahedron*, **24**, 1083 (1968).
[d] J. E. Baldwin and W. B. Fogelsong, *J. Am. Chem. Soc.*, **90**, 4311 (1968).
[e] C. Trindle and O. Sinanoglu, *J. Am. Chem. Soc.*, **91**, 4054 (1969)—Pople parameters; "planar in-plane" and "planar out-plane" structures of relative energies +59.8 and +36.1 kcal mole^{-1} were also considered.
[f] Idem, Wiberg parameters.
[g] V. Buss; unpublished calculations by V. Buss reported by P. von R. Schleyer and V. Buss, *J. Am. Chem. Soc.*, **91**, 5880 (1969).
[h] H. Kollmar and H. O. Smith, *Tetrahedron Lett.*, 3133 (1970).

Table 4. Ab Initio (STO-3G) Molecular Orbital Calculations on the Relative Energies (kcal mole^{-1}) of the 2-Phenylethyl Cation[46]

Bridged ethylenebenzenium (20)	0
Orthogonal perpendicular unbridged (21)	35.4
Orthogonal staggered unbridged (22)	42.3
Planar perpendicular unbridged (23)	46.5
Planar staggered unbridged (24)	48.8

Table 5. Molecular Orbital Calculations of the Relative Energies (kcal mole^{-1}) of Different States of the $C_2H_4F^+$ and $C_2H_4Cl^+$ Cations[47]

	$C_2H_4F^+$		$C_2H_4Cl^+$	
	STO-3	4-31G	STO-3G	4-31G
Staggered 1-haloethyl	0	—	0	—
Eclipsed 1-haloethyl	0	0	0	0
Halogen bridged	16.7	29.8	−1.1	4.53
Hydrogen bridged	39.0	23.1[a]	27.2	19.3[a]
Eclipsed 2-haloethyl	36.4	18.3	18.0	13.7

[a] It was thought that these values were overestimates.

Energies (kcal mole^{-1}) of Different States of the $C_4H_7^+$ Cation

NNDO[a]	CNDO[h]	Ab initio[i] STO-3G	MINDO/2[j]	MINDO/2[j]	INDO[k]	INDO[l]	Ab initio[m] STO-3G	Ab initio[m] 4-31G
0	0	0	0	0	0	0	0	0
22		17.54	9.5	16.0	47.3[n]	41.15		
	20[o]							
					92.8			
					258			
								9.4
	−40						32.7–35	20–23.2

[a] L. Radom, J. A. Pople, V. Buss, and P. von R. Schleyer, *J. Am. Chem. Soc.*, **92**, 6380 (1970).
[j] M. Shanshal, *Theor. Chim. Acta.* **20**, 405 (1971).
[k] C. U. Pittman, C. Dyas, C. Engelman and L. D. Kispert, *J. Chem. Soc. Faraday Trans. 2*, **68**, 345 (1972).
[l] W. C. Danen, *J. Am. Chem. Soc.*, **94**, 4835 (1972).
[m] W. J. Hehre and P. C. Hiberty, *J. Am. Chem. Soc.*, **94**, 5917 (1972); **96**, 302 (1974).
[n] Value considered too large by the authors.
[o] Drawn as puckered cyclobutyl cation but called bicyclobutyl.
[p] No minimum found; molecule has an orbitally degenerate ground state.
[q] Cross-ring interaction; see also K. B. Wiberg and K. Szeimies, *J. Am. Chem. Soc.*, **92**, 571 (1970).

1-fluoroethyl cation, but the energy of the bridged chloronium ion was only 4.53 kcal mole^{-1} greater than that of the 1-chloroethyl cations. The hydrogen-bridged ions and the 2-haloethyl cations were much less stable, but it was thought that the 4-31G calculations underestimated the stability of the former. Calculations were also performed on the effects of methyl substituents.[47]

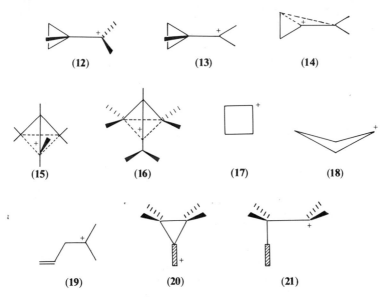

(12)　(13)　(14)
(15)　(16)　(17)　(18)
(19)　(20)　(21)

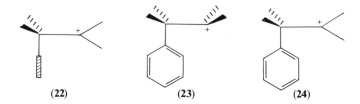

(22) (23) (24)

There have been several calculations on the structure of the 2-norbornyl cation. Trahanovsky, who used the extended Hückel method, calculated the carbon-bridged structure to be 51 kcal mole^{-1} more stable than the unbridged one,[48] and Klopman, using a PNDDO$^+$ method, calculated the relative energies of edge-protonated nortricyclene, corner-protonated nortricyclene, and the unbridged 2-norbornyl cation to be 0, 3, and 44.7 kcal mole^{-1}, respectively.[49] In contrast, MINDO/3 calculations give the unbridged structure to be slightly more stable than the carbon- and hydrogen-bridged structures (see Scheme 1).[50] The unbridged 2-norbornenyl cation was calculated to be 11.5 kcal mole^{-1} less stable than the bridged form (π complex).[50]

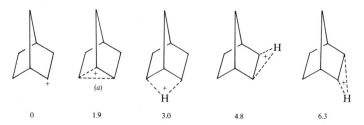

    0   1.9   3.0   4.8   6.3

Scheme 1. Relative energies of different states of the 2-norbornyl cation (kcal mole^{-1}). (a) is written as a π-complex structure.

As pointed out by Hoffman,[51] one of the problems of modern molecular orbital calculations is the difficulty of building a "bridge of understanding" between the computational results and the current ways of thinking of experimentalists. In this section we have reviewed many of the results of calculation of energies of bridged and unbridged carbocations. It is unfortunate that few of them are capable of giving an answer that can be translated into the descriptive language of the organic chemist as to why a certain structure is more stable than another. In Section 2.2 we shall discuss some of the more qualitative descriptive theories of the chemical structures of these and other species.

2.2. Hyperconjugation—An Alternative Explanation?

It has been said that "It is in ionic species that hyperconjugation comes to the fore."[52] With this in mind, it is useful to consider the possibility that the properties of certain ionic species can be accounted for by hyperconjugation, without the need for σ-bridging.

As early as 1952 Winstein considered the possibility that participation and hyperconjugation may merge but came to no definite conclusion,[53] and in 1965 Shiner and Jewett stated that they viewed neighboring hydrogen participation "as an extreme manifestation of a type of interaction also associated with hyperconjugation."[54] On the other hand, Jensen and Smart have stated that "Hyperconjugation should be considered as a phenomenon by which electrons are only partially delocalized to an electron deficient center and that these 'participating electrons' formally remain in the σ bond" and have regarded hyperconjugation as a quite separate phenomenon from participation.[55] A similar view has been taken by Harris,[2] and Olah et al. considered that they had "direct experimental justification for distinction between trivalent carbenium ('classical') and pentacoordinate ('non-classical') ions."[56] Hoffman et al.[52] have emphasized "that on theoretical grounds there is no dichotomy between participation with and without bridging." Presumably this only means that both processes occur with nuclear movement since the authors clearly regard the $F\text{---}CH_2\text{---}CH_2^+$ cation as a distinct species from the H- or F-bridged ions as these are not mentioned in their paper. Also other papers by some of these authors contain discussions in which bridged and unbridged cations are treated as distinct and separate species (see Section 2.1). The view of Hoffmann et al. has been reiterated by Olah and Liang, who, in contrast to the earlier opinion of Olah et al.[56] stated that "it is rather meaningless to overemphasize the so-called limiting 'classical' and 'non-classical' nature of carbocations."[57]

It seems to us that hyperconjugative stabilization of a cationic transition state should in principle be distinguishable experimentally from a bridging stabilization if more than one stabilizing group could be introduced,[58] because of the difficulty of two groups bridging simultaneously, which would require six groups coordinated about the central carbon atom in the transition state (**25**), but two groups should be able to interact hyperconjugatively simultaneously (**26**). Hence if Δ is the rate acceleration that results from one group, the rate acceleration to be expected from n groups is approximately $n\Delta$ for a bridging stabilization and approximately $(\Delta)^n$ for a hyperconjugative stabilization.[58] This does mean that both effects cannot

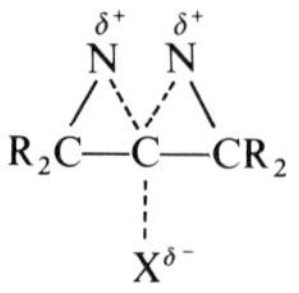

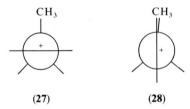

(25) 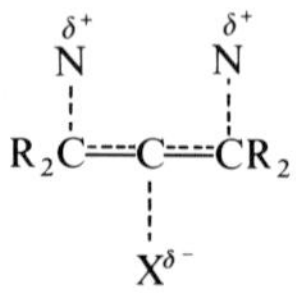   (26)

act at once but that they are in principle distinguishable experimentally from one another. It is therefore valid to consider them as separate effects.

The general picture that emerges from molecular orbital calculations on the effect of hyperconjugation on the stabilities of carbocations is, at present, the following. C—H hyperconjugation is not an important stabilizing or destabilizing factor for planar cations since the rotational barrier in the ethyl cation is virtually zero.[11,12,52,59,60] C—C hyperconjugation by a methyl group is larger but still fairly small, < 3 kcal mole^{-1}. For example, conformation (27) of the 1-propyl cation was calculated to be 2.52 kcal

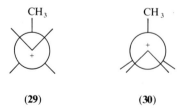

(27)   (28)

mole^{-1} more stable than conformation (28).[59] If the geometry of the CH_2^+ group is changed from planar to tetrahedral, there is an increase in the rotational barrier with conformation (29) favored over conformation (30) by 6.04 kcal mole^{-1}.[60] This conclusion is interesting because certain bridge-

(29)   (30)

head carbocations cannot attain a planar geometry. Hyperconjugative stabilization by cyclopropane rings is large and leads the bisected conformation of the cyclopropylcarbinyl cation to be more stable than the perpendicular conformation. The energy difference was given as 17.54 kcal mole^{-1} by *ab initio* calculations using an STO-3G basis set (see p. 29). Hyperconjugative stabilization by a cyclobutane ring is much smaller, and *ab initio* calculations (STO-3G basis set) gave the bisected conformation of the cyclobutylcarbinyl cation to be about 4 kcal mole^{-1} more stable than the perpendicular conformation.[60] If the group X in a cation XCH$_2$CH$_2^+$ is less electronegative than H, then conformation (**31**) is favored over (**32**);[52] thus when X = BH$_2$ the energy difference between them was calculated to be + 10.4 kcal mole^{-1}. When X is more electronegative than H, conformation (**32**) is favored over (**31**). Thus when X = F and OH conformation (**32**) is 9.31 and 7.67 kcal mole^{-1} more stable than conformation (**31**).[52] It is

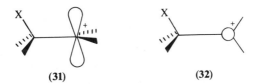

(**31**) (**32**)

interesting to note that with these ions the favored conformation is the one that is least favorable for bridging by the F or OH groups. This presumably precludes a "merging" of bridging and hyperconjugation. When the electronegativity difference between X and H is smaller than for X = BH$_2$, F or OH, the energy difference between the two conformations is also smaller; thus, as mentioned earlier, when X = CH$_3$ conformation (**31**) was calculated to be favored by only 2.52 kcal mole^{-1}.[59]

The possibility that strained-ring systems other than cyclopropyl may stabilize carbocations by hyperconjugation is a matter of considerable interest. In 1969, Jensen and Smart reported that the partial rate factors for benzoylation at the *para* positions of norbornyl benzenes were greater than for simple alkyl benzenes and that, in contrast to the latter, the compound with the tertiary substituent 1-phenylnorbornane reacted faster than those with a secondary substituent, 7-phenylnorbornane and *exo*- and *endo*-2-phenylnorbornane.[55] It was also found that *exo*-2-phenylnorbornane reacted faster than *endo*-2-phenylnorbornane and isopropylbenzene (relative reactivities, 2.89:1.85:1). These results were attributed to enhanced C—C hyperconjugation in the phenylnorbornanes. There was already evidence from ^{13}C—H coupling constants that the C—H bonds in norbornane possess enhanced *s* character, and this presumably would be accompanied by the C—C bonds having enhanced *p* character, to which the enhanced C—C hyperconjugation was attributed. It was argued that the effect of C—C hyperconjugation should be greater in reactions that occur at the carbon atoms of the norbornyl structure, and Jensen and Smart suggested that the difference in reactivity between *exo*- and *endo*-norbornyl halides and arenesulfonates arose because in the transition state for formation of the carbocation from *exo*-2-norbornyl derivatives the developing *p* orbital has the optimum orientation for hyperconjugative stabilization from the electrons in the C_6—C_1 bond, whereas in the transition state from *endo*-2-norbornyl derivatives the developing *p*-orbital is less favorably situated for interaction with the electrons in the C_1—C_7 bond, "while overlap with the C_6—C_1 bond is virtually impossible." An attempt to show a similar effect in the solvolyses of cumyl chlorides, a reaction that is less electron demanding, was, perhaps understandably, less conclusive. The relative reactivities of the *p*-isopropyl-, *p*-2-*endo*-norbornyl- and *p*-2-*exo*-norbornyl cumyl chlorides on solvolysis in 90% aqueous acetone at 25°C were 1 : 1.16 : 1.34.[61]

The trouble with both these pieces of work is that they are concerned with the relative stabilities of transition states that resemble the cationic species (**33**) and (**34**) (and their *endo* isomers), and this may or may not be

(**33**)   (**34**)

relevant to the structure of 2-norbornyl cations and the transition states for their formation. The results seem to be of more relevance to the relative reactivities of *exo*- and *endo*-2-norbornylmethyl derivatives, but there appear to be no experimental kinetic measurements on the solvolyses of these compounds.

That hyperconjugation is an important stabilizing factor for carbocations with strained rings has found its strongest advocate in Traylor in his theory of vertical and nonvertical stabilization.[58,62,63,64] As defined by Traylor, a substituent Y provides *vertical stabilization* if Y does not change its geometry or its distance from the positively charged carbon as the transition state is approached and *nonvertical stabilization* if the substituent Y changes its geometry or its distance from the reaction site as the transition state is approached.[62] Thus, hyperconjugation (termed σ–π conjugation by Traylor) in a transition state (**35**) was considered to be a form of vertical

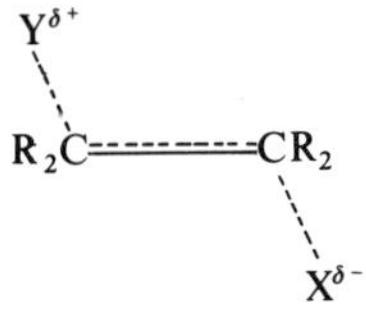

$$Y^{\delta+} \qquad \qquad Y^{\delta+}$$
$$R_2C\text{-------}CR_2 \qquad R_2C\text{------}CR_2$$
$$X^{\delta-} \qquad \qquad X^{\delta-}$$

$$(\mathbf{35}) \qquad \qquad (\mathbf{36})$$

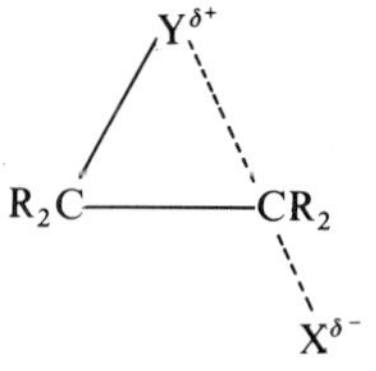

stabilization and bridging (internal displacement) in a transition state (**36**) a form of nonvertical stabilization. It was realized that the C—Y bond length in (**35**) may be slightly changed from what it is in the initial state so that the description of hyperconjugative stabilization of transition states as a form of vertical stabilization is not quite exact. It was also emphasized that (**35**) and (**36**) symbolize limiting behavior and that a third possibility, formulated as (**37**), is possible, where the transition state is stabilized by

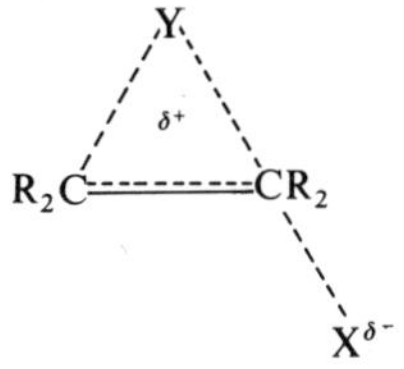

$$Y$$
$$\delta+$$
$$R_2C\text{-------}CR_2$$
$$X^{\delta-}$$

$$(\mathbf{37})$$

hyperconjugation and bridging. This stabilization was considered to be nonvertical, however, since Y has moved.[58] To obtain evidence on whether a carbocation (and the transition state for its formation) was stabilized by hyperconjugation or bridging Traylor and his co-workers investigated the stabilizing effects of various groups R in an authentic vertical process, the charge-transfer interaction between the hydrocarbons R—C_6H_5 and tetracyanoethylene (TCNE). This was thought to involve the change

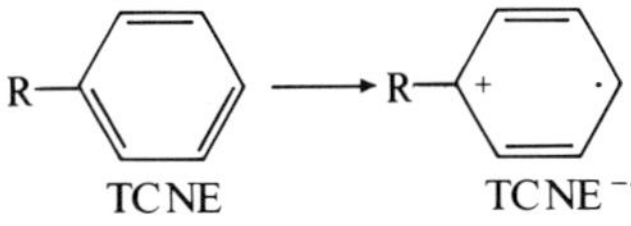

R—⟨⟩ ⟶ R—⟨+⟩
TCNE   TCNE⁻·

and to occur without nuclear movement (Franck–Condon principle). They plotted the logarithm of the rate constants for the solvolyses of a number of compounds R—CH$_2$—X against the frequency of the charge-transfer band (see Table 6) and obtained a straight line. They conclude that "this correlation implies that acceleration of the reaction is primarily due to a vertical electronic effect involving σ–π conjugation of one or more of the strained bonds."[63] There are several problems with this conclusion. The first is that there is a second obvious potential source of the variation of rates of solvolysis of the compounds listed in Table 6; this is relief of steric strain on going to a bridged transition state. The reactions of most of the compounds listed in Table 6 are what Traylor terms nonvertical *processes*; i.e., the products are formed with skeletal rearrangement. It is therefore difficult to be certain that the skeletal rearrangement has not started before the rate-determining transition state is reached. If the relief of strain on going to a bridged transition state were proportional to the total strain energy of the initial state, and if the change in frequency of the charge-transfer band were also proportional to this, there would be a proportionality between the logarithm of the rate enhancement arising from release of steric strain and the frequency of the charge-transfer band, as found. Traylor and his co-workers considered this possibility and argued against it on the grounds that the charge-transfer results showed that the groups R are capable of providing substantial vertical stabilization for an excited state (see Table 6) and that it is not unreasonable that they should do so for the transition states of the solvolysis reactions. In our view this argument may have some validity since the density of positive charge on the carbon α to the group R may be even higher in the transition state for the solvolyses than for the corresponding excited states. Also, on going to the excited state maximum stabilization will be obtained only with those molecules that have the correct rotational angle between the R and Ph groups in the initial state since electronic excitation is a much faster process than molecular rotation.[65] However, on passing to the transition state rotation could occur around the bond between R and Ph groups to achieve the most favorable conformation for hyperconjugative stabilization. This means that hyperconjugative stabilization could be greater in the latter than in the former of these processes. Thus in the rapid formation of an excited state hyperconjugative stabilization is restricted to molecules that have the correct ground-state conformation, whereas in the slower formation of the transition state for solvolyses all molecules have sufficient time to achieve the most favorable conformation. These arguments show that hyperconjugative stabilization should be considered seriously as a stabilizing factor in the formation of transition states from compounds with strained rings. They also illustrate

Table 6. Charge-Transfer Frequencies for RC_6H_5 with Tetracyanoethylene (TCNE) Compared with Relative Rates of Solvolysis of RCH_2X[63]

	R	γ (cm^{-1} × 10^{-3})	$\delta\Delta E$ (kcal mole^{-1})	$k_{rel}{}^a$	$\delta\Delta G^{\ddagger}$ (100°C) (kcal mole^{-1})
1.	t-Bu	22.8	0	1	0
2.	(cyclohexyl)	22	−1.7	8	−1.5
3.	(cyclopentyl)	21.5	−3.8	190	−3.89
4.	(cyclopropyl)	21[b]	−5.1	5 × 10^4	−8.02
5.	(norbornyl)	20.6	−6.3	—	—
6.	(bicyclic)	19.8[c]	−8.7	1 × 10^7	−12.0
7.	(bicyclic)	19.6	−9.2	3 × 10^7	−12.8
8.	(cubyl)	19.2	−10.3	—	—

[a] The way in which these relative rates were arrived at is not immediately obvious. The following is a possible method. Reaction 2: R. L. Bixler and C. Niemann, J. Org. Chem., 23, 742 (1958), give $k_{99.7}$ as 11.69 × 10^{-6} sec^{-1} for acetolysis of the tosylate and 1.66 × 10^{-6} sec^{-1} for acetolysis of neopentyl tosylate. K. B. Wiberg and B. R. Lowry, J. Am. Chem. Soc., 85, 3188 (1963), give k_{80} as 1.4 × 10^{-6} sec^{-1} for acetolysis of the tosylate and 0.17 × 10^{-6} sec^{-1} for acetolysis of neopentyl tosylate. The product is mainly 1-bicyclo[2.2.2]octyl acetate. Reaction 3: K. B. Wiberg and B. R. Lowry, loc. cit., give k_{80} as 33 × 10^{-6} sec^{-1} as the total rate constant for acetolysis and rearrangement. The product is 90% 1-norbornyl tosylate, 8% 1-norbornyl acetate, and 2% unrearranged acetate. Reaction 4: D. D. Roberts, J. Org. Chem., 29, 294 (1964), quotes rate constants for the acetolysis of cyclopropylcarbinyl tosylate at 15°–30°C and activation parameters that lead to an extrapolated value of k_{100} of 1.09 × 10^{-1} sec^{-1}. The values of k_{rel} for reactions 6 and 7 also appear to have been determined via the value of k_{rel} for reaction 4. Reaction 6: P. von R. Schleyer and G. W. Van Dine, J. Am. Chem. Soc., 88, 2321 (1966), report k_{100} as 9.65 × 10^{-5} sec^{-1} for solvolysis of the 3,5-dinitrobenzoate in 60% aqueous acetone at 100° and 4.30 × 10^{-7} sec^{-1} for solvolysis of cyclopropylcarbinyl 3,5-dinitrobenzoate. No products were reported. Reaction 7: W. D. Closson and G. T. Kwiatkowski, Tetrahedron, 21, 2779 (1965), give k_{100} as 2.17 × 10^{-5} sec^{-1} for solvolysis in 60% aqueous acetone and ca. 0.003 sec^{-1} for cyclopropylcarbinyl p-nitrobenzoate. Rearranged products were obtained.
[b] The plotted value appears to be 20.85. Reference 62 gives 21.3.
[c] Reference 62 gives 20.

what in our view is a serious semantic problem introduced by Traylor and his co-workers' use of the term vertical stabilization in connection with nonphotochemical processes. It was admitted by them that the use of the term vertical stabilization in connection with thermal processes is not exact,[58] and it seems to us that in such processes there is always the possibility that bond lengths, bond angles, and torsional angles will be adjusted to their optimum values, whereas this is not possible in a photochemical process. Therefore it is probably better to drop the term vertical stabilization in this connection and use a term such as hyperconjugative stabilization, which arises from a fairly well-defined process[52] and does not carry overtones that cannot be justified or that have to be heavily qualified.

Traylor and his co-workers made two interesting predictions on the basis of their theory, namely, the relative rates of solvolysis of compounds (**38**) and (**39**).[63] Unfortunately these predictions do not appear to have been

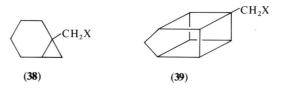

(**38**)   (**39**)

tested experimentally. It would also be expected that the homocubylbenzene would undergo electrophilic substitution in the *para* position faster that cyclopropylbenzene, but this also appears to have never been tested.

Hyperconjugative stabilization has also been considered to be important for bridgehead carbocations since 10-tricyclo[5.2.1.0^{4,10}]decyl tosylate (**40**) reacts nearly 10^9 times slower than expected on the basis of molecular mechanics calculations.[66] The carbocation from this compound has none of the β–γ C—C or C—H bonds antiperiplanar to the vacant p orbital, and so hyperconjugative stabilization should be absent, whereas with most

(**40**)

other bridgehead cations this is not so. Two factors may lead to hyperconjugative stabilization of bridgehead carbocations being particularly favorable: increased p character of C—C bonds as a result of increased strain and the cationic carbon having a tetrahedral geometry that was calculated to lead to enhanced hyperconjugative stabilization in the *anti-* conformation of the 1-propyl cation (see p. 32).

Another compound for which there is evidence that hyperconjugative stabilization may be an important factor in its solvolysis reaction is 8-vinyl-*exo*-8-bicyclo[3.2.1]octyl 3,5-dinitrobenzoate (**41**), which reacts 515 times faster than its *endo* isomer (**42**) in aqueous acetone.[67] This ratio is

(41)   (42)

reduced to 122 if correction is made for the difference in energy of the initial states. Previously the 8000-fold difference in the acetolysis rates between the *exo-* and *endo*-8-bicyclo[3.2.1]octyl toluene-*p*-sulfonates was attributed to σ participation by the C_1—C_2 bond in the reaction of the *exo* isomer,[68] but it seems that at least part of this rate difference may arise from hyperconjugation since a substantial rate difference persists with the 8-vinyl-substituted compounds, which react without migration of the 1,2 bond, and which would be expected to react without significant participation, since they yield allyl cations that would be expected to be formed without nucleophilic assistance.

The group for which evidence for stabilization of a carbocationic center by hyperconjugation appears to be best is, however, the cyclopropyl group. Indeed there is much spectroscopic evidence that a cyclopropane ring can interact with various unsaturated systems, as well as cationic centers, and that the most favorable conformation for this interaction is the one in which the plane of the cyclopropane ring lies perpendicular to the plane of the adjacent unsaturated system.[69] As has often been described before,[70,71] there are two simple, mathematically equivalent[72] MO descriptions of a cyclopropane ring that have been used to describe this behavior. In Walsh's description[73] the cyclopropane ring is considered to be made up of three sp^2 hybridized CH_2 groups that are brought together to occupy the corners of a triangle with one of the sp^2 orbitals of each group

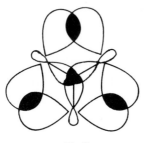 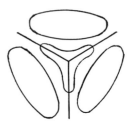

Fig. 2. Fig. 3.

that is not bonded to H directed toward the center. The axes of the sp orbitals lie on the plane of the triangle so that overlap occurs to give regions of electron density outside the triangle; the AO's are shown in Fig. 2 and the resulting MO in Fig. 3. It was pointed out by Walsh[73] that "For conjugation the groups must set themselves such that the plane of the ring is parallel to the axis of the neighboring $2p$ AOs." This appears to be the first suggestion that the bisected conformation should be the most stable for compounds in which the cyclopropyl group interacts (hyper) conjugatively with an adjacent sp^2 hybridized carbon. In Coulson and Moffitt's description[74] the C—C bonds of the ring are made up of orbitals with $sp^{4.12}$ hybridization[75] and the C—H bonds made up of orbitals that are $sp^{2.28}$ hybridized. Again maximum overlap between the $sp^{4.12}$-hybridized orbitals and an adjacent p orbital on an sp^2-hybridized carbon is obtained in the bisected conformation.

There is very good evidence that if the cyclopropyl group of a cyclopropylcarbinyl derivative can interact with the developing cationic center in a bisected conformation a very large rate enhancement results, whereas if this conformation is prevented sterically, there is no rate enhancement. Thus on the basis of the rate of solvolysis of 8,9-dehydro-2-adamantyl 3,5-dinitrobenzoate in aqueous acetone it was estimated that the corresponding toluene-p-sulfonate (**43**) would undergo acetolysis about 8 powers of 10 faster than 2-adamantyl toluene-p-sulfonate (**44**).[76] It was also estimated that nortricyclyl toluene-p-sulfonate (**45**) undergoes acetolysis about 10^8 times faster than 7-norbornyl toluene-p-sulfonate.[77] Both compounds (**43**) and (**45**) are constrained to give cyclopropylcarbinyl cations in a bisected conformation. In contrast, compound (**46**), which is constrained to a cyclopropylcarbinyl cation in the perpendicular conformation, reacts about 150 times more slowly than compound (**47**).[78]

(**46**)    (**47**)

Although Roberts originally[79] ascribed the stability of cyclopropylcarbinyl cations to hyperconjugation, he later[80] proposed that it arose from a bonding interaction between one of the rear carbon atoms of the cyclopropane ring and the CH_2^+ group to form a bicyclobutonium ion, (**48**) and (**49**).

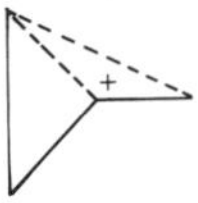

(**48**) 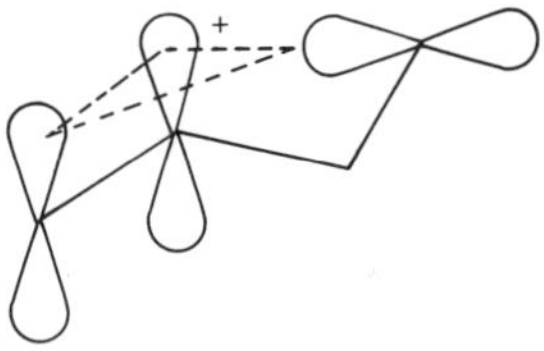   (**49**)

An important difference between the bicyclobutonium ion and the hyperconjugatively stabilized cyclopropylcarbinyl cation is their symmetry, and attempts to distinguish between them and transition states leading to them have utilized this. Thus Schleyer and Van Dine showed that the effect of methyl substituents in the 3 and 4 positions were additive on the rate of solvolysis of cyclopropylcarbinyl 3,5-dinitrobenzoate and concluded that the transition state was symmetrical.[81] In another series of revealing experiments Hart and his co-workers showed that each successive replacement of one of the isopropyl groups of tris-isopropylcarbinyl p-nitrobenzoates by a cyclopropyl group caused a similar enhancement of the rate of solvolysis.[82] This is of course a $(\Delta)^n$ effect expected for hyperconjugative stabilization of the transition state, not an $n(\Delta)$ effect expected for bridging (see p. 32).

There is also much other evidence that supports the bisected hyperconjugativity stabilized structure.[83] The one piece of evidence that has been taken to support a bicyclobutonium ion structure is the NMR spectrum of the species obtained from cyclopropylcarbinol and cyclobutanol in SbF_5–SO_2ClF.[44] The NMR spectra of the 1-methylcyclopropylcarbinyl and 1,1-dimethylcyclopropylcarbinyl cations indicate that they have bisected conformations. However, the spectra of the species obtained from cyclopropylcarbinol and cyclobutanol were considered to be best interpreted as arising from a rapidly equilibrating set of bicyclobutonium ions. In particular, if the ^{13}C chemical shifts to be expected from a set of rapidly equilibrating bisected cations are calculated by extrapolation from the measured shifts of the 1-methyl and 1,1-dimethylcyclopropylcarbinyl cations, they do not agree with the observed shifts. Therefore it was considered that there was a difference in structure between the unsubstituted and methyl-substituted ions and that the former were bicyclobutonium ions. However, the extrapolation used was a linear one based on a straight line defined by two points and, as pointed out by Richey,[84] it is possible that the differences between the calculated and experimental chemical shifts "result only from dramatically increased absorption of charge by the cyclopropyl ring in a structure otherwise similar to those of the more stable ions." It is also clear that the relationship between ^{13}C chemical shifts of carbocations and their stability is not simple.[85] On the other hand, as Richey also pointed out, the interconversion of different structures occurs very rapidly, and hence the differences in energies between ground states and transition states are small.[84] Therefore there is always the possibility that changing a substituent may reverse the relative stabilities of different types of structures. It is interesting that the interconversion of cyclopropylcarbinyl cations occurs rapidly even if they are constrained to a bisected conformation. Thus the NMR spectra of the dehydro-2-adamantyl cation were interpreted in terms of a set rapidly equilibrating bisected ions. On the basis of the estimated ^{13}C shifts of C_1 and C_8, 235 and 125 ppm (downfield from external TMS), and the fact that a single signal is observed for these nuclei at $-120°C$, the energy barrier for interconversion must be less than 6.6 kcal mole^{-1}.

In conclusion, most of the evidence and molecular orbital calculations (see p. 27) support the view that cyclopropylcarbinyl cations have a bisected conformation that is stabilized because of "the increased *p*-character of the 'bent' C—C cyclopropane bonds which increases greatly their C—C hyperconjugative ability."[86] The most important evidence that has been interpreted as arguing against this formulation is the NMR spectra of the cyclopropylmethyl cation itself, and hence there must be certain reservations about the above conclusion concerning this ion. It must also be remembered that the energies of intermediate structures must often lie fairly close to that of the bisected cyclopropylcarbinyl structure since these sometimes undergo very rapid interconversions.

If this conclusion is correct, it means that the effect by which a cyclopropyl group stabilizes an adjacent cationic center in cyclopropylcarbinyl cations is not neighboring group participation as defined initially in Chapter 1 and that the rate enhancement should not be referred to as anchimeric assistance.

2.3. The Effect of Intramolecularity—Ring Size and gem-Dialkyl Effects

The factors that lead intramolecular reactions frequently to occur more rapidly than intermolecular reactions have been widely discussed in the last five to six years.[87-91] What we consider to be the correct explanation is that given by Page and Jencks,[90] namely, that the entropy change on passing to the transition state for an intramolecular reaction is more favorable than for an analogous intermolecular one. An excellent discussion of the differences in the entropy changes between inter- and intramolecular reactions has recently been given by Page,[91] and the reader is referred to his review for a full discussion of this topic.

It was pointed out by Wiberg[92] that in the gas phase there is a large unfavorable free-energy change (> 9 kcal mole^{-1}) when two molecules come together to form a third as a result of loss of three degrees of translational freedom and a smaller one from the loss of three degrees of external rotational freedom. These unfavorable free-energy changes are partly offset by a favorable entropy change associated with the creation of three new degrees of vibrational freedom [see Eq. (1)]. As the number of degrees of translational and external rotational freedom remain unchanged in the analogous intermolecular process [Eq. (2)], it should be favored. Page and

	A	+	B	⇌	A—B	
translation:	3		3		3	(1)
rotation:	3		3		3	
vibration:	$3n - 6$		$3n' - 6$		$3n + 3n' - 6$	

	A⌣B	⇌	A—B	
translation:	3		3	(2)
rotation:	3		3	
vibration:	$3n - 6$		$3n - 6$	

Jencks showed that it would be expected that intramolecular reactions would be favored entropically in solution as well as in the gas phase.[90]

Despite these more favorable entropy changes, intramolecular reactions are not always favored relative to analogous intermolecular reactions owing to the effect of ring strain in the product, if equilibrium constants are being considered, or to a lesser extent in the transition state, if rate constants are being considered. This is illustrated by the figures given in Tables 7, 8, and 9. Here the equilibrium constants and thermodynamic parameters for cyclization of a series of alkanes to cycloalkanes, diols to cyclic ethers, and hydroxy thiols to cyclic sulfides in the gas phase are given along with analogous intermolecular reactions. The intramolecular reactions have more favorable $\Delta G°$'s and equilibrium constants only when five- and six-membered rings are formed. With rings of other sizes the unfavorable $\Delta H°$ values more than offset the favorable $T\Delta S°$ values. This behavior is different from the kinetic behavior of reactions in solution described in Section 2.3.1, where three-membered rings are often formed as rapidly (see Table 13, reactions 10–14, 23–27) or more rapidly than five-membered rings. It seems likely that this is because in the transition states for cyclizations the ring is only partly formed, and hence the strain energies in the transition states are smaller than in the fully formed rings. The factor that makes formation of three-membered rings unfavorable relative to five-membered ones is therefore reduced. The figures in Tables 7–9 also illustrate how strongly intramolecular reactions are favored entropically and suggest that if the unfavorable $\Delta H°$ terms could be reduced the equilibrium and rate constants for intramolecular reaction would be increased further.

Some examples of changes in structure that cause an increase in the rate of cyclization for formation of rings of the same size are shown in Table 10. Undoubtedly several factors are important in determining the variation in rates. One may be that on passing to the transition state for cyclization of the acyclic compounds (**50**) and (**53**) there is an increase in torsional strain since in the initial state the bonds are all staggered, whereas in the five-membered ring some are eclipsed. In the initial state for compounds (**51**), (**52**), and (**54**) some bonds are already eclipsed and so the initial-state free energy is increased, but there is not a corresponding increase in the transition-state free energy. It is difficult to estimate what this effect will be, but the strain energies of five-membered rings are ca. 5 kcal mole^{-1} (cf. Table 12) and so it may be substantial. It has also been estimated that there should be a substantial relief of steric strain on cyclization of (**51**) and (**52**).[93] The high rates of reaction of compounds (**51**) and (**52**) therefore

Table 7. Comparison of the Thermodynamic Properties for the Cyclization of Hydrocarbons (298°K)[a]

Reaction	$\Delta G°$ (kcal mole^{-1})	$\Delta H°$ (kcal mole^{-1})	$\Delta S°$ (cal °K^{-1})	K_p	K_c (mole liter^{-1})
 ⌇ → △ + H$_2$	30.56	37.56	23.45	3.56×10^{-23}	6.02×10^{-26}
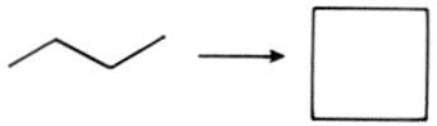 ⌇ → □ + H$_2$	30.40	36.52	20.52	4.66×10^{-23}	7.88×10^{-26}
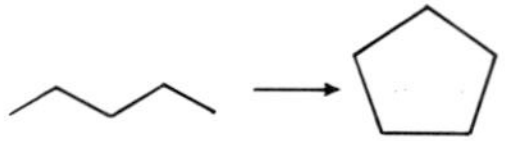 ⌇ → ⬠ + H$_2$	11.23	16.54	17.81	5.63×10^{-9}	9.52×10^{-12}
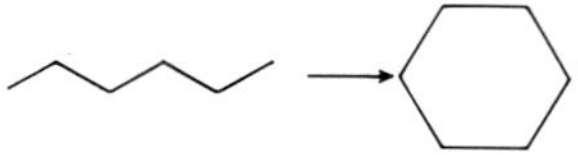 ⌇ → ⬡ + H$_2$	7.65	10.53	9.66	2.40×10^{-6}	4.06×10^{-9}
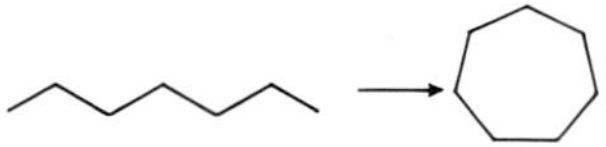 ⌇ → ⬣ + H$_2$	13.15	16.36	10.76	2.19×10^{-1}	3.70×10^{-13}
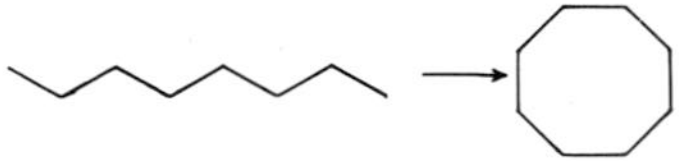 ⌇ → ⬢ + H$_2$	17.57	19.76	7.32	1.24×10^{-13}	2.10×10^{-16}
CH$_3$—CH$_2$ + CH$_3$—CH$_2$—CH$_3$  → ⌇ + H$_2$	11.48	10.06	−4.75	3.69×10^{-9}	3.69×10^{-9b}

[a] Data from D. R. Stull, E. F. Westrum, and G. C. Sinke, *The Chemical Thermodynamics of Organic Compounds*, John Wiley & Sons, Inc., New York (1969).
[b] Dimensionless.

Table 8. Comparison of the Thermodynamic Properties for the Cyclization of Diols (298°K)[a]

Reaction	ΔG (kcal mole^{-1})	ΔH (kcal mole^{-1})	ΔS (cal K^{-1})	K_p	K_c (mole liter^{-1})
 HO–CH₂–CH₂–OH ⟶ (epoxide) + H₂O	15.01	22.67	25.72	9.41×10^{-12}	1.59×10^{-14}
 HO–(CH₂)₃–OH ⟶ (oxetane) + H₂O	13.57	20.66	23.79	1.08×10^{-10}	1.83×10^{-13}
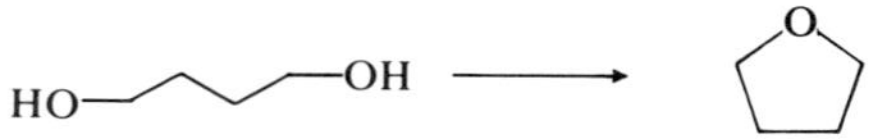 HO–(CH₂)₄–OH ⟶ (THF) + H₂O	−5.34	0.83	21.17	8.37×10^{3}	1.42×10^{1}
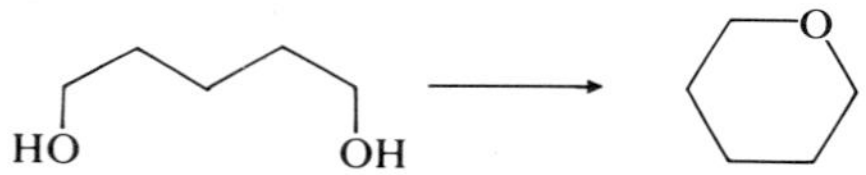 HO–(CH₂)₅–OH ⟶ (THP) + H₂O	−7.79	−3.69	13.75	5.28×10^{5}	8.93×10^{2}
CH₃–CH₂–OH + CH₃OH ⟶ CH₃CH₂OCH₃ + H₂O	−3.70	−5.33	−5.48	5.22×10^{2}	5.22×10^{2b}

[a] Data from D. R. Stull, E. F. Westrum, and G. C. Sinke, *The Chemical Thermodynamics of Organic Compounds*, John Wiley & Sons, Inc., New York (1969), and D. W. Scott, *J. Chem. Thermodyn.*, **2**, 833 (1970), or estimated using the additivity rules of S. W. Benson, F. R. Cruickshank, D. M. Golden, G. R. Haughen, H. E. O'Neal, A. S. Rodgers, R. Shaw, and R. Walsh, *Chem. Rev.*, **69**, 279 (1969).
[b] Dimensionless.

Table 9. Comparison of the Thermodynamic Properties for the Cyclization of Hydroxy Thiols (298° K)[a]

Reaction	$\Delta G°$ (kcal mole^{-1})	$\Delta H°$ (kcal mole^{-1})	$\Delta S°$ (cal °K^{-1})	K_p	K_c (mole liter^{-1})
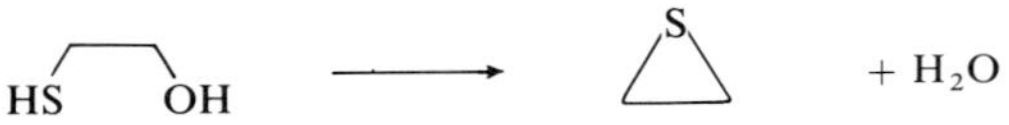 HS⌒OH → ▵S + H$_2$O	2.07	9.26	24.14	3.01×10^{-2}	5.09×10^{-5}
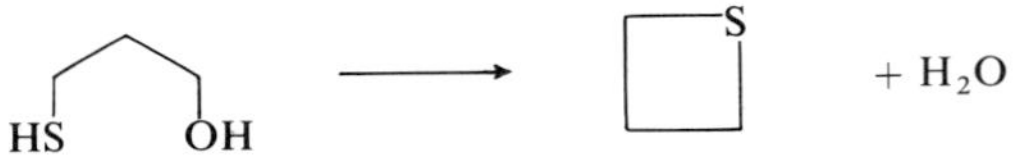 HS⌒⌒OH → ☐S + H$_2$O	2.62	9.17	21.98	1.19×10^{-2}	2.01×10^{-5}
 HS⌒⌒⌒OH → ⬠S + H$_2$O	−14.03	−8.57	18.33	2.02×10^{10}	3.42×10^7
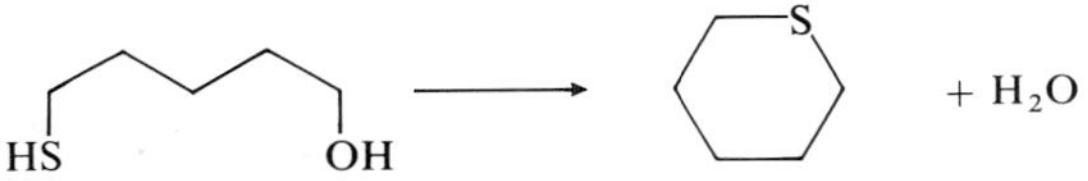 HS⌒⌒⌒⌒OH → ⬡S + H$_2$O	−14.30	−10.66	12.23	3.20×10^{10}	5.41×10^7
$CH_3CH_2OH + CH_3SH \rightarrow CH_3CH_2-S-CH_3 + H_2O$	−9.32	−10.44	−3.77	7.02×10^6	7.02×10^{6b}

[a] Data from D. R. Stull, E. F. Westrum, and G. C. Sinke, *The Chemical Thermodynamics of Organic Compounds*, John Wiley & Sons, Inc., New York (1969) or estimated using the additivity rules of S. W. Benson, F. R. Cruickshank, D. M. Golden, S. R. Haughen, H. E. O'Neal, A. S. Rodgers, R. Shaw, and R. Walsh, *Chem. Rev.*, **69**, 279 (1969).
[b] Dimensionless.

Table 10. Some Structural Changes that Favor Cyclization

	Lactonization[a]		Anhydride formation.[b]
	$k_{H^{-1}}$ (liter mole^{-1} min^{-1}) (25°C, 20% aqueous ethanol)	K_{eq}	$10^2 k_1$ (min^{-1})
(50) [structure with CO_2H and OH]	0.086	6.15	
(53) [structure with CO_2Ar and CO_2^-]			4.5
(51) [norbornane with CO_2H and CH_2OH]	7.23	2810	
(54) [structure with O, CO_2Ar, CO_2^-]			1030
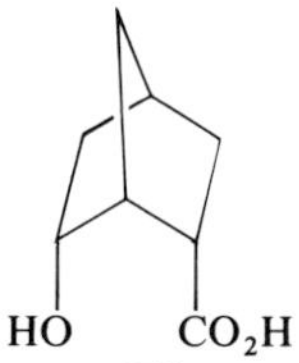 (52) [norbornane with HO and CO_2H]	1120	12,740	

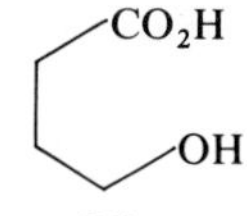

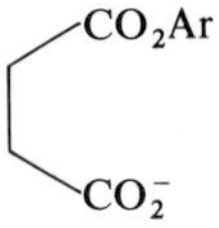

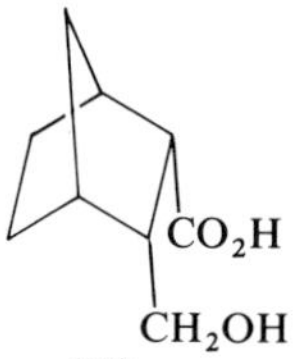

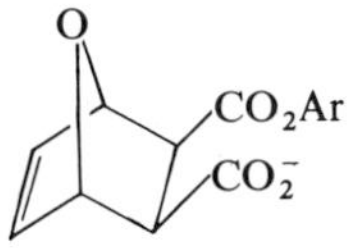

Ar = p-MeOC$_6$H$_4$

[a] D. R. Storm and D. E. Koshland, *J. Am. Chem. Soc.*, **94**, 5804 (1972).
[b] T. C. Bruice and U. K. Pandit, *J. Am. Chem. Soc.*, **82**, 5858 (1960).

probably arise mainly from these two factors plus a smaller entropy effect arising from the fact that cyclization of (50) involves loss of rotational freedom around four bonds whereas cyclization of (51) and (52) involves loss of rotational freedom around two bonds and one bond, respectively (see also Chapter 4, Section 4.2.4).

When the size of the ring that is being formed is varied there is of course a large variation in the anchimeric assistance. Moreover, the ring size associated with the fastest rate varies from one reaction to another. The degree of substitution of the chain also affects the rate, and generally the rate increases with increasing number of substituents. These effects have been most widely studied for reactions in which the neighboring group acts

2.3.1. Ring Size

The dependence of the rate of an intramolecular nucleophilic displacement on the size of the ring that is being formed is determined by the interplay of at least three quantities[94]: (1) the entropy loss on formation of the transition state, (2) the increase in strain energy on formation of the transition state, and (3) the electronic effects of the neighboring group and reaction group on one another.

2.3.1.1. Entropy Effects

Page has tabulated the standard entropy loss on cyclization of a series of terminal olefins (Table 11).[91] The simple expectation that these would

Table 11. Entropy Changes Accompanying Cyclization at $298°K$ $(cal°K^{-1}\ mole^{-1})$

Cyclization	$-\Delta S°$
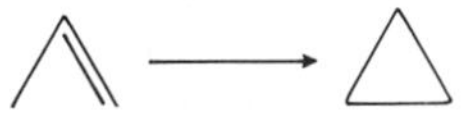 △ → △	7.0
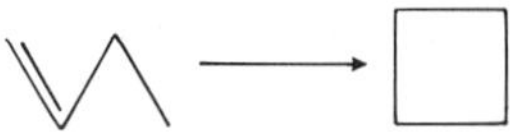 ⋁⋀ → □	10.30
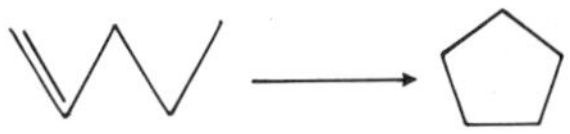 ⋁⋀⋁ → ⬠	13.10
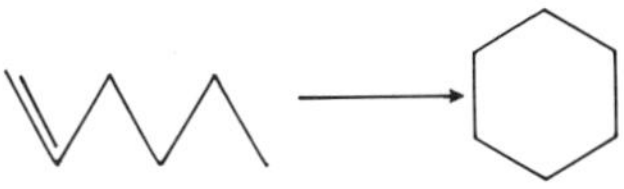 ⋁⋀⋀ → ⬡	21.01
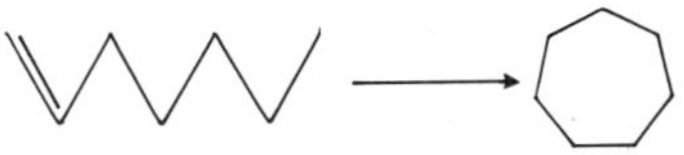 ⋁⋀⋀⋁ → ⬠(7)	19.60
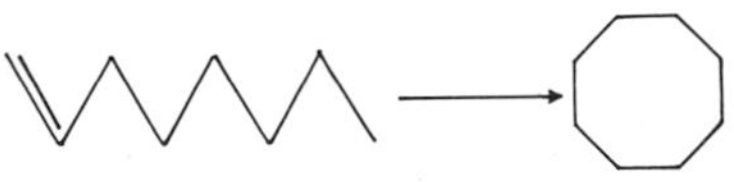 ⋁⋀⋀⋀ → ⬠(8)	18.79

Table 12. Strain Energy of Cycloalkanes in Saturated Heterocyclic Rings (kcal mole^{-1})[95]

Ring size	Alkane	Ether	Imine	Sulfide
3	27.43	22.78	26.87	19.78
4	26.04	25.51	—	19.64
5	6.05	5.63	5.80	1.97
6	−0.02	1.16	−0.15	−0.27

ether (3) − ether (5) = 21.65
imine (3) − imine (5) = 21.07
sulfide (3) − sulfide (5) = 17.81

increase with increasing length of the chain is not followed completely, presumably because the entropies of the rings do not vary in a regular manner with ring size. Thus low-frequency torsional motions of the methylene groups of the eight- and nine-membered rings probably compensate for the additional loss of internal rotational entropy compared to the six-membered ring. Nevertheless, on passing along the series from three- to six-membered rings the loss of entropy is in the order $6 > 5 > 4 > 3$, and hence if this were the only factor, the relative rates of ring closure would be $3 > 4 > 5 > 6$.

2.3.1.2. Ring Strain

Pell and Pilcher have tabulated the strain energies of cyclic alkanes, ethers, imines, and sulfides (Table 12), which all decrease with ring size in the order $3 > 4 > 5 > 6$.[95] Thus if the same fraction of the strain energy is always developed in the transition state, it would be expected that if this were the only controlling factor the rates of reactions that form rings of this type would be in the order $6 > 5 > 4 > 3$.

2.3.1.3. The Electronic Effects of the Neighboring and Reaction Groups on One Another

Since in this class of reaction most of the leaving and neighboring groups are electron withdrawing, inductively these effects act to decrease the nucleophilicity of the neighboring group and to decrease the tendency of the leaving group to depart. They therefore tend to decrease the rate of ring formation, the decrease being the greatest for three-membered rings, when the neighboring and leaving groups are attached to adjacent carbon atoms, and decreasing with increasing ring size.

A comparison of many different types of reactions (Table 13) indicates that the balance of these factors may result in formation of three-, five-, or six-membered rings being most rapid, depending on the type of reaction and the substituents present.

The first generalization that can be made is that cyclizations of unsubstituted polymethylene chains involving nucleophilic attack on saturated carbon by an oxygen (Table 13, reactions 1–4), nitrogen (Table 13, reactions 6–8), or chlorine (Table 13, reaction 9) neighboring group occur in the order 5 > 6 > 3. If the element of the neighboring group is changed to sulfur and participation by an arylthio group is considered, the relative rate of formation of three-membered rings increases markedly (Table 13, reactions 10–14), and in all the examples known three-membered rings are formed more rapidly than five-membered ones. This can be correlated with the relative strain energies of cyclic thioethers compared to cyclic imines and cyclic ethers. Thus the difference in strain energies between three- and five-membered cyclic sulfides is 3–4 kcal mole^{-1} less than between the corresponding oxygen and nitrogen compounds (Table 12). It is not clear if this order for participation by thioether groups refers only to arylthio groups or whether it is different for alkylthio groups.[123] That this is possible is suggested by the observation of Bennett et al.[104] that 4-chlorobutyl ethyl sulfide is very reactive, and the report of Böhme and Sell[124] that the rate constant for solvolysis of 2-chloroethyl ethyl sulfide is 0.17 min^{-1} at 100°C in aqueous dioxane.[125] Hence it is possible that with the ethylthio group five-membered rings are formed faster than three-membered ones, but there is no exact kinetic comparison. Three-membered rings are formed faster than five-membered ones with carbon (Table 13, reactions 26 and 27) and carbanion (Table 13, reactions 23–25) neighboring groups. As pointed out by Knipe and Stirling,[112,113] this may arise from the conjugating ability of the cyclopropane ring. In reactions 23–29 another substituent is present that could conjugate with the developing cyclopropane ring in the transition state, hence leading to a lowering in the energy of the transition state for formation of the three-membered ring relative to that for formation of the five-membered ring. It is possible that this conjugative stabilization outweighs the increased strain energy for formation of a three-membered ring compared to a five-membered ring in the transition state. However, it probably does not do so in the final state. Generally three-membered rings are thermodynamically less stable than five-membered ones even if they are formed faster (cf. p. 45), so it is possible that in the final state the extra strain energy of the three-membered ring outweighs any conjugative stabili-

Table 13. Rates of Ring Closure as a Function of Ring Size

Reaction					Relative rates[a]		
No.	Type	Starting material	Conditions (°C)	$n = 3$	5	6	Ref.
1	MeO-n	MeO$(CH_2)_{n-1}$·OBs	HOAc (25°)	0.00043	1.0	0.187[b]	96
2	MeO-n	o-MeO·C_6H_4·CMe$_2$$(CH_2)_{n-4}$·OTs	HOAc (75°)	—	1.0	1.2	96
3	HO-n	HO·$(CH_2)_{n-1}$·Cl	H_2O (70.5°)	0.0010	1.0	0.41	97, 98
4	HO-n	HO·$(CH_2)_{n-1}$·Cl	50% aq. acetone (100°)	0.00135	1.0	0.0354	99
5	O$^-$-n	$^-$O·$(CH_2)_{n-1}$·Cl	H_2O, $^-$OH (18°)	0.2	1.0	0.001	98, 100
6	H_2N-n	H_2N·$(CH_2)_{n-1}$·Br	H_2O (25°)	0.0012	1.0	0.017[c]	101
7	ArHN-n	PhNH$(CH_2)_{n-1}$·Cl	50% aq. dioxane (25°)	0.00120	1.0	—	102
8	ArHN-n	PhNH$(CH_2)_{n-1}$·Br	60% EtOH–K_2CO_3 (25°)	0.105	—	1.0	102
9	Cl-n	MeCHCl·$(CH_2)_{n-2}$ONs	TFA (30°)	ca. 0.002	1.0	0.028[d]	103
10	ArS-n	PhS·$(CH_2)_{n-1}$·Cl	50% aq. acetone (80°)	—	1.0	0.0103	104
11	ArS-n	PhS·$(CH_2)_{n-1}$·Cl	20% aq. dioxane (100°)	5.0	1.0	—	104
12	ArS-n	PhS·$(CH_2)_{n-1}$·Cl	MeOH (?)	1.15	1.0	0.033	105
13	ArS-n	p-Me·C_6H_4S$(CH_2)_{n-2}$·CH(Ph)Cl	80% aq. EtOH (30°)	114	1.0	—	106
14	ArS-n	p-Me·C_6H_4S$(CH_2)_{n-1}$·Cl	80% aq. EtOH (55°)	1.36	1.0	—	106
15	CO_2^--n	^-O_2C·$(CH_2)_{n-2}$·Cl	H_2O, $^-$OH (37.5°)	—	1.0	0.9	99
16	CO_2^--n	^-O_2C·$(CH_2)_{n-3}$CHBrMe	H_2O, $^-$OH (25°)	0.00008	1.0	—	100
17	O-n	PhCO$(CH_2)_{n-2}$Cl	80% aq. EtOH–AgClO$_4$ (56.2°)	0.0017	1.0	0.028	107
18	O-n	PhCO$(CH_2)_{n-2}$OBs	TFA (30°)	—	1.0	0.0255	108
19	O-n	PhSO$(CH_2)_{n-2}$CMe$_2$Cl	80% aq. sulfolane (35°)	—	1.0	3.6	109

Some Factors that Influence Anchimeric Assistance

20	O-n	PhNHCO$_2$(CH$_2$)$_{n-3}$Br	80% aq. EtOH (75°)	—	1.0	110
21	N$^-$-n	PhNHCO$_2$(CH$_2$)$_{n-3}$Br	EtONa–EtOH (0°)	—	1.0	110
22	N$^-$-n	p-Me·C$_6$H$_4$SO$_2$NH(CH$_2$)$_{n-1}$Cl	EtONa–EtOH (25°)	22.7	—	102, 111
23	R$_3$C$^-$-n	p-Me·C$_6$H$_4$SO$_2$(CH$_2$)$_n$Br	t-BuOK–t-BuOH (55°)	100	1.0	112
24	R$_3$C$^-$-n	(EtOCO)$_2$CH·(CH$_2$)$_{n-1}$Cl	t-BuOK–t-BuOH (25°)	>100	1.0	113
25	Ar$_1^-$-n^e	$^-$O·C$_6$H$_4$·(CH$_2$)$_{n-1}$Br	MeOH (25°)	1100	1.0	114
26	Ar$_1$-n	p-MeO·C$_6$H$_4$·(CH$_2$)$_{n-1}$OBs	HOAc (75°)	147	1.0	115
27	Ar$_1$-n	2,4-(MeO)$_2$·C$_6$H$_3$·(CH$_2$)$_{n-1}$OBs	HOAc (75°)	262	1.0	115
28	Ar$_2$-n	3,5-(MeO)$_2$·C$_6$H$_3$·(CH$_2$)$_{n-2}$OBs	HCO$_2$H (75°)	—	1.0	116
29	Ar$_2$-n	m-MeO·C$_6$H$_4$·(CH$_2$)$_{n-2}$OBs	HCO$_2$H	—	1.0	116
30	HO-n	HO·(CH$_2$)$_{n-2}$CO$_2$H	H$^+$, H$_2$O (25°)	—	1.0	117
31	HO-n	HO·(CH$_2$)$_{n-2}$CO·NH$_2$	H$^+$, H$_2$O (20°)	—	1.0	118
32	HO-n	HO·(CH$_2$)$_{n-2}$CO$_2$Ph	H$^+$, H$_2$O (29.8°)	—	1.0	119
33	HO-n	o-HO·C$_6$H$_4$·(CH$_2$)$_{n-4}$CONH$_2$	H$^+$, H$_2$O (90°)	—	1.0	120
34	HO-n	o-HO·C$_6$H$_4$·(CH$_2$)$_{n-4}$CONH$_2$	H$_2$O (90°)	—	1.0	120
35	O$^-$-n	HO·(CH$_2$)$_{n-2}$CO$_2$Ph	H$_2$O, pH 7.47 (29.8°)	—	1.0	119
36	O$^-$-n	HO·(CH$_3$)·CH(CH$_2$)$_{n-3}$CONH$_2$	H$_2$O, NaOH (50°)	—	1.0	120
37	O$^-$-n	o-HO·C$_6$H$_4$·(CH$_2$)$_{n-4}$CO$_2$β·C$_{10}$H$_7$	20% aq. dioxane, pH = 4.79 (25°)	—	1.0	119
38	R$_2$N-n	Me$_2$N·(CH$_2$)$_{n-2}$CO$_2$Ph	H$_2$O (25°)	—	1.0	121
39	R$_2$N-n	H$_2$N·(CH$_2$)$_{n-3}$·CH(NHTs)·CO$_2$Me	H$_2$O, $^-$OH (25°)	—	1.0	122

Rate-ratio values (column before refs): 0.0424, 0.0072, 1.0, 0, —, —, —, 0.014, 100, 17, 6.4, 2.6, 2.13, 0.24, 0.79, 0.055, 1.39, 0.22, 0.389, 3.03

^{a}The slower reactions may follow alternative mechanisms; thus the cyclization rates may be even less than the values given here.
^{b}For $n = 7$, rel. $k = 0.0018$.
^{c}For $n = 7$, rel. $k = 0.00003$.
^{d}Calculated on the basis of 90% k_Δ reaction. For $n = 7$, rel. $k = 0.001$.
^{e}This terminology is described in Chapter 1.

zation. This is another indication that the strain developed in the transition states for cyclizations is less than in the final states (see p. 44).

When unsaturation is introduced into the chain the rate of formation of the six-membered ring relative to the five-membered ring increases (cf. Table 13, reactions 2, 28, and 29). The introduction of unsaturation increases the rigidity of the chain, and hence the entropy loss on ring closure is reduced, but ring formation is now probably accompanied by a greater increase in strain. These two factors probably favor formation of six-membered rings relative to five-membered ones. The relative rate of formation of six-membered rings is also increased when the hybridization of the carbon at which the nucleophilic substitution takes place is changed from sp^3 to sp^2 (Table 13, reactions 30–39).

Rings with more than six atoms are formed very slowly. Their formation was studied qualitatively by Ziegler[126,127] and Ruzicka,[94] and more recently Illuminati and his co-workers have published kinetic measurements for the formation of lactones from ω-bromoalkanoate ions[128] and the cyclization of O-ω-bromoalkoxyphenoxides.[129] The former reactions show a rate minimum at ring sizes 7 and 8, and the rates of the latter decrease as the ring size is increased from 6 to 10. This difference in behavior is presumably related to the relative strains of these different types of ring.

When the leaving group is part of a three-membered ring the relative rates of formation of rings of different sizes may be changed from what it is in more simple situations, and four-membered ring formation becomes particularly favorable. Danishevsky and his co-workers have studied the *bis*-methoxycarbonyl carbanion in a three-membered ring as leaving group and have defined two modes of ring opening–ring formation: the spiro and the fused mode.[130] There is a strong preference for the spiro mode,

Spiro mode   Fused mode

which is readily rationalized as arising from the greater ease of achievement of a collinear transition state in this mode. This this is so is seen very easily from a consideration of Dreiding models. Thus, compound (**55**) forms preferentially a five-membered ring, and (**56**) a six-membered ring. These results are unexceptional in that five-membered rings are normally formed faster than six-membered ones, and six-membered ones faster than seven-

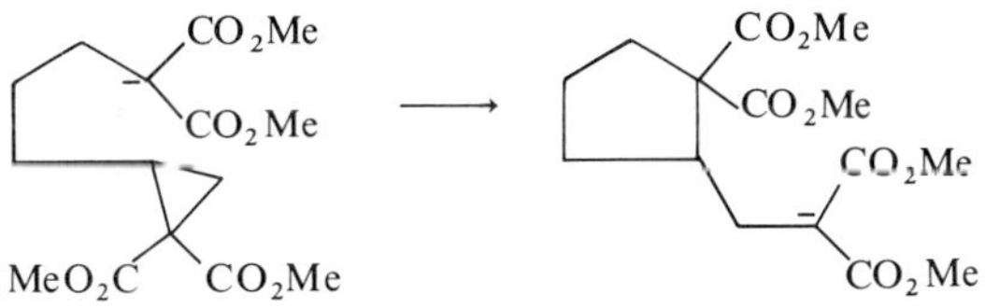

(55) →

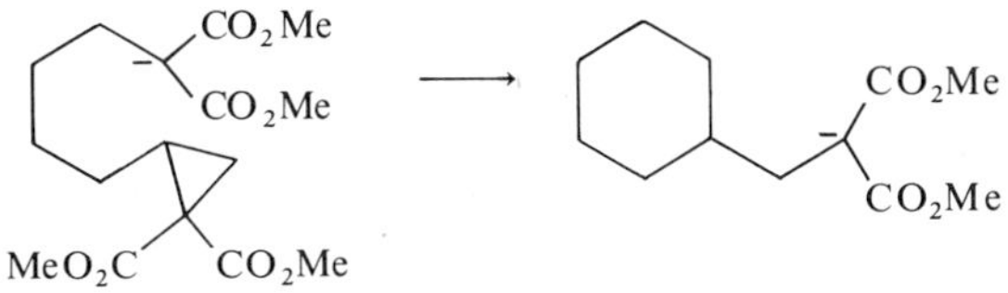

(56) →

membered ones. However, what is at first sight surprising is that the four-membered ring is formed faster than the five-membered one from compound (**58**). The former (**57**) is formed reversibly, and so after a short reaction

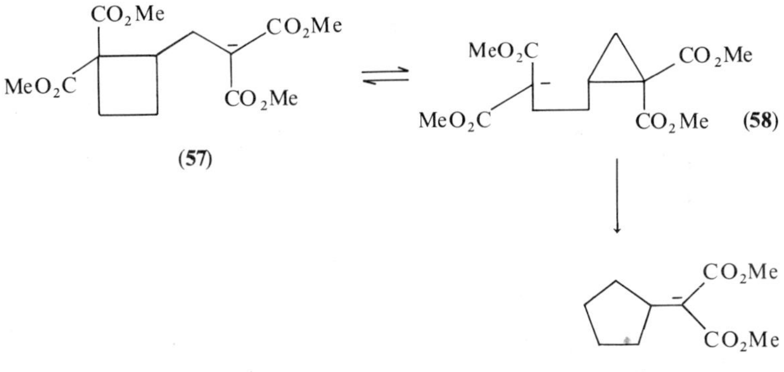

(57) ⇌ (58)

↓

(59)

time it is the major product, whereas after a long reaction time the five-membered ring compound (**59**) predominates. It was estimated that the ratio of the rate constants for formation of four- and five-membered rings from (**58**) was ca. 2.5:1.[130] It seems that here the requirement for a linear transition state[131] is a more important factor in determining the rate than the ring strain of the product.

With the next lower member of this series with which the spiro mode could lead to a three-membered ring and the fused mode to a four-membered

(3)

(4)

(5)

ring, the former occurs exclusively. The esters **(60)** and **(61)** can be equilibrated without formation of any of the cyclobutane isomer.[132]

Similar behavior is found in the opening of epoxide rings by the neighboring carbanion [cf. Eq. (3)–(5)].[133] Although five-membered rings are formed in preference to four-membered ones with the *trans*-epoxide **(63)**, the *cis*-epoxide **(62)** yields only cyclobutane products.[134] When the epoxide is fused to a six-membered ring opening occurs only in the fused mode, and the compound that forms the six-membered ring **(64)** reacts much

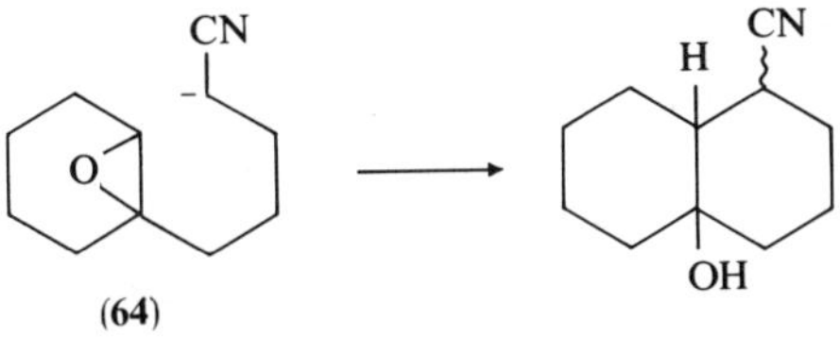

faster than the one that forms the five-membered ring **(65)**.[133] The slowness of opening in the spiro mode may arise from an unfavorable steric interaction between the entering carbanion and the hydrogen on C_6, which is *trans* to the epoxide group. Examination of models makes it very clear that achievement of a collinear transition state in the fused mode is much easier for formation of a six- than a five-membered ring. Similar behavior is found in epoxide migrations.[135]

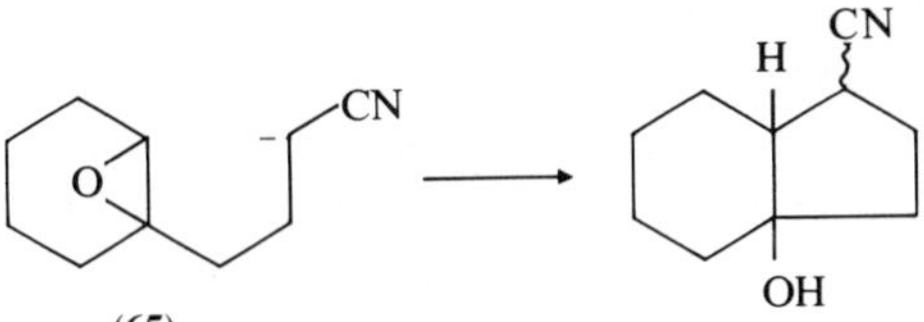

The spiro mode is also favored in the ring opening of cyclic halonium ions. Thus the kinetically controlled products of the reactions of the 3-butenoate ion with bromine and iodine under carefully controlled conditions are β-lactones, which are presumably formed from cyclic bromonium and iodonium ions [Eq. (6)].[136]

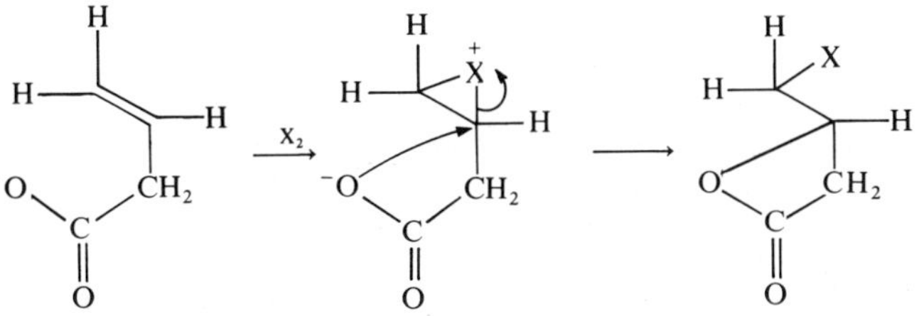

(6)

There is some evidence that an eight-membered ring is the most favorable ring size for reactions that involve an intramolecular proton transfer.[137,138,139] Thus intramolecular deuteration of the ω-dimethylaminoalkylimines of acetone-d_6 (**66**) occurs most rapidly when $n = 8$.[137] In the

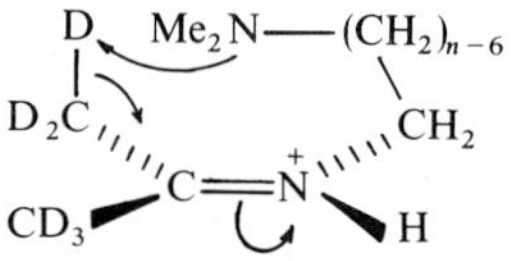

(**66**)

initial state of these reactions four of the atoms that are going to be part of the cyclic transition state are coplanar, and three of these remain coplanar in the product. This, and the requirement that the bond to the deuteron that is transferred lie in a perpendicular plane, places certain steric restrictions on the transition state such that a linear N—D—C bond and complete staggering of the methylene groups can be achieved only if the ring is nine membered. In fact, maximum velocity is obtained when it is eight membered. Thus presumably the additional loss of internal rotational entropy on having a nine-membered ring is less favorable than having a nonlinear N—D—C bond or some eclipsing in the cyclic transition state with eight members. However, in reactions where the developing double bond is not part of the ring, (**67**)[140] and (**68**),[141] a six-membered ring appears to be most highly favored.

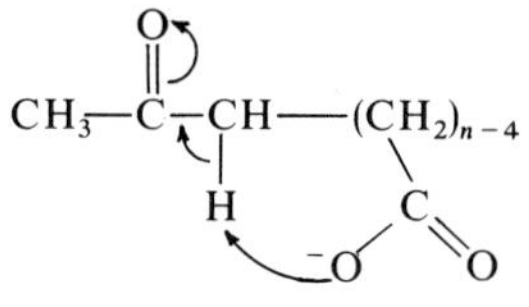

(**67**) 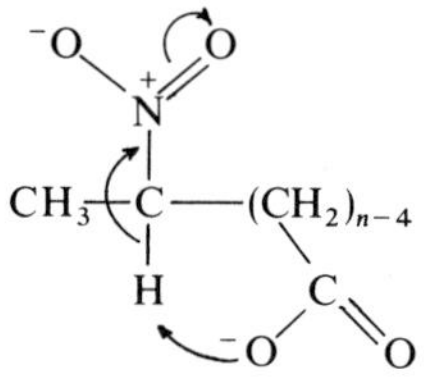   (**68**)

2.3.2. The gem-Dialkyl Effect[142]

It has been known for a long time that when *gem*-dialkyl substituents are introduced into a polymethylene chain the equilibrium and rate constants for cyclization are frequently increased. There is not a lot of data available for reactions in the gas phase, but the effect of *gem*-dimethyl substituents on the cyclization of cyclopentane and cyclohexane are known (Table 14).

Table 14. *Effect of Dimethyl Substitution on the Equilibrium Constants and Thermodynamic Properties for the Cyclization of Cyclopentane and Cyclohexane (298°K)*

Reaction	$\Delta G°$ (kcal mole^{-1})	$\Delta H°$ (kcal mole^{-1})	$\Delta S°$ (cal °K^{-1})	K_p
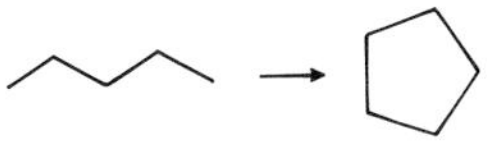 pentane → cyclopentane + H$_2$	11.23	16.54	17.81	5.63×10^{-9}
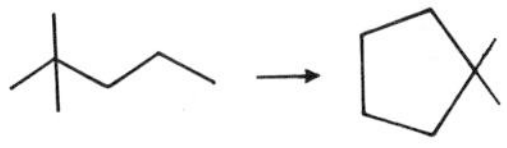 2,2-dimethyl → + H$_2$	9.31	16.22	23.18	1.45×10^{-7}
 3,3-dimethyl → + H$_2$	8.70	15.12	21.55	4.07×10^{-7}
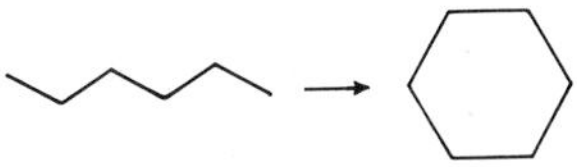 hexane → cyclohexane + H$_2$	7.65	10.53	9.66	2.40×10^{-7}
 dimethyl → + H$_2$	5.86	10.45	15.39	4.96×10^{-5}
 dimethyl → + H$_2$	5.23	9.35	13.85	1.44×10^{-4}

^aData from D. R. Stull, E. F. Westman, and G. C. Sinke, *The Chemical Thermodynamics of Organic Compounds*, John Wiley & Sons Inc., New York (1969).

The values of K_p are increased by 10^1–10^2, and this increase is associated with more favorable values of $\Delta H°$ and $\Delta S°$. There are many examples of this effect for reactions in solution. The rate constants for anhydride formation from substituted succinic acids, succinanilic acids, and the anions of *p*-bromophenyl hydrogen succinates and the equilibrium constants for anhydride formation from the acids are given in Table 15. The increase in the rate and equilibrium constants with increasing methyl substitution is not completely regular, which no doubt signifies that the effects are complex, with more than one factor being important. Nevertheless, the general trend is that methyl substituents increase the rate and equilibrium constants. The effect is greatest on the equilibrium constant for anhydride formation from the acid and for the one compound for which datum is available and is relatively small on anhydride formation from the monoanion of *p*-bromophenyl hydrogen succinate. Probably the rate-limiting step for formation

Table 15. *Effect of Methyl Substitution on the Rate and Equilibrium Constants for the Formation of Alkyl-1-Substituted Succinic Anhydrides from the Corresponding Acids and Mono-N-Phenylamides*

	$10^7 k_1{}^a$	$10^6 K^b$	$10^5 k^c$	$10^2 k^d$
Unsubstituted	1.5	7	2.1	1.7
Methyl	9.8	42	5.55	—
meso-2,3-Dimethyl	—	—	11.8	—
d,l-2,3-Dimethyl	53	250	49.2	—
2,2-Dimethyl	18	100	24.3	5.3
Trimethyl	—	—	273	—
Tetramethyl	3000	170,000	805	—

[a] Rate constant (sec^{-1}) for formation of anhydride from acid at 60°C; L. Eberson and H. Welinder, *J. Am. Chem. Soc.*, **93**, 5821 (1971).
[b] Equilibrium constant for formation of anhydride from the acid at 60°C, idem.
[c] Rate constant (sec^{-1}) for the formation of the anhydride from the mono-N-phenylamide at 60°C; T. Higuchi, L. Eberson, and J. D. McRae, *J. Am. Chem. Soc.*, **89**, 3001 (1967).
[d] Rate constant (sec^{-1}) for the formation of the anhydride from the anion of mono-p-bromophenyl esters in 50% (v/v) dioxane–water at 30°C; T. C. Bruice and U. K. Pandit, *J. Am. Chem. Soc.*, **82**, 5858 (1960).

of the anhydride from the acids and anilic acids is decomposition of the tetrahedral intermediate [Eq. (7)], whereas in the cyclization of the esters,

$$\text{(7)}$$

X = OH or NHAr

$$\text{(8)}$$

where the leaving group is much better, it is attack of carboxylate ion [Eq. (8)]. Therefore in the former reactions the ring is fully closed in the rate-limiting transition state, whereas in the latter it is only partially closed, and so any steric factors that favor ring closure should be less important with the esters than with the anilides and acids.

When the size of the alkyl group is increased there is a tendency for the rate and equilibrium constants to increase, but the increase is not completely regular (cf. Table 16) and is greater for the equilibrium than for the rate constants. Similar effects are found on formation of glutaric anhydride

Table 16. Effect of Alkyl Substituents on the Rate and Equilibrium Constants for the Formation of Alkyl-Substituted Succinic Anhydrides at 60°C[a]

	$10^4 k\ \text{sec}^{-1}$	K_{eq}
Tetramethyl	3.00	0.17
dl-2,3-Diethyl-2,3-dimethyl	2.58	3.4
meso-2,3-Diethyl-2,3-dimethyl	1.67	1.0
Tetraethyl	23.5–24.9	10.0
dl-2,3-Di-t-butyl	—	6.0[b]

[a] L. Eberson and H. Welinder, J. Am. Chem. Soc., 93, 5821 (1971).
[b] At 20°C.

from the anion of p-bromophenyl monoglutarate (Table 17). The gem-dialkyl effect has also been shown to operate in lactonization reactions (Tables 18 and 20). Again the effects on the equilibrium constants are greater than on the rate constants.

The effect of gem-dialkyl substituents on the rates of formation of fully saturated three- and five-membered rings in reactions that involve substitution at saturated carbon are usually greater than in anhydride and lactone formation (see Tables 19 and 21), but in the one example where the effect and 19). Again the effects on the equilibrium constants are greater than on was reported. Insufficient data are available to make any generalizations, but it seems that the effect on the rate of introducing a gem-dimethyl group, for example, is not a constant factor. Thus whereas 4-bromo-2,2-dimethylbutylamine undergoes cyclization 158 times faster than the parent compound, 4-bromo-1,1-dimethylbutylamine reacts only 2.19 times as fast and 4-bromo-3,3-dimethylbutylamine reacts more slowly. The latter compound has a neopentyl-type structure, and the low rate may arise from an unfavorable steric interaction in the transition state. A similar factor may be responsible for the relatively small increase in the rate of cyclization of 3-chloro-2,2-dimethylpropyl dimethylamine compared to the parent compound (Table 19). On the other hand, all the gem-dimethyl compounds listed in Table 21 have neopentyl-type structures, but they cyclize to three-membered rings 40–937 times as fast as the unsubstituted compounds.

Another interesting manifestation of the gem-dialkyl effect appears in the series of epoxide migrations studied by Payne[143] [see Eqs. (9)–(13)] in which the most highly substituted epoxides were found to be most stable at equilibrium. The epoxides were equilibrated in a deficit 0.5 M sodium

Table 17. *Effect of Alkyl Substituents on the Rate of Formation of Glutaric Anhydrides from the Anions of 3-Substituted and 3,3-Disubstituted p-Bromophenyl Monoglutarates*[a]

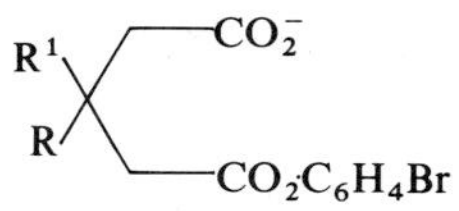

$$\text{R}^1\text{R}\text{C}(\text{CO}_2^-)(\text{CO}_2\text{C}_6\text{H}_4\text{Br})$$

	$10^5 k (\text{sec}^{-1})$		$10^5 k (\text{sec}^{-1})$		$10^5 k (\text{sec}^{-1})$		$10^5 k (\text{sec}^{-1})$
$R = R^1 = H$	5.65	$R = H, R^1 = C_6H_5$	23.2	$R = R^1 = CH_3$	131	$R = CH_3, R^1 = i-C_3H_7$	1243
$R = H, R^1 = CH_3$	27.0	$R = CH_3, R^1 = CH_6H_5$	182	$R = R^1 = C_2H_5$	1010	$R = CH_3, R^1 = t-C_4H_9$	5983
$R = H, R^1 = C_3H_7$	73.0	$R = C_2H_5, R^1 = C_6H_5$	1213	$R = R^1 = C_3H_7$	1008		
$R = H, R^1 = i-C_3H_7$	189.0	$R = C_3H_7, R^1 = C_6H_5$	1130	$R = R^1 = C_6H_5$	1528		

[a] T. C. Bruice and W. C. Bradbury, *J. Am. Chem. Soc.*, **87**, 4846 (1965).

Table 18. *Effect of Methyl Substituents on the Equilibrium and Rate Constants for Formation of γ-Butyrolactone at 25°C*

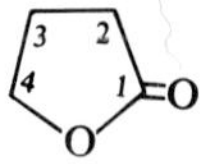

	Unsubstituted	2-Me	3-Me	4-Me	2,2-Me$_2$	3,3-Me$_2$	4,4-Me$_2$
K^a	2.68	20.7		13.3			545
k (min^{-1})^a	0.0377	0.2		0.010			0.465
K^b	2.88	4.08	20.8	12.8	>100	>100	35.7
k^b (liter mole^{-1} min^{-1})^b	0.0634			0.202			

^{a}In 1 M, nitric acid; H. Sebelius, Inaugural dissertation, Lund (1927), quoted by W. Hückel, *Theoretical Principles of Organic Chemistry* (trans. W. H. Rathmann), Elsevier Publishing Company, Amsterdam (1958), p. 895.
^{b}In 0.025 M hydrochloric acid; O. H. Wheeler and E. E. Granell de Rodriguez, *J. Org. Chem.*, **29**, 1227 (1964).

Table 19. *Effect of gem-Dialkyl Groups on the Cyclization of Four- and Five-Membered Rings*

	Formation of four-membered rings in 80% aq. ethanol at 56° 10$^5 k$(sec^{-1})^a		Formation of five-membered rings in aq. acetate buffer at 30°C, rel. rates^b
Me$_2$N · CH$_2$CH$_2$CH$_2$Cl	10.5	H$_2$N · CH$_2$CH$_2$CH$_2$CH$_2$Br	1
Me$_2$N · CH$_2$CMe$_2$ · CH$_2$Cl	98.0	H$_2$N · CH$_2$CMe$_2$CH$_2$CH$_2$Br	158
		H$_2$H · CH$_2$CEt$_2$ · CH$_2$CH$_2$Br	594
		H$_2$N · CH$_2$CPr$_2^i$ · CH$_2$CH$_2$Br	9190
		H$_2$N · CH$_2$CPh$_2$ · CH$_2$CH$_2$Br	5250
		H$_2$N · CMe$_2$CH$_2$CH$_2$CH$_2$Br	2.19
		H$_2$N · CH$_2$CH$_2$CMe$_2$CH$_2$Br	0.158

^{a}C. A. Grob and F. A. Jenny, *Tetrahedron Lett.*, No. 23, 25 (1960).
^{b}R. F. Brown and N. M. van Gulick, *J. Org. Chem.*, **21**, 1046 (1956).

Table 20. *Effect of Methyl Substituents on the Equilibrium and Rate Constants for the Formation of δ-Valerolactone at 25°C*

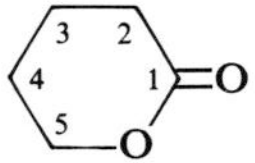

	Unsubstituted	2-Me	3-Me	5-Me	3,3-Me$_2$	5,5-Me$_2$	3-Me–5,5-Me$_2$	4-Me–5,5-Me$_2$
K^a	0.099	0.198		0.269		0.335	21.2	3
$k(\min^{-1})^a$	0.24	0.36		0.56		0.052	3.02	0.399
K^b	0.0613		1.08	0.379	11.1	0.433		
k(liter mole^{-1} min^{-1})^b	0.133		2.08	0.439	8.16	0.0541		

[a] In 1 M nitric acid: H. Sebelius, Inaugural dissertation, Lund (1927), quoted by W. Hückel, *Theoretical Principles of Organic Chemistry* (trans. W. H. Rathmann), Elsevier Publishing Company, Amsterdam (1958), p. 895.
[b] In 0.02 M hydrochloric acid: O. H. Wheeler and E. E. Granell de Rodriguez, *J. Org. Chem.*, **29**, 1227 (1964).

Table 21. Effect of Methyl Substituents on the Formation of Three-Membered Rings

Epoxide formation in aq. alkali at 18°C

	$10^3 k_2$ (liter mole^{-1} sec^{-1})[a]	Estimated k_1 (sec^{-1}) for ring closure of the anion[b]
$Cl \cdot CH_2 \cdot CH_2 \cdot OH$	5	5
$Cl \cdot CH_2 \cdot CHMe \cdot OH$	110	130
$Cl \cdot CH_2 \cdot CMe_2 \cdot OH$	1300	2500

Ethylene imine formation

	25°C, $10^6 k$ (sec^{-1})[c]	75°C, $10^6 k$ (sec^{-1})[d]
$Cl \cdot CH_2 CH_2 \cdot NH_2$	8	
$Cl \cdot CH_2 \cdot CHMe \cdot NH_2$	250	
$Cl \cdot CH_2 \cdot CMe_2 \cdot NH_2$	7500	
$^-O_3SO \cdot CH_2 \cdot CH_2 \cdot NH_2$		8.0
$^-O_3SO \cdot CH_2 \cdot CHMe \cdot NH_2$		51.7
$^-O_3SO \cdot CH_2 \cdot CMe_2 \cdot NH_2$		325

[a] H. Nilsson and L. Smith, Z. Phys. Chem. Abt. A, **166**, 136 (1933).
[b] S. Winstein and E. Grunwald, J. Am. Chem. Soc., **70**, 829 (1948).
[c] H. Freundlich and G. Salomon, Z. Phys. Chem. Abt. A, **166**, 179 (1933).
[d] C. S. Dewey and R. A. Bafford, J. Org. Chem., **32**, 3108 (1967).

Table 22. Rate and Equilibrium Constants for the Lactonization of 3-(2-Hydroxylphenyl) propionic Acids at 30°C in 20% Aqueous Dioxane[145]

$k_{H_3O^+}$ (mole^{-1} sec^{-1})	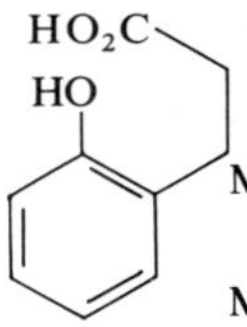 5.9×10^{-6}	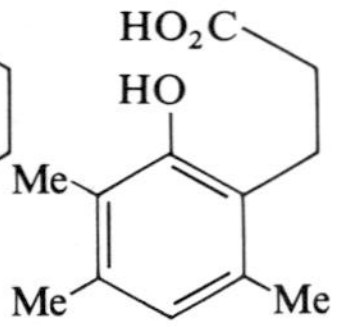 4.0×10^{-5}	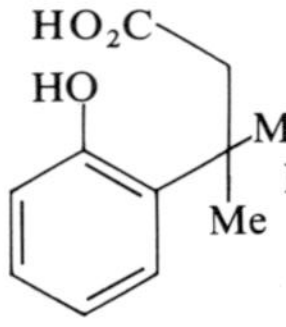 2.62×10^{-2}	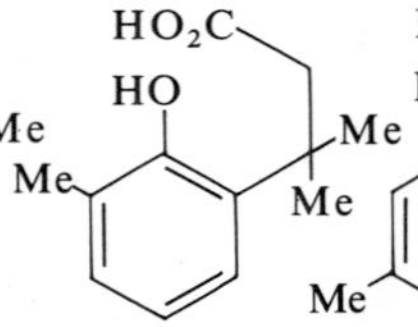 9.85×10^{-2}	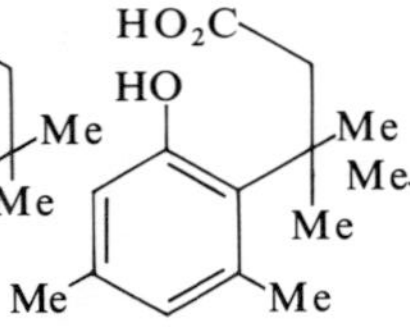 5×10^5	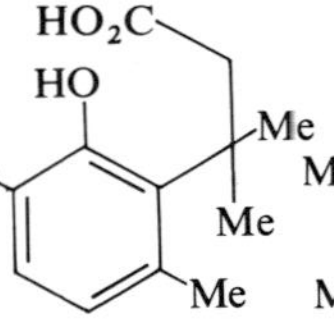 1.5×10^6	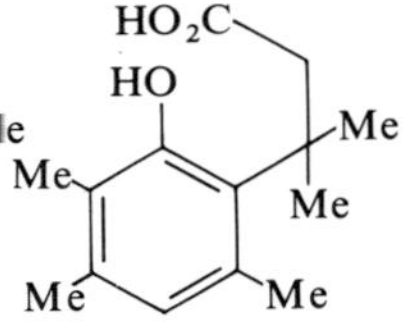 2×10^6
K_{eq}	0.0373	0.621	25.67	—	—	—	>99

hydroxide, and so the equilibria were probably set up between mixtures of un-ionized and ionized epoxides.

$$H_2C\text{—}CHCMe_2\text{(OH)} \rightleftharpoons HOCH_2CH\text{—}CMe_2\text{(O)} \quad (9)$$
(8%)   (92%)

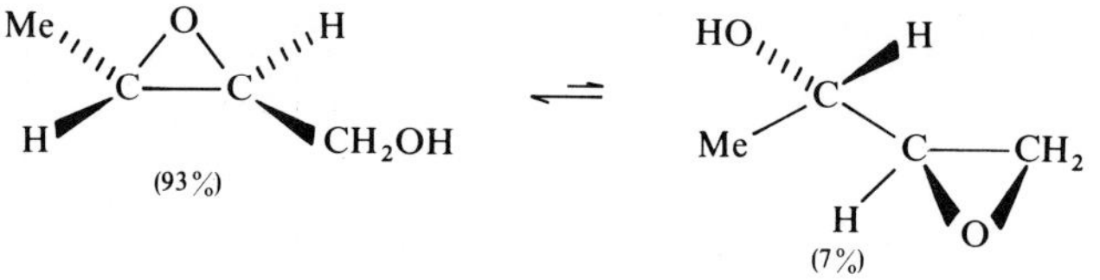

(10) (93%) ⇌ (7%)

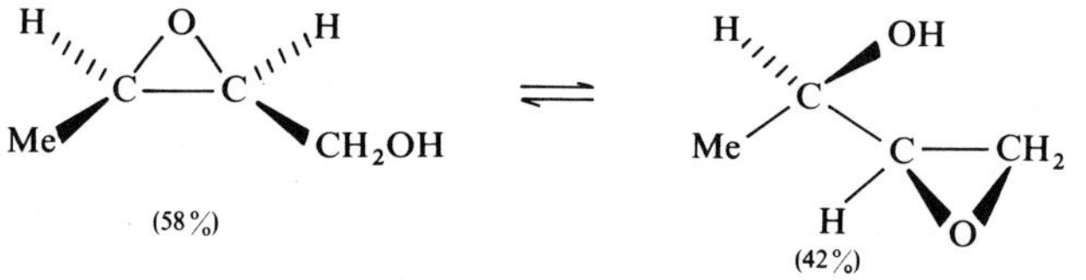

(11) (58%) ⇌ (42%)

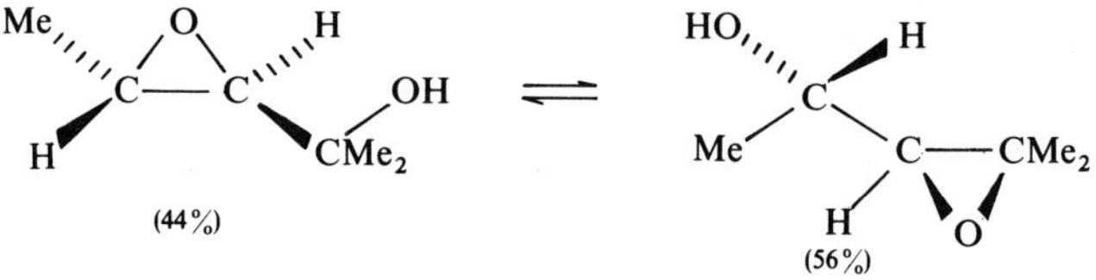

(12) (44%) ⇌ (56%)

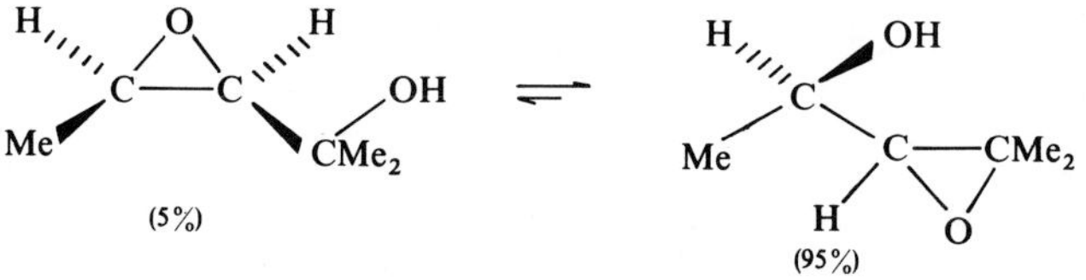

(13) (5%) ⇌ (95%)

Much larger enhancements of rate and equilibrium constants are found on *gem*-dimethyl substitution in more sterically restricted situations.[144,145] Thus the hydronium-ion-catalyzed lactonization of acid (**70**) occurs 1.1×10^4 faster than that of acid (**69**), and it appears that both react with acyl oxygen fission.[144] The most striking examples, however, are those

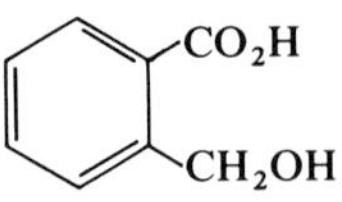

(**69**) o-HOCH$_2$-C$_6$H$_4$-CO$_2$H

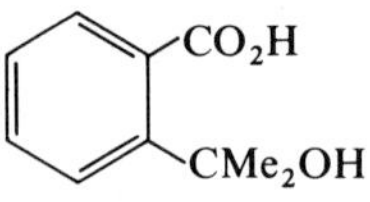

(**70**) o-HOCMe$_2$-C$_6$H$_4$-CO$_2$H

Table 23. Rate Constants for Cyclization of the Anions of 3-(2-Hydroxyphenyl)propyl Mesylates at 30°C in Aqueous Solution[146]

k (sec^{-1})	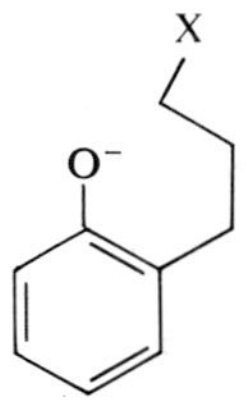 0.237	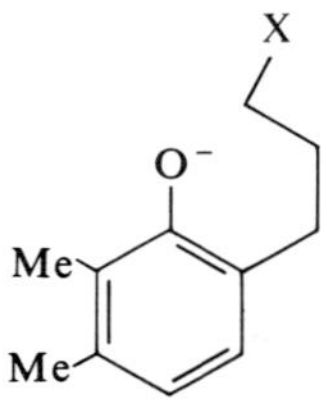 0.840	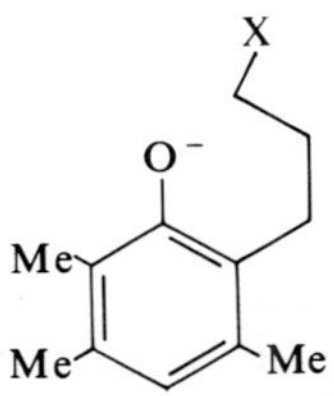 1.65	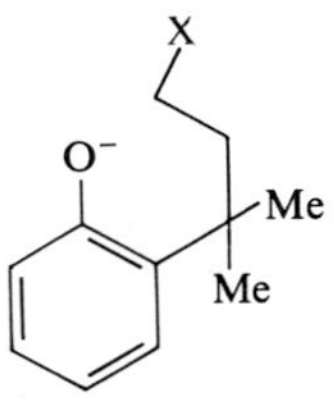 736	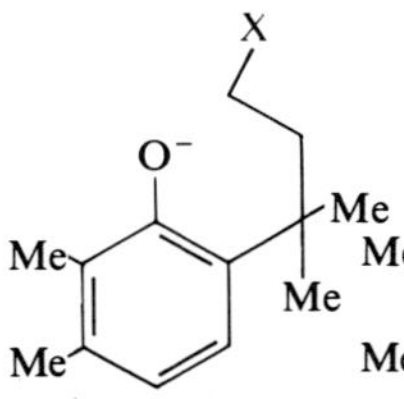 2480	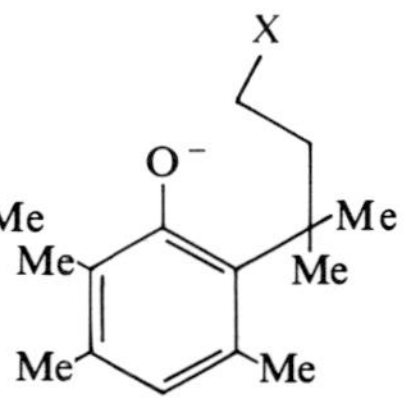 2.05×10^8

$X = OSO_2Me$.

Table 24. ΔpK_a of 3,3-Disubstituted Glutaric Acids at 25°C[149]

3,3 Substituents	H, H	Me, Me	Et, Et	Pr^i, Pr^i
ΔpK_a	1.02	2.60	3.75	4.05

reported by Cohen and his co-workers on the lactonization of 3-(2-hydroxyphenyl)propionic acids (Table 22)[145] and the cyclization of the anions of 3-(2-hydroxyphenyl)propyl mesylates (Table 23).[146] In contrast to the reactions mentioned above, larger effects were observed in the lactonizations than in the nucleophilic substitution reactions at saturated carbon.

The above results suggest that the origin of the *gem*-dialkyl effect is not simple and that probably several factors are involved. It was suggested by Ingold that substituents caused a decrease in the bond angle between the remaining valencies to bring them closer to that found in the ring.[147] While this may be a factor for small rings, it seems unlikely that it could be a factor for five- and six-membered rings where the bond angles are close to the normal values. Bruice and Pandit suggested that "the ring closure proceeds at a higher rate on geminal (or alkyl) substitution because of the resultant decrease in unprofitable rotamer distribution."[148] Bruice and Bradbury produced good evidence that *gem*-dialkyl substituents affect the rotamer distribution of glutaric acids by showing that the difference in the pK_a's of glutaric acid increased when alkyl and especially when 3,3-dialkyl substituents were introduced (see Table 24), which suggests that the substituents bring the carbonyl groups closer together.[149] It is therefore reasonable to suppose that alkyl substituents would also bring the carboxylate and ester groups of the anion of *p*-bromophenyl hydrogen glutarate closer together, and the rate enhancement (Table 15) may be viewed as arising through this.

It seems unlikely, however, that the very large rate enhancements (ca. 10^{11}) observed with some systems can arise solely from a change in the

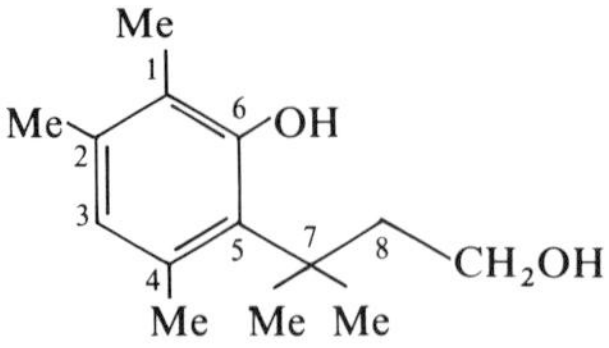

(71)

rotamer populations. A crystal structure determination on the alcohol (71), which is structurally related to the acid of Table 22 and the mesylate of Table 23, showed that whereas the conformation around the bond C_5—C_7 is one that would be highly favorable for cyclization, the conformation around the bond C_7—C_8 is not, and that a rotation of ca. 120° around the bond would be necessary for ring closure.[150] This result implies that if the

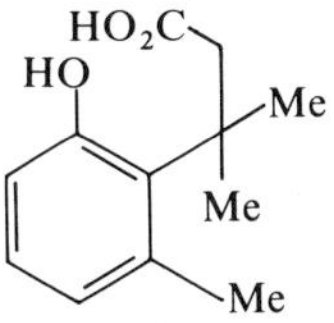

(72) 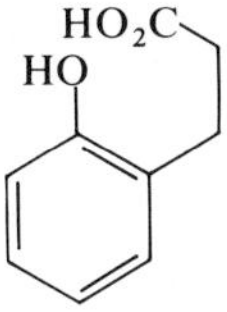   (73)

rate difference for lactonization between (72) and (73), which was estimated[145] to be 5×10^{10}, arises solely from an effect on the rotamer population, the difference in energies between unfavorable and favorable rotamers around the C_5—C_7 bond would be $2.303 RT \log 5 \times 10^{10} = 14.8$ kcal mole^{-1}, which seems to be very large. It seems more likely that relief of steric strain on passing to the transition state is an important contributing factor.[151] In a sterically crowded acid such as (72) there should be unfavorable non-bonding van der Waals' interactions that are partly converted to bonding interactions in the transition state, with a resulting lowering of the energy difference between it and the initial state. That other factors still may be important is indicated by Milstein and Cohen's observation that the ratio of the rate constants for acid-catalyzed lactonization of (72) to (73) is larger than for the base-catalyzed lactonization.[145] As suggested by them, it is possible that there is a conformational requirement in the rate-limiting decomposition of the tetrahedral intermediate in the acid-catalyzed reaction that is made difficult to achieve by the three methyl groups of (72). This could arise if interaction between the loan pair on the oxygen of the phenolic hydroxyl group and the developing carbonium ion were important (cf. 74).

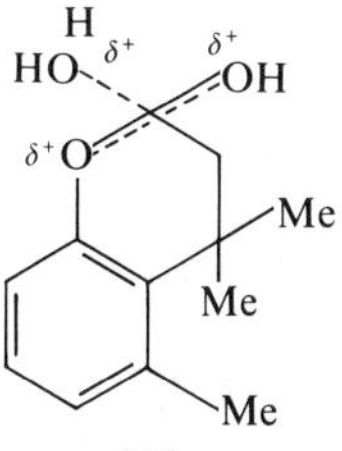

(74)

It is interesting that Kirby and his co-workers have shown that alkyl substituents have a very large rate enhancing effect on the formation of maleic anhydride from N-alkyl maleamic acids[152] and that Eberson and Welinder have shown that they have a large effect on the equilibrium constant for the formation of maleic anhydride from maleic acid.[151] In these compounds there is no question of a change in the rotamer population, and the effect must arise from a relief of nonbonding interactions on cyclization.

The most satisfactory general discussion of the *gem*-dialkyl effect is still that of Allinger and Zalkow.[153] It was shown that alkyl substitution causes a decrease in $\Delta H°$ and an increase in $\Delta S°$ for the reaction hexane → cyclohexane + hydrogen (cf. Table 14). This was explained by the fact that on formation of cyclohexane from hexane there are six additional gauche interactions but on ring closure of an alkyl-substituted cyclohexane there are less than six. Hence $\Delta H°$ for the formation of a six-membered ring and $\Delta H^{\ddagger}$ for the formation of a six-membered cyclic transition state should be decreased on alkyl substitution. In addition, alkyl substituents restrict rotation in the acyclic form and thus decrease the rotational entropy. Hence, loss of rotational entropy on ring closure of an alkyl-substituted chain, or going to the transition state for such ring closure, is less than for an unsubstituted chain. Therefore alkyl substitution causes a decrease in the free energy of formation of a six-membered ring from an open chain through more favorable $\Delta H°$ and $T \Delta S°$ terms. The effect of alkyl substituents on $\Delta H°$ and $T \Delta S°$ for the cyclization of n-pentane to cyclopentane + hydrogen is similar to the effect on cyclization of n-hexane (see Table 14), and it seems likely that it would be similar for the formation of rings of other sizes as well. Thus even in these relatively simple gas-phase equilibria at least two factors are involved in the *gem*-dialkyl effect. There may be additional ones in more complex reactions in solution.

References

1. S. Winstein and E. Grunwald. *J. Am. Chem. Soc.*, **70**, 828 (1948); S. Winstein, B. K. Morse, E. Grunwald, K. C. Schreiber, and J. Corse, *ibid.*, **74**, 1113 (1952); P. G. Gassman, J. Zeller, and J. T. Lumb, *Chem. Commun.* 69 (1968); P. G. Gassman and A. F. Fentiman, *J. Am. Chem. Soc.*, **91**, 1545 (1969); **92**, 2549, 2551 (1970); R. K. Lustgarten, J. Lhomme, and S. Winstein, *J. Org. Chem.*, **37**, 1075 (1972); P. G. Gassman and J. M. Pascone, *J. Am. Chem. Soc.*, **95**, 7801 (1973); E. N. Peters and H. C. Brown, *ibid.*, **95**, 2397 (1973); H. C. Brown, M. Ravindranathan, and E. N. Peters, *ibid.*, **96**, 7351 (1974).
2. J. M. Harris, *Prog. Phys. Org. Chem.*, **11**, 132 (1974).

3. B. N. McMaster, *Specialist Periodical Reports, Mass Spectrometry*, Vol. 3, The Chemical Society, London (1975), p. 2.
4. V. Gold, *J. Chem. Soc. Faraday Trans. 1*, **68**, 1611 (1972).
5. R. Hoffmann, *J. Chem. Phys.*, **40**, 2480 (1964).
6. R. E. Davis and A. S. N. Murthy, *Tetrahedron*, **24**, 4595 (1968).
7. T. Yonezawa, H. Nakatsuji, and H. Kato, *J. Am. Chem. Soc.*, **90**, 1239 (1968).
8. J. Dannenberg and T. D. Berke, *Abstracts of Papers, 158 ACS National Meeting, Sept. 1969*, The American Chemical Society, Washington, D.C. (1969), Phys. 163.
9. R. Sustmann, J. E. Williams, M. J. S. Dewar, L. C. Allen, and P. von R. Schleyer, *J. Am. Chem. Soc.*, **91**, 5350 (1969).
10. F. Fratev, R. Janoschek, and H. Preuss, *Int. J. Quantum Chem.*, **3**, 873 (1969).
11. J. E. Williams, V. Buss, L. C. Allen, P. von R. Schleyer, W. A. Latham, W. J. Hehre, and J. A. Pople, *J. Am. Chem. Soc.*, **92**, 2141 (1970).
12. G. V. Pfeiffer and J. G. Jewett, *J. Am. Chem. Soc.*, **92**, 2143 (1970).
13. D. T. Clark and D. M. J. Lilley, *Chem. Commun.*, 549 (1970).
14. H. Kollmar and H. O. Smith, *Angew. Chem. Int. Ed. Engl.*, **9**, 462 (1970); *Theor. Chim. Acta*, **20**, 65 (1971).
15. W. A. Latham, W. J. Hehre, and J. A. Pople, *J. Am. Chem. Soc.*, **93**, 808 (1971).
16. J. E. Williams, V. Buss, and L. C. Allen, *J. Am. Chem. Soc.*, **93**, 6867 (1971).
17. P. C. Hariharan, W. A. Latham, and J. A. Pople, *Chem. Phys. Lett.*, **14**, 385 (1972).
18. D. A. Dixon and W. N. Lipscomb, *J. Am. Chem. Soc.*, **95**, 2853 (1973).
19. B. Zurawski, R. Ahlrichs, and W. Kutzelnigg, *Chem. Phys. Lett.*, **21**, 309 (1973).
20. P. Ausloos, R. E. Rebbert, L. W. Sieck, and T. O. Tiernan, *J. Am. Chem. Soc.*, **94**, 8939 (1972).
21. J. D. Roberts and J. A. Yancey, *J. Am. Chem. Soc.*, **74**, 5943 (1952).
22. C. C. Lee and M. K. Frost, *Can. J. Chem.*, **43**, 526 (1965).
23. I. L. Reich, A. Diaz, and S. Winstein, *J. Am. Chem. Soc.*, **91**, 5635 (1969).
24. P. C. Myhre and K. S. Brown, *J. Am. Chem. Soc.*, **91**, 5639 (1969); P. C. Myhre and E. Evans, *ibid.*, **91**, 5641 (1969).
25. G. A. Olah, J. R. DeMember, R. H. Schlosberg, and Y. Halpern, *J. Am. Chem. Soc.*, **94**, 156 (1972).
26. J. H. Vorachek, G. G. Meisels, R. A. Geanangel, and R. H. Emmel, *J. Am. Chem. Soc.*, **95**, 4078 (1973).
27. J. D. Petke and J. L. Whitten, *J. Am. Chem. Soc.*, **90**, 3338 (1968).
28. H. Fisher, H. Kollmar, H. O. Smith, and K. Miller, *Tetrahedron Lett.*, 5821 (1968).
29. H. Kollmar and H. O. Smith, *Tetrahedron Lett.*, 1833 (1970).
30. P. C. Hariharan, L. Radom, J. A. Pople, and P. von R. Schleyer, *J. Am. Chem. Soc.*, **96**, 599 (1974).
31. P. K. Bischof and M. J. S. Dewar, *J. Am. Chem. Soc.*, **97**, 2278 (1975).
32. M. Saunders and E. L. Hagen, *J. Am. Chem. Soc.*, **90**, 6881 (1968).
33. G. A. Olah and A. M. White, *J. Am. Chem. Soc.*, **91**, 5801 (1969).
34. M. Saunders, P. Vogel, E. L. Hagen, and J. Rosenfeld, *Acc. Chem. Res.*, **6**, 53 (1973).
35. C. J. Collins, *Chem. Rev.*, **69**, 543 (1969).
36. J. L. Fry and G. J. Karabatsos, in: *Carbonium Ions* (G. A. Olah and P. von R. Schleyer, eds), Vol. II, p. 527, John Wiley & Sons, Inc. (Interscience Division), New York (1970).
37. C. C. Lee, *Prog. Phys. Org. Chem.*, **7**, 129 (1970).
38. M. Simonetta and S. Winstein, *J. Am. Chem. Soc.*, **76**, 18 (1954); cf. M. Simonetta, in: *Chemical Reactivity and Reaction Paths* (G. Klopman, ed.), p. 16, John Wiley & Sons, Inc. (Interscience Division), New York (1974).
39. C. A. Coulson and M. J. S. Dewar, *Discuss. Faraday Soc.*, **2**, 54 (1947); M. J. S. Dewar, *Rev. Mod. Phys.*, **35**, 586 (1963); M. J. S. Dewar, *Chem. Soc. Spec. Publ.*, **21**, 181 (1967); J. C. Slater, *Quantum Theory of Molecules and Solids*, Vol. 1, McGraw-Hill Book Com-

pany, New York (1963), p. 108; M. J. S. Dewar, *The Molecular Orbital Theory of Organic Chemistry*, McGraw-Hill Book Company, New York (1969), p. 361); T. F. W. McKillop and B. C. Webster, *Tetrahedron*, **26**, 1879 (1970).
40. R. J. Piccolini and S. Winstein, *Tetrahedron, Suppl. 2*, **19**, 423 (1963).
41. M. E. H. Howden and J. D. Roberts, *Tetrahedron, Suppl. 2*, **19**, 403 (1963); cf. E. A. C. Lucken *ibid.*, **19**, 413 (1963).
42. S. Winstein, *Quart. Rev. Chem. Soc.*, **23**, 141 (1969).
43. D. S. Kabakoff and E. Namanworth, *J. Am. Chem. Soc.*, **92**, 3234 (1970).
44. G. Olah, D. P. Kelly, C. L. Jeuell, and R. D. Porter, *J. Am. Chem. Soc.*, **92**, 2544 (1970); **94**, 146 (1972); see also D. P. Kelly and H. C. Brown, *ibid.*, **97**, 3897 (1975).
45. E. I. Snyder, *J. Am. Chem. Soc.*, **92**, 7529 (1970).
46. W. J. Hehre, *J. Am. Chem. Soc.*, **94**, 5919 (1972).
47. W. J. Hehre and P. C. Hiberty, *J. Am. Chem. Soc.*, **96**, 2665 (1974).
48. W. S. Trahanovsky, *J. Org. Chem.*, **30**, 1666 (1965).
49. G. Klopman, J. Am. Chem. Soc., **91**, 89 (1969).
50. M. J. S. Dewar, *Chem. Brit.*, **11**, 97 (1975).
51. R. Hoffmann, *Acc. Chem. Res.*, **4**, 1 (1971).
52. R. Hoffmann, L. Radom, J. A. Pople, P. von R. Schleyer, W. J. Hehre, and L. Salem, *J. Am. Chem. Soc.*, **94**, 6221 (1972).
53. S. Winstein, B. K. Morse, E. Grunwald, K. C. Schreiber, and J. Corse, *J. Am. Chem. Soc.*, **74**, 1113 (1952), footnote 19.
54. V. J. Shiner and J. G. Jewett, *J. Am. Chem. Soc.*, **87**, 1382 (1965).
55. F. R. Jensen and B. E. Smart, *J. Am. Chem. Soc.*, **91**, 5688 (1969); see also D. E. Sunko and S. Borčić *in: Isotope Effects in Chemical Reactions*, ACS Monograph 167 (C. J. Collins and N. S. Bowman, eds.), p. 171, Van Nostrand Reinhold Company, New York (1970).
56. G. A. Olah, C. L. Jeuell, D. P. Kelly, and R. D. Porter, *J. Am. Chem. Soc.*, **94**, 146 (1972).
57. G. A. Olah and G. Liang, *J. Am. Chem. Soc.*, **97**, 1920 (1975).
58. T. G. Traylor, W. Hanstein, H. J. Berwin, N. A. Clinton, and R. S. Brown, *J. Am. Chem. Soc.*, **93**, 5715 (1971); see also C. J. Lancelot and P. von R. Schleyer, *ibid.*, **91**, 4296 (1969), and P. Laszlo and Z. Welvart, *Bull. Soc. Chim. France*, 2412 (1966).
59. L. Radom, J. A. Pople, and P. von R. Schleyer, *J. Am. Chem. Soc.*, **94**, 5935 (1972).
60. L. Radom, J. A. Pople, V. Buss, and P. von R. Schleyer, *J. Am. Chem. Soc.*, **92**, 6380 (1970).
61. H. C. Brown, B. G. Gnedin, K. Takeuchi, and E. N. Peters, *J. Am. Chem. Soc.*, **97**, 610 (1975).
62. W. Hanstein, H. J. Berwin, and T. G. Traylor, *J. Am. Chem. Soc.*, **92**, 829 (1970).
63. N. A. Clinton, R. S. Brown, and T. G. Traylor, *J. Am. Chem. Soc.*, **92**, 5228 (1970).
64. D. F. Eaton and T. G. Traylor, *J. Am. Chem. Soc.*, **96**, 1226 (1974).
65. B. Capon and N. B. Chapman, *J. Chem. Soc.*, 600 (1957).
66. R. C. Bingham and P. von R. Schleyer, *J. Am. Chem. Soc.*, **93**, 3189 (1971).
67. G. D. Sargent and T. J. Mason, *J. Am. Chem. Soc.*, **96**, 1063 (1974).
68. C. S. Foote and R. B. Woodward, *Tetrahedron*, **20**, 687 (1964).
69. J. F. Music and F. A. Matsen, *J. Am. Chem. Soc.*, **72**, 5256 (1950); W. W. Robertson, J. F. Music, and F. A. Matsen, *ibid.*, **72**, 5260 (1950); N. H. Cromwell and M. A. Graff, *J. Org. Chem.*, **17**, 414 (1952); N. H. Cromwell and G. V. Hudson, *J. Am. Chem. Soc.*, **75**, 872 (1953); A. L. Goodman and R. H. Eastman, *ibid.*, **86**, 908 (1964); E. Kosower and M. Ito, *Proc. Chem. Soc. London*, 25 (1962); L. S. Bartell, B. L. Carroll, and J. P. Guillory, *Tetrahedron Lett.*, 705 (1964); L. S. Bartell and J. P. Guillory, *J. Chem. Phys.*, **43**, 647 (1965); L. S. Bartell, J. P. Guillory, and A. T. Parks, *J. Phys. Chem.*, **69**, 3043 (1965); G. J. Karabatsos and N. Hsi, *J. Am. Chem. Soc.*, **87**, 2864 (1965); G. L. Closs and H. B. Klinger, *ibid.*, **87**, 3265 (1965); J. L. Pierre and P. Arnaud, *Bull Soc. Chim.*

France, 1690 (1966); W. Lüttke, A. de Meijere, H. Wolff, H. Ludwig, and H. W. Schrötter, *Angew. Chem. Int. Ed. Engl.,* **5**, 123 (1965); O. Bastiansen and A. de Meijere, *ibid.,* **5**, 124 (1965); W. Lüttke and A. de Meijere, *ibid.,* **5**, 512 (1966); H. Günter and D. Wendisch, *ibid.,* **5**, 251 (1966); G. R. De Mare and J. S. Martin, *J. Am. Chem. Soc.,* **88**, 5033 (1966); S. A. Monti, D. J. Buchek, and J. C. Shepard, *J. Org. Chem.,* **34**, 3080 (1969); R. C. Hahn, P. H. Howard, S.-M. Kong, G. A. Lorenzo, and N. L. Miller, *J. Am. Chem. Soc.,* **91**, 3558 (1969);

70. M. Yu Lukina, *Usp. Khim,* **31**, 901 (1962); *Russ. Chem. Rev.,* **31**, 419 (1962).
71. R. Hoffmann and R. B. Davidson, *J. Am. Chem. Soc.,* **93**, 5699 (1971).
72. W. A. Bernett, *J. Chem. Educ.,* **44**, 17 (1967).
73. A. D. Walsh, *Trans. Faraday Soc.,* **45**, 179 (1949).
74. C. A. Coulson and W. E. Moffitt, *J. Chem. Phys.* **15**, 151 (1947); *Philos. Mag.,* **40**, 1 (1949).
75. L. L. Ingraham, *in: Steric Effects in Organic Chemistry* (M. S. Newman, ed.), p. 518, John Wiley & Sons, Inc., New York (1956).
76. J. E. Baldwin and W. D. Fogelsong, *J. Am. Chem. Soc.,* **90**, 4303 (1968).
77. H. G. Richey and N. C. Buckley, *J. Am. Chem. Soc.,* **85**, 3057 (1963).
78. B. R. Ree and J. C. Martin, *J. Am. Chem. Soc.,* **92**, 1660 (1970); V. Buss, R. Gleiter, and P. von R. Schleyer, *ibid.,* **93**, 3927 (1971).
79. J. D. Roberts, W. Bennett, and R. Armstrong, *J. Am. Chem. Soc.,* **72**, 3329 (1950); J. Roberts and R. H. Mazur, *ibid.,* **73**, 2509 (1951).
80. R. H. Mazur, W. N. White, D. A. Semenow, C. C. Lee, M. S. Silver, and J. D. Roberts, *J. Am. Chem. Soc.,* **81**, 4390 (1959); M. C. Caserio, W. H. Graham, and J. D. Roberts, *Tetrahedron,* **11**, 171 (1960).
81. P. von R. Schleyer and G. W. Van Dine, *J. Am. Chem. Soc.,* **88**, 2321 (1966).
82. H. Hart and J. M. Sandri, *J. Am. Chem. Soc.,* **81**, 320 (1959); H. Hart and P. A. Law, *ibid.,* **84**, 2462 (1962).
83. K. B. Wiberg, B. A. Hess, and A. J. Ashe, *in: Carbonium Ions* (G. A. Olah and P. von R. Scheyler, eds), Vol. III, p. 1295, John Wiley & Sons, Inc. (Interscience Division). New York (1972).
84. H. G. Richey, *in: Carbonium Ions* (G. A. Olah and P. von R. Schleyer, eds.), Vol. III, p. 1284, John Wiley & Sons, Inc. (Interscience Division), New York (1972).
85. H. Volz, J. H. Shin and H.-J. Streicher, *Tetrahedron Lett.,* 1297 (1975); G. A. Olah and D. A. Forsyth, *J. Am. Chem. Soc.,* **97**, 3137 (1975).
86. C. U. Pittman, C. Dyas, C. Engelman, and L. D. Kispert, *J. Chem. Soc. Faraday Trans. 2,* **68**, 345 (1972).
87. W. P. Jencks, *Catalysis in Chemistry and Enzymology,* McGraw-Hill Book Company, New York (1969), p. 15; B. Capon, *Essays Chem.,* **3**, 127 (1972); T. C. Bruice, *in: The Enzymes,* 3rd ed. (P. D. Boyer, ed.), Vol. II, p. 217, Academic Press, Inc., New York (1970).
88. D. R. Storm and D. E. Koshland, *Proc. Natl. Acad. Sci. U.S.A.,* **66**, 445 (1970); G. A. Dafforn and D. E. Koshland, *Bioorg. Chem.,* **1**, 129 (1971); D. R. Storm and D. E. Koshland, *J. Am. Chem. Soc.,* **94**, 5805, 5815 (1972); G. A. Dafforn and D. E. Koshland, *Biochem Biophys. Res. Commun.,* **52**, 779 (1973).
89. B. Capon, *J. Chem. Soc. B,* 1207 (1971); T. C. Bruice, A. Brown, and D. O. Harris, *Proc. Natl. Acad. Sci. U.S.A.,* **68**, 658 (1971); J. Reuben, *ibid.,* **68**, 653 (1971); G. N. J. Port and W. G. Richards, *Nature (London),* **231**, 312 (1971); C. Delisi and D. M. Crothers, *Biopolymers,* **12**, 1689 (1972); J. W. Larsen, *Biochem. Biophys. Res. Commun.,* **50**, 839 (1973); C. R. Partrick, *Int. J. Chem. Kinet.,* **5**, 769 (1973).
90. M. I. Page and W. P. Jencks, *Proc. Natl. Acad. Sci. U.S.A.,* **68**, 1678 (1971); M. I. Page, *Biochem. Biophys. Res. Commun.* **49**, 940 (1970).
91. M. I. Page, *Chem. Soc. Rev.,* **2**, 295 (1973).
92. K. B. Wiberg, *Physical Organic Chemistry,* John Wiley & Sons, Inc., New York (1964), p. 221.

93. D. F. De Tar, *J. Am. Chem. Soc.*, **96**, 1254, 1255 (1974).
94. L. Ruzicka. *Chem. Ind. (London)*, **54**, 2 (1934).
95. A. S. Pell and G. Pilcher, *Trans. Faraday Soc.*, **61**, 71 (1965).
96. S. Winstein, E. Allred, R. Heck, and R. Glick, *Tetrahedron*, **3**, 1 (1958).
97. H. W. Heine, A. D. Miller, W. H. Barton, and R. W. Greiner, *J. Am. Chem. Soc.*, **75**, 4778 (1953).
98. B. Capon and I. Farazmand, unpublished observation; I. Farazmand, Ph.D. thesis, University of Birmingham, Birmingham (1964).
99. G. Kohnstam and M. Penty, in: *Hydrogen-Bonded Solvent Systems* (A. K. Covington and P. Jones, eds.), p. 275, Taylor and Francis, London (1968).
100. H. W. Heine and W. Siegfried, *J. Am. Chem. Soc.*, **76**, 489 (1954).
101. H. Freundlich and H. Kroepelin, *Z. Phys. Chem. (Leipzig)*, **122**, 39 (1926).
102. R. Bird, A. C. Knipe, and C. J. M. Stirling, *J. Chem. Soc. Perkin Trans. 2*, 1215 (1973).
103. P. E. Peterson and J. F. Coffey, *J. Am. Chem. Soc.*, **93**, 5208 (1971); P. E. Peterson, *Acc. Chem. Res.*, **4**, 406 (1971).
104. G. M. Bennett, F. Heathcoat, and A. N. Mosses, *J. Chem. Soc.*, 2567 (1929).
105. F. G. Bordwell and W. T. Brannen, *J. Am. Chem. Soc.*, **86**, 4645 (1964).
106. R. Bird and C. J. M. Stirling, *J. Chem. Soc. Perkin Trans. 2*, 1221 (1973).
107. D. J. Pasto and M. P. Serve, *J. Am. Chem. Soc.*, **87**, 1515 (1965).
108. H. R. Ward and P. D. Sherman, *J. Am. Chem. Soc.*, **90**, 3812 (1968).
109. F. Montanari, R. Danieli, H. Hogeveen, and G. Maccagnani, *Tetrahedron Lett.*, 2685 (1964); M. Cinquini, S. Colonna, and F. Montanari, *ibid.*, 3181 (1966).
110. F. L. Scott and D. F. Fenton, *Tetrahedron Lett.*, 1681 (1964); 685 (1970); F. L. Scott, E. J. Flynn, and D. F. Fenton, *J. Chem. Soc. B*, 277 (1971).
111. F. L. Scott and E. Flynn, *Tetrahedron Lett.*, 1675 (1964).
112. A. C. Knipe and C. J. M. Stirling, *J. Chem. Soc. B*, 808 (1967).
113. A. C. Knipe and C. J. M. Stirling, *J. Chem. Soc., B*, 67 (1968).
114. R. Baird and S. Winstein, *J. Am. Chem. Soc.*, **84**, 788 (1962); **85**, 567 (1963).
115. R. Heck and S. Winstein, *J. Am. Chem. Soc.*, **79**, 3105 (1957).
116. R. Heck and S. Winstein, *J. Am. Chem. Soc.*, **79**, 3114 (1957).
117. C. A. Matusak, Ph.D. thesis, Ohio State University, Columbus (1957); *Diss. Abstr.*, **18**, 792 (1958).
118. L. Zürn. *Justus Liebigs Ann. Chem.*, **631**, 56 (1960); R. B. Martin, R. Hedrick, and A. Parcell, *J. Org. Chem.* **29**, 158 (1964).
119. B. Capon, S. McDowell, and W. V. Raftery, *J. Chem. Soc. Perkin Trans. 2*, 1118 (1973).
120. A. Tsuji, T. Yamana, and Y. Mizukami, *Chem. Pharm. Bull.*, **20**, 2528 (1972); T. Yamana, A. Tsuji, and Y. Mizukami, *ibid.*, **21**, 721 (1973).
121. T. C. Bruice and S. J. Benkovic, *J. Am. Chem. Soc.*, **85**, 1 (1963).
122. E. F. Curragh and D. T. Elmore, *J. Chem. Soc.*, 2948 (1962).
123. C. J. M. Stirling, *J. Chem. Educ.*, **50**, 844 (1973).
124. H. Böhme and K. Sell, *Chem. Ber.*, **81**, 123 (1948).
125. B. Capon, *Quart. Rev. Chem. Soc.*, **18**, 108 (1964).
126. K. Ziegler, *Ber.*, **67A**, 139 (1934).
127. K. Ziegler, A. Lüttringhaus, and K. Wohlgemuth, *Justus Liebigs Ann. Chem.*, **528**, 162 (1937).
128. C. Galli, G. Illuminati, and L. Mandolini, *J. Am. Chem. Soc.*, **95**, 8374 (1973).
129. G. Illuminati, L. Mandolini, and B. Masci, *J. Am. Chem. Soc.*, **96**, 1422 (1974).
130. S. Danishefsky, J. Dynak, E. Hatch, and M. Yamamoto, *J. Am. Chem. Soc.*, **96**, 1256 (1974).
131. L. Tenud, S. Farooq, J. Seibl, and A. Eschenmoser, *Helv. Chim. Acta*, **53**, 2059 (1970).
132. S. Danishefsky, J. Dynak, and M. Yamamoto, *Chem. Commun.*, 81 (1973).
133. G. Stork and J. F. Cohen, *J. Am. Chem. Soc.*, **96**, 5270 (1974); also see ref. 154.

134. J. Y. Lallemand and M. Onanga, *Tetrahedron Lett.*, 585 (1975).
135. J. G. Buchanan and H. Z. Sable, *Selective Organic Transformations* (B. S. Thyagarajan, ed.), Vol. 2, pp. 53–54, John Wiley & Sons, Inc. (Interscience Division), New York (1972).
136. W. E. Barnett and J. C. McKenna, *Tetrahedron Lett.*, 2595 (1971); W. E. Barnett and W. H. Sohn, *ibid.*, 1777 (1972).
137. J. Hine, M. S. Cholod, and J. H. Jensen, *J. Am. Chem. Soc.*, **93**, 2321 (1971); J. Hine, M. S. Cholod, and R. A. King, *ibid.*, **96**, 835 (1974).
138. R. D. Gandour and R. L. Schowen, *J. Am. Chem. Soc.*, **96**, 2231 (1974); R. D. Gandour, *Tetrahedron Lett.*, 295 (1974).
139. B. Capon, *in: Proton Transfer Reactions* (E. F. Calden and V. Gold, eds.), p. 339, Chapman & Hall., London (1975).
140. R. P. Bell and M. A. D. Fluendy, *Trans. Faraday Soc.*, **59**, 1623 (1963); R. P. Bell and P. de Maria, *ibid.*, **66**, 930 (1970).
141. H. Wilson and E. S. Lewis, *J. Am. Chem. Soc.*, **94**, 2283 (1972).
142. G. S. Hammond, *in: Steric Effects in Organic Chemistry* (M. S. Newman, ed.), p. 468, John Wiley & Sons, Inc., New York (1956) (a review of earlier work).
143. G. B. Payne, *J. Org. Chem.*, **27**, 3819 (1962).
144. D. P. Weeks and X. Creary, *J. Am. Chem. Soc.*, **92**, 3418 (1970).
145. S. Milstein and L. A. Cohen, *Proc. Natl. Acad. Sci. U.S.A.*, **67**, 1143 (1970); *J. Am. Chem. Soc.*, **94**, 9158 (1972).
146. R. T. Borchardt and L. A. Cohen, *J. Am. Chem. Soc.*, **94**, 9166 (1972).
147. C. K. Ingold, *J. Chem. Soc.*, 308 (1921).
148. T. C. Bruice and U. K. Pandit, *J. Am. Chem. Soc.*, **82**, 5858 (1960).
149. T. C. Bruice and W. C. Bradbury, *J. Am. Chem. Soc.*, **87**, 4846 (1965).
150. J. M. Karle and I. L. Karle, *J. Am. Chem. Soc.*, **94**, 9182 (1972).
151. L. Eberson and H. Welinder, *J. Am. Chem. Soc.*, **93**, 5821 (1971).
152. A. J. Kirby and P. W. Lancaster, *J. Chem. Soc. Perkin Trans. 2*, 1206 (1972); M. F. Aldersley, A. J. Kirby, P. W. Lancaster, R. S. McDonald, and C. R. Smith, *J. Chem. Soc. Perkin Trans 2*, 1487 (1974).
153. N. L. Allinger and V. Zalkow, *J. Org. Chem.*, **25**, 701 (1960).
154. P. A. Cruickshank and M. Fishman, *J. Org. Chem.*, **34**, 4060 (1969), describe an exception.

3
Some Experimental Methods Used in the Study of Neighboring Group Participation

The study of neighboring group participation involves, for the most part, the same experimental techniques required in the study of other chemical mechanisms. A discussion of common techniques is beyond the scope of this monograph, and a general knowledge of physical organic chemistry is assumed; for appropriate general and special treatments of specific physical organic methods, see Ref. 1. The purpose of this chapter is to describe the use of various experimental techniques in defining the presence (and extent) or absence of neighboring group participation.

Just how much experimental information is required before a neighboring group mechanism can be said to be involved? This, as will become evident, is a variable that cannot be generalized. Often one may learn much from a single experimental observation, such as the formation of a cyclic

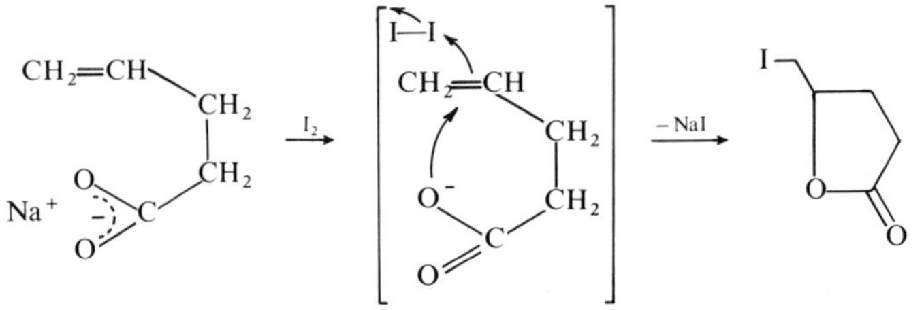

(1)

(1)   (2)

product. For example, the sodium salt of 4-pentenoic acid [Eq. (1)] reacts in aqueous solution with iodine to give the iodolactone (**2**) as the sole product,[2*] an obvious neighboring group process. Is the reaction anchimerically assisted or does the carboxylate anion capture a carbocationic intermediate after the rate-determining step? A comparison with other iodine addition experiments is informative. Under the condition that (**1**) is converted to (**2**), a similar olefinic acid, 7-octenoic acid, fails to give a measurable reaction rate. Thus, an anchimerically assisted synchronous process for the reaction in Eq. (1) may be suspected, but this is not the only possibility. Addition of iodine to an olefinic double bond is a reversible reaction,[4] and the failure of octenoic acid to give a measurable reaction rate with iodine may arise from an unfavorable equilibrium constant under the reaction conditions. An alternative reaction pathway is shown in Scheme I, and the reason that 4-pentenoic acid reacts with iodine whereas 7-octenoic

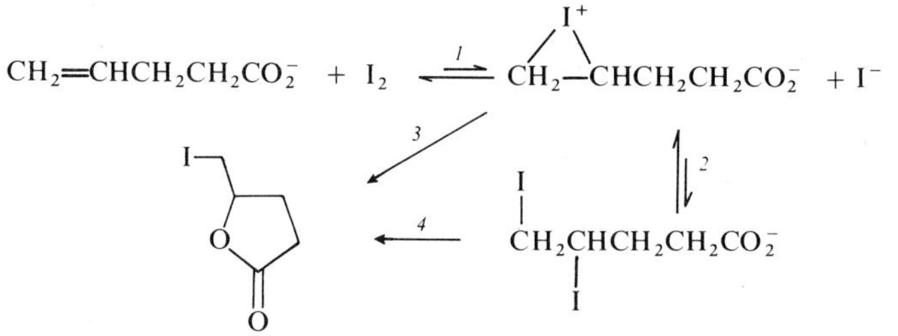
CH2=CHCH2CH2CO2- + I2
1
I+
CH2—CHCH2CH2CO2- + I-
3
2
I
4
O
O
CH2CHCH2CH2CO2-
I

(**2**)

Scheme 1

acid does not may be that it has step *3* or *4* available and that analogous steps with 7-octenoic acid are unfavorable. Thus the reaction rate and product indicate the occurrence of neighboring group participation, but it is difficult to be certain in which step it occurs.‡

* This type of mechanism was proposed by Lucas *et al.*[3] and is still used (Williams *et al.*[3]).

‡ Similarly, in nucleophilic displacement reactions it is possible for anchimeric assistance to be observed because a neighboring group process circumvents ion pair return, i.e.,

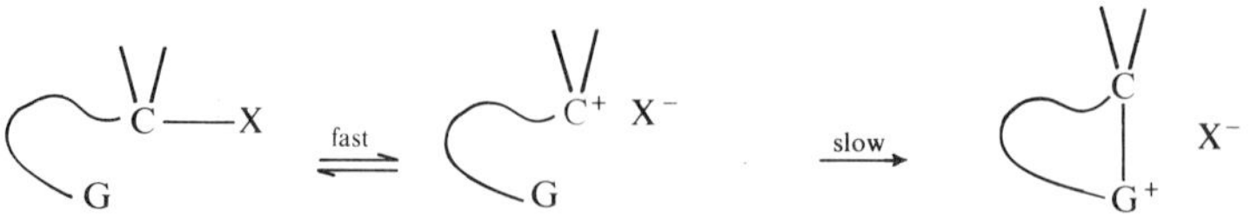
C—X
G
fast
C+ X-
G
slow
C
G+
X-

The occurrence of the ion-pair mechanism, rather than neighboring group displacement at a neutral carbon, is a controversial topic (see pp. 248–249 and 255–258).

In the above example a product analysis suggested a kinetic analysis. There are also as many examples of the opposite approach, that is, a kinetic result suggesting careful product studies. In one of the least controversial examples of anchimeric assistance derived from a kinetic approach, Winstein et al.[5] showed that the p-toluenesulfonate ester of anti-7-norbornenol (3) undergoes acetolysis with a half-life of 11 min at 25°C. This is more than 10^{11} times faster than the corresponding saturated derivative [(5), X = OTs], which undergoes acetolysis under the same conditions with a half-life of 3.5 million years. The presence of the suggested intermediate (4) is confirmed by the isolation, in a series of careful methanolysis experiments,[6] of the ether (6), the kinetically favored product. Under typical solvolysis conditions substitution without rearrangement and with retention of configuration is obtained; e.g., acetolysis of (3) gives (7), the

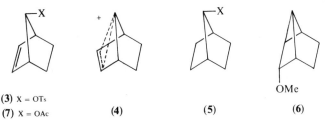

(3) X = OTs
(7) X = OAc   (4)   (5)   (6)

thermodynamically favored product. This alone is evidence of a neighboring group mechanism since complete retention is uncommon in solvolytic reactions.[1] The isolation of (6), however, removes doubt about the intermediacy of (4).

In both the above examples anchimeric assistance was easily detected from kinetic experiments. There are a sufficient number of examples, however, where there is little noticeable rate increase relative to similar model systems, yet anchimeric assistance appears to be occurring from product stereochemistry. In such cases, many approaches may be required to escape the dilemma. We shall discuss several methods below.

3.1. Kinetic Methods

3.1.1 Estimation of Reaction Rates

To measure anchimeric assistance it is necessary to estimate the reaction rate of the hypothetical reaction in the absence of anchimeric assistance.

An intermolecular reaction as a model for an intramolecular reaction is often worthless since the intermolecular reaction may not occur at all, or it may occur by an altogether different mechanism. Because of this, one normally must resort to rate comparisons obtained by using a similar molecule as a basis for comparison; for example, a molecule containing the neighboring group in a different configuration may be used. In using such comparisons, one must recognize the potential pitfalls of the particular comparison. We shall now discuss some of the available methods.

3.1.1.1. Rate Comparisons of Isomers

For the establishment of anchimeric assistance in *trans*-substituted cyclic compounds, the *cis* isomer contains essentially the same σ-inductive effects and may, from a cursory inspection, seem to be a reasonable model. There are some problems with this model, however: (1) Field effects are difficult to determine, and they probably affect many rate comparisons[7]; (2) the conformation for reaction may differ from the ground-state-preferred conformation, and hence estimates based on the wrong conformation may lead to errors; and (3) hydrogen participation (or facile elimination) in the *cis* model may introduce errors. We shall now discuss this method and its potential pitfalls with the use of examples.

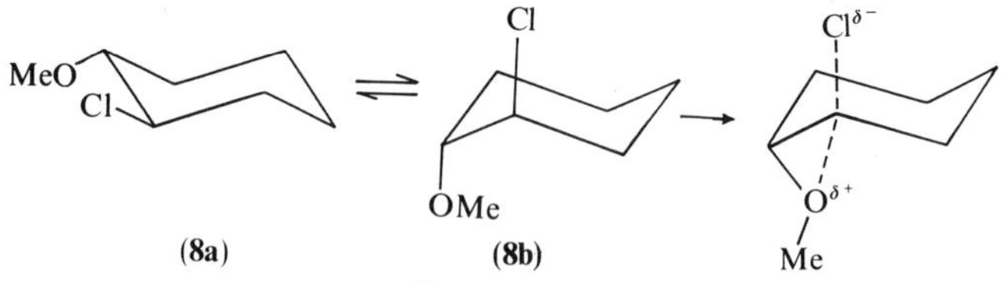

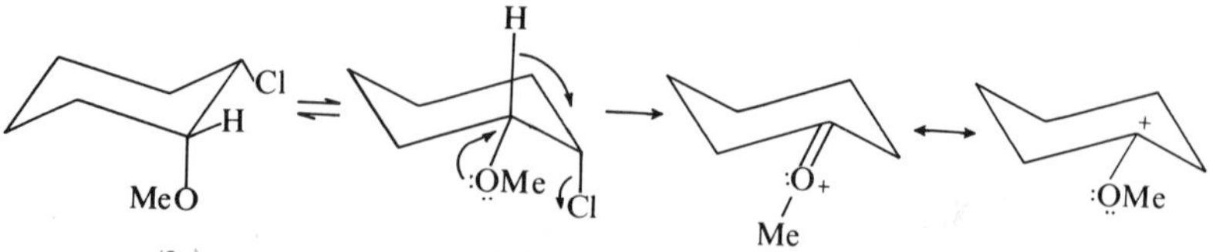

In comparing *cis*- and *trans*-2-methoxycyclohexyl chloride, the more stable ground state for the *trans* isomer (**8**) may be the diequatorial conformation (**8a**) when steric considerations are weighed, but the diaxial conformation (**8b**) may be preferred owing to dipole repulsion. Only one conformation, the diaxial conformation (**8b**), allows optimum neighboring group participation.[8] On the other hand, the *cis* isomer may exist in either conformation (**9a**) or (**9b**), yet it probably solvolyzes only from (**9b**).[8,9] Such factors could affect a comparison of rates.

Another potential problem that may arise is that the isomer used as the model may undergo reaction by a mechanism different from that expected. For example, although rate-limiting elimination has not been rigorously excluded hydrogen participation [e.g., (**9b**) → (**10**)] has been claimed in *cis* substrates such as (**9**) (see Chapter 4, p. 167). Thus, what may appear to be a good model (i.e., *cis* versus *trans*) may in fact be a complex model yielding data of little quantitative significance unless one thoroughly treats all the above complicating factors.

Another widely used comparison, especially in bicyclic systems, is the *exo/endo* rate ratio. This comparison is also fraught with pitfalls. Problems arise because the *endo* and *exo* isomers of a polycyclic system have unique configurations and hence different steric requirements. This can affect anchimeric assistance by neighboring groups. In addition there are inherent differences in field effects and nucleophilic solvent assistance. The 350-fold difference in rate constant for acetolysis of *exo-* and *endo-*2-norbornyl *p*-bromobenzenesulfonates [(**11**) and (**12**), respectively] was taken as evidence for σ participation in the solvolysis of (**11**) by Winstein and Trifan.[10] Brown,[11] on the other hand, has maintained that the difference results primarily, if not wholly, from steric retardation to the ionization of (**12**). There is, however, considerable internal return with (**11**) but none measurable with (**12**).[10] Furthermore, (**12**) undergoes solvolysis with a considerable amount of nucleophilic solvent assistance, while (**11**) apparently has little.[10,12] Thus the basis for a direct comparison is tenuous at best.

Finally, when the neighboring group is acting from the *ortho* position of an aryl system, an approach used is to compare the *ortho/para* rate ratio. Steric differences in the isomers are obvious and may contribute to error; some improvement to the comparison is provided by multiplying by the *ortho/para* rate ratio found for similar nonparticipating substituents.[13]

3.1.1.2. Rate Comparisons of Similar Compounds

a. Relative Rate Comparisons. When a substituent such as hydroxyl or carboxyl is providing assistance by acting as an intramolecular catalyst, the rate of the methylated derivative can be used to give an estimate of the unassisted rate. As with the use of *ortho/para* rate ratios, discussed above, electronic and steric differences are neglected in making these comparisons. Generally, these errors are often negligible compared to the large differences in rates measured if anchimeric assistance is occurring.

Homomorphs can also be useful for relative rate comparisons in certain aliphatic reactions. In a search for participation by the remote dimethylamino group in compounds such as (**13**), the corresponding compound with an isopropyl group. i.e., (**14**), has been used[14] as a model in order to maintain similar steric requirements (see Chapter 6).

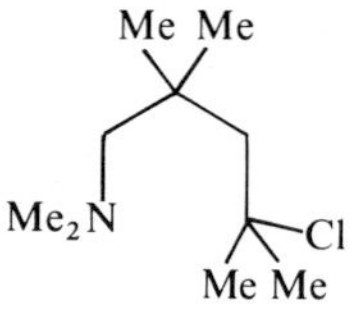

(**13**) 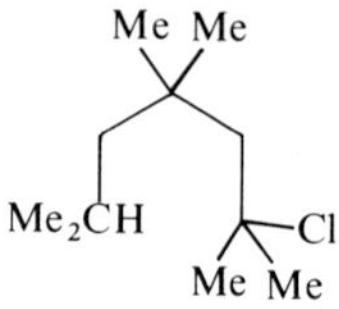   (**14**)

$k_{13}/k_{14} \equiv k/k_H = 7.2$

In a study of sulfur participation in perester decomposition[15] it was of interest to eliminate steric acceleration as a factor. Hence the rates of thermolysis of (**15**) and (**16**) were compared as evidence that anchimeric assistance by sulfur rather than steric acceleration was responsible for the high relative rates observed for the *o*-phenylthioperesters (see Chapter 5).

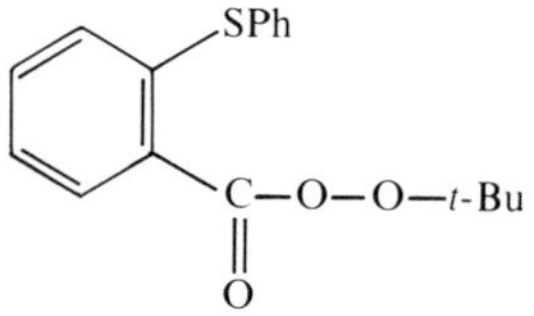

(**15**)   (**16**)

$k_{rel} = 2.72 \times 10^4$ 1.0

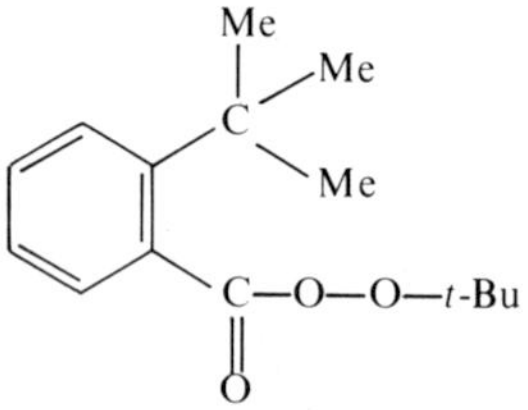

b. Linear Free-Energy Relationships. Although some estimates of anchimeric assistance by the use of comparable models are generally accepted, as discussed above, even with the neighboring group removed sufficiently from the reaction center (e.g., 4-methoxy-1-butyl chloride), the rate of the unsubstituted derivative (e.g., 1-butyl chloride) does not give a true value of the polar factors. When the neighboring group is yet closer to the reaction center, polar and steric effects generally become more important. In such cases the Taft equation,[16] $\log k/k_0 = \rho^* \sigma^*$, can be applied with reasonable success in estimating rates of unassisted processes. Streitwieser[17] first advocated this method for use in detecting anchimeric assistance. For example, aryl participation in the solvolysis of 1-aryl-2-propyl derivatives is indicated when a plot of the logarithm of the titrimetric rate constant, k_t, versus σ^* is constructed. For the acetolysis of secondary alkyl p-toluenesulfonates at 100°C, deactivated β-arylalkyl derivatives fall on the same line that is defined by purely alkyl derivatives (see Fig. 1).

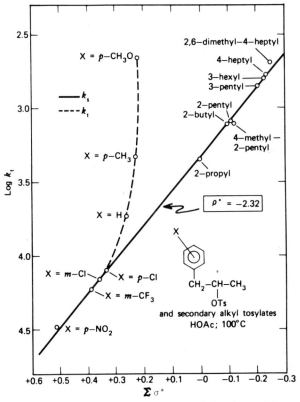

Fig. 1. Taft treatment of the acetolysis of secondary alkyl and 1-aryl-2-propyl p-toluenesulfonates.[18,19] Reproduced by permission of John Wiley and Sons.

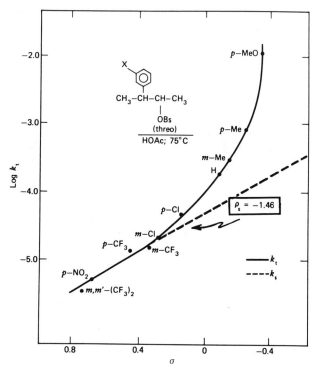

Fig. 2. Hammett treatment of the acetolysis of *threo*-3-aryl-2-butyl *p*-bromobenzenesulfonates.[18,19] Reproduced by permission of John Wiley and Sons.

Since the slope of this line is ρ^* for the solvent-assisted reaction (k_s), anchimeric assistance (k_Δ) is indicated by positive deviation from the line.[18,19] The amount of anchimeric assistance is then taken to be the ratio of k_t/k_s.[18] The Hammett treatment has also been applied to β-arylalkyl derivatives[19] (see Fig. 2), but the number of points available for the construction of the k_s line is normally smaller.

Peterson et al.[20,21] have shown that substituents remotely situated may affect the reactivity of a substrate in highly ionizing but nonnucleophilic solvents such as trifluoroacetic acid. When a substituent is not nucleophilically involved in a reaction the rate data are correlated by assuming that the inductive effect, defined as $\log k_H - \log k_X \equiv \Delta \log k$, falls off by a constant factor (ε) per methylene group, where ε is called the attenuation factor and k_H and k_X are the rate constants for the unsubstituted and substituted substrates undergoing reaction.[20,22] For application to addition to a homologous series of alkenes, Eq. (4) was formulated and transformed into Eq. (5), where n, m $(n > m)$ are the numbers of carbons in the alkenes

$$\Delta \log k_n = \varepsilon^{n-m} (\Delta \log k_m) \tag{4}$$

$$\log \Delta \log k_n = \log \varepsilon \, (n - m) + \log \Delta \log k_m \tag{5}$$

whose inductive effects are compared. When $\log \Delta \log k_n$ is plotted against the number of carbons in the alkene carbon skeleton, a straight line is defined for a homologous series of alkenes bearing a neighboring group that does not participate in the rate-determining step of addition.[20] Figure 3 illustrates this method. Anchimeric assistance is indicated when points deviate substantially from the line (on the fast side) defined by a homologous series.[20] For example, Peterson et al.[20] have used this method to demonstrate the presence of anchimeric assistance by the methoxy and chloro groups in addition of trifluoroacetic acid to homologous series of ω-substituted 1-alkenes (see Fig. 1 of Chapter 4) and Williams et al.[23] used the method to show the absence of anchimeric assistance by the bromo group but the presence of assistance by the hydroxyl group upon bromination of ω-substituted 1-alkenes.

c. *Other Empirical Methods.* Direct application of the Taft treatment to some systems, including the controversial cyclic and bicyclic systems, has been difficult, if not impossible, owing to the availability of σ^* values only for noncyclic substrates.[24] The problems associated with model com-

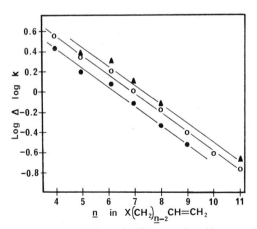

Fig. 3. Plot of rate data for the addition of trifluoroacetic acid to ω-substituted 1-alkenes having cyano (▲), trifluoroacetoxy (○), and acetoxy (●) substituents. Slopes of the lines correspond to an attenuation factor, ε, of 0.65.[20]

parisons are discussed above. Foote[25] and Schleyer[26] devised a method for estimating acetolysis rates of certain cyclic substrates that predicts with remarkable success solvolysis rates of many cyclic systems. Their empirical relationship between acetolysis rates of secondary *p*-toluenesulfonates and the carbonyl absorption frequency in the infrared of the derived ketone correlates acetolysis rates of "normal" substrates, as shown in Fig. 4. It is, however, bothersome to see known k_c and k_s substrates correlated by the same line. It is now known that the Foote–Schleyer relationship fails

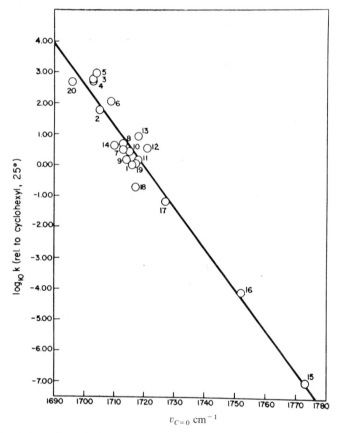

Fig. 4. Correlation of acetolysis rates of arenesulfonates and carbonyl frequencies of the corresponding ketones.[25] The numbers represent the following: 1–10, cyclohexyl through cyclopentadecyl, respectively; 11, 2-propyl; 12, 2-butyl; 13, 3-methyl-2-butyl; 14, 3,3-dimethyl-2-butyl; 15, 7-norbornyl; 16, *endo*-8-bicyclo[3.2.1]octyl; 17, 2-adamantyl; 18, α-nopinyl; 19, β-nopinyl; 20, 1,4-α-5,8-β-dimethanperhydro-9-anthracyl. Reprinted by permission of the American Chemical Society.

when other solvents are used [27,28] and when there is steric retardation to ionization.[29] The relationship predicts that the rates of some systems are slower than observed. Thus, it is said that these particular substrates are anchimerically assisted by the difference between the amount calculated and that observed. [25,26]

Another method[24] for predicting the rates of cyclic secondary systems (**17**) has recently become available. The method is a refinement of the α-Me/H ratio method[30] of estimating the magnitude of solvent and anchimeric assistance. Since a linear σ^* plot (Taft plot) is observed for the solvolysis of acyclic tertiary derivatives (**18**) in various solvents, by placing

(**17**)   (**18**)

the rates of cyclic tertiary derivatives (**18**) on a plot where a ρ^* value has previously been determined, a substituent constant for R in (**18**) can be determined. These constants, called σ_t^* values, will accurately reflect inductive, hyperconjugative, and steric effects for solvolysis of a secondary acyclic alkyl system (**17**) if there are no major steric differences between the secondary and tertiary models.[24] Exceptions are systems that adopt different ground-state conformations as secondary (**17**) or tertiary (**18**) systems (e.g., the medium ring systems) and systems where converting (**17**) to (**18**) adds ground-state steric strain [e.g., (**19**) and (**20**)]. In the exceptions noted, the

(**19**)   (**20**)

tertiary models would not reflect the steric conditions of the secondary systems and would result in erroneous predictions about the secondary system. In other cases, the σ_t^* values can be used to calculate the rates of solvolysis of secondary substrates using Eq. (6), where ρ^* is the reaction constant for the solvolysis of acyclic substrates in a particular solvent and $\log k_{2\text{-pr}}$ is the rate of solvolysis of the 2-propyl derivative in the same

$$\log k_{\text{calc}} = \rho^* \sigma_t^* + \log k_{2\text{-pr}} \qquad (6)$$

solvent. This particular method has been used[24,27] to determine σ_t^* values from the solvolysis of tertiary p-nitrobenzoates in 60% dioxane and tertiary chlorides in 80% aqueous ethanol. These σ_t^* values have been used to predict rates of acetolysis and ethanolysis of p-toluenesulfonates at 25°C with reasonable success.[24] The method works less well for formolysis[27] and predicts poorly for reactions in trifluoroacetic acid.[27] The reasons for the failure in highly ionizing solvents are not known, but it is possible that the electronic contributions of the methyl group in the tertiary model is not

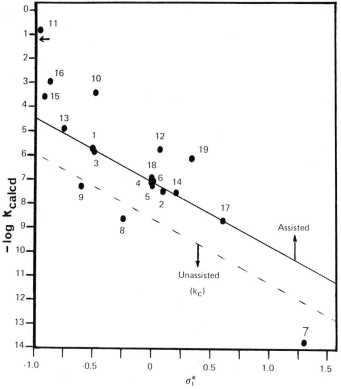

Fig. 5. Plot of actual (●) and calculated values (solid line) of acetolysis (25°C) rate constants versus calculated substituent constants for cyclic and polycyclic secondary alkyl p-toluenesulfonates. The solid line is defined by actual acyclic secondary acetolysis results and is for k_s substrates. The solid circles are 1, cyclopentyl; 2, cyclohexyl; 3, cycloheptyl; 4, cyclotridecyl; 5, cyclopentadecyl; 6, cycloheptadecyl; 7, 7-norbornyl; 8, 2-adamantyl; 9, 7,7-dimethyl-endo-2-norbornyl; 10, anti-7-norbornenyl; 11, 7-norbornadienyl; 12, exo-2-benzonorbornenyl; 13, exo-2-norbornyl; 14, endo-2-norbornyl; 15, 7,7-dimethyl-exo-2-norbornyl; 16, 1-methyl-exo-2-norbornyl; 17, 6,6-dimethyl-endo-2-norbornyl; 18, 1-methyl-endo-2-norbornyl; 19, cyclobutyl.

a constant.[27,28] Figure 5 shows a plot of calculated σ_I^* values versus log k for the acetolysis of several cyclic compounds for which the method is applicable. Since the solid line is defined by solvent-assisted acyclic systems, all compounds that fall on the line are either solvent or anchimerically assisted. Those compounds that fall significantly below the line are not assisted (i.e., k_c substrates), while those that are significantly above the line are strongly assisted.[24,27]

   d. *Calculation of Steric Effects.* There has been considerable debate[29,31,32] over the origin of rate accelerations in certain systems where steric factors may play a large role. Is the measured rate acceleration for 2,2,2-triphenylethyl derivatives[31] (**21**) a consequence of steric acceleration (Brown, *Chem. Soc. Spec. Publ.*[31]), i.e., the release of nonbonding com-

$$Ph_3C-CH_2X \longrightarrow Ph_2C\overset{Ph}{\underset{X^-}{\cdots\cdots}}CH_2 \longrightarrow Ph_2C=CHPh \qquad (7)$$

(**21**)

pression on going to the transition state for rearrangement, or of anchimeric assistance (Ingold[31]), i.e., stabilization of the transition state by bonding of the phenyl group to C-1? This question could be answered if one could quantitatively predict steric acceleration. There has been recent success[33] in treating steric factors in some kinds of reactions based on modification to a method first advocated by Westheimer.[34] The further development of this method may aid in resolving many long-standing controversies.

3.1.2. Participation as a Function of Electron Demand

   It is now well established that the extent of participation by a neighboring group is a function of several factors. Some of these factors were discussed at length in Chapter 2. We know, for example, that all nucleophilic neighboring groups do not participate to the same extent based on a certain electron demand at the center undergoing ionization. Instead, each neighboring group has its own unique ability to participate. Some groups are so powerful that, despite the lack of a recognized need, participation will occur. Other groups might be so weak that participation is difficult to recognize. The groups in the latter category are normally the source of controversy.

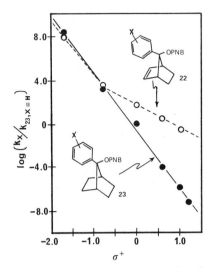

Fig. 6. ρ–σ^+ Plot of 7-aryl-7-norbornyl and syn-7-aryl-anti-7-norbornenyl p-nitrobenzoates in 70% aqueous dioxane at 25°C.

Gassman and co-workers[35] pitted the aryl delocalization of the p-anisyl group against the π-electron participation of the double bond in syn-7-p-anisyl-anti-7-norbornenyl p-nitrobenzoate (**22**) and found that (**22**) solvolyzes only ca. three times faster than the saturated ester (**23**). Since a rate difference of 10^{11} is observed for the parent secondary tosylates [i.e., (**3**) versus (**5**)], the π participation was reduced or leveled by the p-anisyl group by a factor of ca. 3×10^{10}. By extending their study over a reactivity range of 15 powers of 10, Gassman and Fentiman[36] demonstrated the linear relationship between electron demand and π participation. Graphically presented (Fig. 6), their results show the dramatic cessation of participation in (**22**) when aryl groups with substituents more capable of stabilizing a carbocation than the p-methoxy group are present.

The absence of participation in (22), X = p-OMe, was further demonstrated by a product analysis.[37] Instead of obtaining only the *anti*-alcohol (e.g., see p. 79), a mixture of the *syn-* and *anti*-alcohols were formed [Eq. (8)].

To test their conclusion that the *p*-anisyl group had a leveling capacity of ca. 3×10^{10}, Gassman and Fentiman[37] employed yet a stronger neighboring group, i.e., the remote cyclopropyl group in (24). The remote cyclopropyl group[38] was estimated[39,40] to fall into the relative reactivity order of secondary *p*-toluenesulfonates as follows:

k_{rel}: 1 10^{11} 10^{14}

Thus, Gassman and Fentiman[37] reasoned that the *p*-anisyl group in (24) should leave a residual factor of ca. 3×10^3 when the rate of solvolysis of (24) is compared with (18), X = *p*-OMe. The observed rate difference, $k_{(24)}/k_{(23),p\text{-OMe}} = 3.8 \times 10^3$, is remarkably close to that predicted. Furthermore, the cyclopropyl group exercises complete product control [Eq. (9)], consistent with participation.[37]

Since the *p*-anisyl group exhibits the same leveling capacity with two different neighboring groups, the use of aryl groups to probe for neighboring group participation seems valid. Brown has extensively used this tool in his quest to establish the presence or absence of π and σ participation in the norbornyl and related systems.[11] For example, π participation in the 6-methoxybenzonorbornenyl (25)[41] and the 5-methyl-2-norbornenyl[42] systems is revealed by a comparison of ρ values [cf. (25) versus (26) and (27) and (29) versus (30) and (31); see Table 1]. The magnitude of $\Delta\rho^+$ is considered a rough indication of the extent of anchimeric assistance.

In the above comparisons, anchimeric assistance in the model compounds (27) and (31) can be considered to be absent since ρ comparisons with their respective isomers (28) and (32) reveal similar electron demand.[41-43] It is not surprising that π participation is insignificant in the tertiary derivatives (27) and (31) if the *exo/endo* rate ratio of 7000 found for solvolysis of the secondary brosylates (38) and (39) has been properly dissected by Peters and Brown[45] into a factor of 350 for the inherent steric contribution and 20 for π participation. If the *p*-anisyl group can level

Table 1. ρ^+ Values for Solvolysis of Tertiary p-Nitrobenzoates in 80% Aqueous Acetone at 25°C

Structure	No.	ρ^+	Ref.
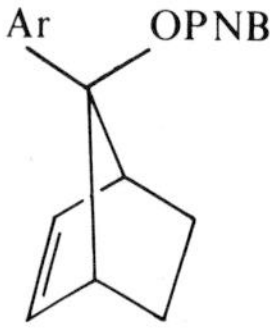 Ar-OPNB (norbornenyl)	(22)	-2.30^a	36
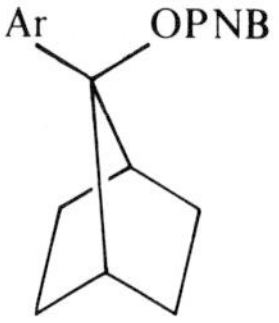 Ar-OPNB (norbornyl)	(23)	-5.27	36
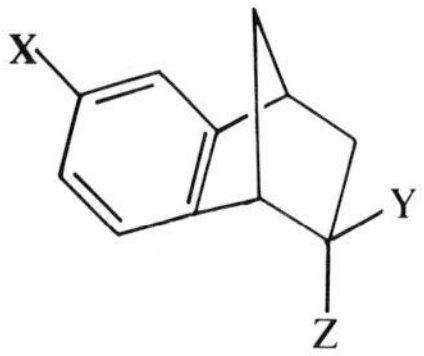 (benzonorbornenyl with X, Y, Z)	(25), X = OMe, Y = OPNB, Z = Ar	-3.72	41
	(26), X = OMe, Y = Ar, Z = OPNB	-4.05	41
	(27), X = H, Y = OPNB, Z = Ar	-4.50	41
	(28), X = H, Y = Ar, Z = OPNB	-4.52	41
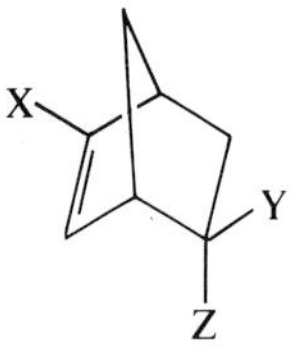 (norbornenyl with X, Y, Z)	(29), X = Me, Y = OPNB, Z = Ar	-3.27	42
	(30), X = Me, Y = Ar, Z = OPNB	-4.19	42
	(31), X = H, Y = OPNB, Z = Ar	-4.21	43
	(32), X = H, Y = Ar, Z = OPNB	-4.17	43
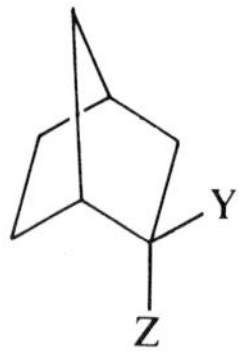 (norbornyl with Y, Z)	(33), Y = OPNB, Z = Ar	-3.83	44
	(34), Y = Ar, Z = OPNB	-3.75	44
1-Aryl-1-cyclopentyl-OPNB	(35)	-3.82	44
1-Aryl-1-cyclopropylethyl-OPNB	(36)	-2.78	43
2-Aryl-3-methyl-2-butyl-OPNB	(37)	-4.65	43

^{a}Measured in 70% dioxane; this value is only for the k_Δ portion of the reaction (see Fig. 6).

participation to the extent of 3×10^{11} (measured in the secondary derivatives), the 2,5-*bis*-(trifluoromethyl)phenyl group should level an anchimeric

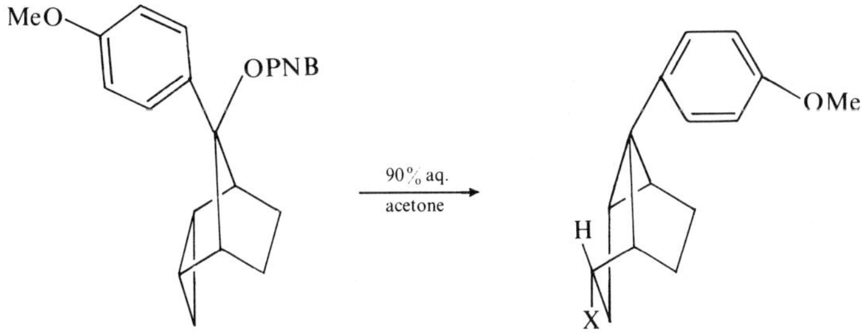

(9)

(24)   X = OH or OPNB

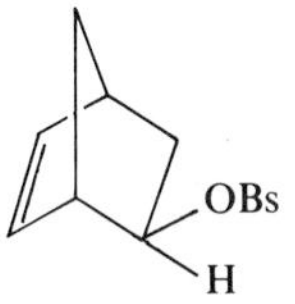

(38) k_{rel}: 7000

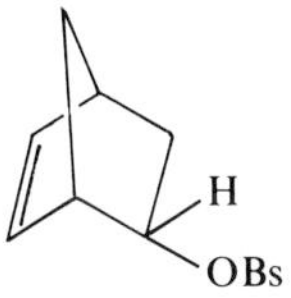

(39) 1

assistance rate factor of > 20. A linear correlation of log k_X/k_H versus σ^+ for (31) verifies that there is no change in mechanism over the range of reactivities measured (*p*-OMe to *m,m'*-di-CF$_3$). Brown *et al.*,[46] however, indicate that an aryl group with a more positive σ^+ than the 3,5-*bis*-(trifluoromethyl)phenyl group might show the onset of π participation.*

The study of a series of *exo*- and *endo*-2-aryl-2-norbornyl *p*-nitrobenzoates is said to offer no evidence favoring σ participation in the secondary 2-norbornyl system[44] [cf. the ρ values of (33), (34), and (35)]. However, this treatment suffers from the same handicap as found in the 5-norbornen-2-yl system; that is, the amount of anchimeric assistance present in the secondary derivative (if any) is small, and hence all the aryl groups commonly used are apparently capable of leveling the proposed σ participation.† Nevertheless, σ contributions can be measured by the

---

* It is possible that aryl groups may exhibit different apparent leveling effects in different substrates. Gassman and Fentiman[36] studied two different neighboring groups in the 7-norbornyl system and found that the *p*-anisyl levels participation of ca. 3×10^{11}. In the 5-methyl-2-norbornenyl system, however, the *p*-anisyl group appears to be leveling π participation of ca. 2.2×10^7; cf. footnote 11 of Ref. 46.
† There may be other factors contributing to a misleading result in the 2-norbornyl system (cf. Ref. 48). Some of these factors may invalidate other conclusions; cf. H. C. Brown and M. Ravindranathan, *J. Am. Chem. Soc.*, **97**, 2895 (1975).

use of this tool as it is unequivocally revealed in (36) when compared to (32).[43]

In summary, it should be emphasized that an intimate understanding of the solvolyses of all the systems under discussion, and especially the secondary derivatives of some of the tertiaries listed in Table 1, is not yet available. When a break in the Hammett plot occurs [as with (17)] there is a transition from a k_Δ to a k_c mechanism. When no such break occurs over a certain range of aryl derivatives the mechanism may be either k_Δ or k_c, and one concludes that all such cases in Table 1 are of the k_c type. If the tertiary norbornyl systems (33) and (34) are of the k_c type, it is doubtful that their behavior has any relevance to the behavior of the complex secondary 2-norbornyl systems.[10,11]

An interpretation of the magnitude of ρ is difficult since many factors are involved. One sees, for example, the effect of inductive electron withdrawal on the ρ values of (31) and (32) by their comparison with the saturated isomers (33) and (38).[45] Hyperconjugation is also said to be a factor[28,47]: Compounds stabilized hyperconjugatively [such as (36)] have low values of ρ, whereas compounds with little hyperconjugative stabilization [such as (23)] have high values of ρ. Yet it is difficult to rationalize from our present knowledge of inductive and hyperconjugative effects why the bicyclic and cyclic derivatives (33) and (35), which have nearly equal ρ values, should have a ρ value so unlike the structurally similar acyclic system (37).[48]

(33) ρ^+: −3.83

(37) −4.65

(35) −3.82

Virtually all authors of reviews on the Hammett treatment caution about the use of the results because of its empirical nature.[49] Hence, a full understanding of these and other complex problems correlated by Hammett plots awaits a method of predicting relative rates and ρ's for a new system before the actual rates are known.

McEwen et al.[106] have interpreted the results of their quarterization studies of o-anisyl-substituted phosphines and arsines in terms of participation as a function of electron demand.

3.1.3. Kinetic Isotope Effects*

In a number of different kinds of neighboring group processes, deuterium isotope effects have proven useful in providing information about the transition state of the rate-determining step. There are several kinds of isotope effects that can be studied; they are solvent isotope effects, primary isotope effects, and secondary isotope effects. Of these types, primary and secondary isotope effects are of more general importance in the study of neighboring group effects.

Kinetic isotope effects are understood[50] from statistical thermodynamics and absolute rate theory to result primarily from the difference in the zero-point energies in the bonding of isotopes to a particular atom. Zero-point energy is one of the factors that determines the effect of isotopic substitution on rate since a weakening of a bond (e.g., C—H) at the transition state implies a lower force constant for bond stretching and a lower zero-point energy for vibration. A C—H bond has a higher zero-point energy than a C—D bond initially; thus more zero-point energy is lost at the transition state for the C—H bond. As a result, the total energy difference between the ground state and the transition state is less (see Fig. 7).

The interpretation of isotope effects is far from straightforward in many cases. In the study of enzyme reactions in aqueous media, D_2O is sometimes used in place of H_2O. Rate differences noted, however, are of no use unless a careful analysis of the system is performed.[50] A change in rate may result from a change in the strength of a strong acid catalyst in an acid-catalyzed reaction. The effect of solvent (H_2O versus D_2O) on acid strength results because of zero-point energy differences in the bonds to oxygen. Solvation (association) is also affected when acidic bonds are changed from O—H to O—D (Melander[50]).

Solvent effects may be primary effects if the solvent is involved directly in the reaction (e.g., proton transfer from water in the rate-determining step). Since water as a solvent can cause rate differences in other ways (e.g., solvation differences), we shall not treat these complex effects here.

There are now many examples of primary isotope effects in chemical reactions.[50] In fact, three basically different kinds of primary isotope effects are measurable with hydrogen and deuterium [Eqs. (10)–(12)], yet there seems to be no certain way to interpret a given k_H/k_D magnitude in terms of one type of hydrogen transfer rather than another.[50]

* For more complete discussions, see Refs. 50 and 51.

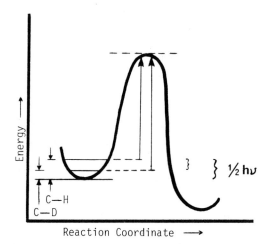

Fig. 7. In a reaction where there is no zero-point energy difference in the transition state, the isotope effect will result from the difference in the zero-point energies of the C—H and C—D bonds in the ground state.

In solvolytic reactions hydride transfer is an important process. If hydride transfer occurs during the rate-determining step, a deuterium

$$\underset{R_2}{\overset{R_1}{>}}\underset{R_3}{\overset{|}{C}}\!-\!\underset{(D)}{\overset{}{H}} \xrightarrow{B^-} \underset{R_2}{\overset{R_1}{>}}\underset{R_3}{\overset{\ominus}{C}} \quad + \quad \underset{(D)}{\overset{}{BH}} \tag{10}$$

$$\underset{R_2}{\overset{R_1}{>}}\underset{R_3}{\overset{|}{C}}\!-\!\underset{(D)}{\overset{}{H}} \xrightarrow{rad \cdot} \underset{R_2}{\overset{R_1}{>}}\overset{}{C}\!-\!R_3 \quad + \quad \underset{(D)}{\overset{}{rad\!-\!H}} \tag{11}$$

$$\underset{R_2}{\overset{R_1}{>}}\underset{R_3}{\overset{|}{C}}\!-\!\underset{(D)}{\overset{}{H}} \xrightarrow{R^+} \underset{R_2}{\overset{R_1}{>}}\overset{+}{C}\!\underset{R_3}{\diagdown} \quad + \quad \underset{(D)}{\overset{}{RH}} \tag{12}$$

isotope effect should be measurable. Although the number of reports of deuterium isotope effects measured in hydride transfers is not large, the k_H/k_D may range up to about 5, which is the value obtained for hydride transfer from cycloheptatriene to the di-p-anisylphenylmethyl cation.[52] The exact value may vary from this, however, since the magnitude of the kinetic isotope effect is a function of changes in zero-point energy in the transition state, which in turn is a function of the symmetry of the transition state.[53] To give a simple illustration, consider the symmetrical transition state (S); there will be no motion of the hydrogen in this vibration; thus there will be no difference between the zero-point energies for hydrogen and deuterium in the transition state, and a normal kinetic isotope effect should be observed. If the transition state is asymmetrical as in (A), the hydrogen atom

$$\overset{\leftarrow}{G_1} \cdots H \cdots \overset{\rightarrow}{G_2} \qquad \overset{\leftarrow}{G_1} \cdots \overset{\leftarrow}{H} \cdots \cdots \vec{B}$$

(S)   (A)

will likely move in the transition state, resulting in a change in zero-point energy upon substituting deuterium for hydrogen. The observed isotope effect will then be reduced from the theoretical maximum to the extent that the difference in zero-point energy of the starting material is maintained in the transition state by this vibration.[53]

Hyperconjugation has been claimed (Shiner[50]) to be the predominant factor in determining the magnitude of β-deuterium isotope (β-d) effects. Accordingly Shiner et al. have found that charge delocalization through resonance in the transition state reduces the magnitude of β-d effects.

	Me	OMe
(40)	(41)	(42)
CHCD$_3$–Cl	CHCD$_3$–Cl	CHCD$_3$–Cl
k_H/k_D: 1.22	1.20	1.11

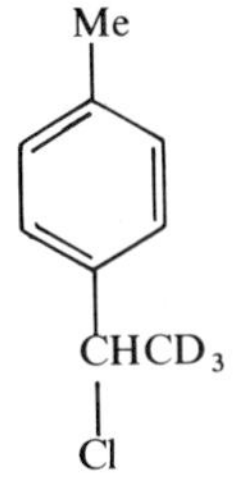

This is clearly indicated by a comparison of k_H/k_D for the chlorides (40), (41), and (42) upon solvolysis in 60% ethanol at 25°C.[54]

Sunko and Borcic[51] conclude that bridging by a neighboring group can also reduce the usual magnitude of both α-d and β-d effects if bridging

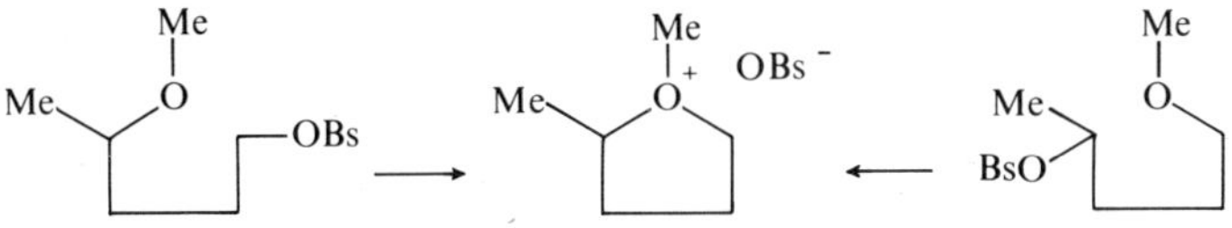

(43) → (13) ← (44)   (13)

is very advanced in the transition state. These conclusions are supported by the noncontroversial examples of *n* participation in the solvolysis of the ethers (**43**) and (**44**), both of which give the same oxonium ion (see pp. 127–129). These anchimerically assisted reactions give very small isotope effects (Table 2), if any, which was taken as evidence of considerable bridging in the transition state.[51]

Unfortunately, a large degree of anchimeric assistance does not automatically lead to a greatly reduced isotope effect. The cholesteryl derivative (**45**),[57] the cholestanyl derivative (**46**),[57] and the 7-norbornenyl derivative (**47**)[55] show the same α-*d* effect, yet (**45**) and (**47**) react with strong anchimeric assistance from the π electrons. Furthermore, (**45**) and (**47**) have greatly

(**45**)   (**46**)   (**47**)
k_H/k_D: 1.13   1.13   1.13
(50 C.96 % EtOH)   (60 C.96 % EtOH)   (30 C.AcOH)

different π-*p* overlap configurations. The saturated derivative is presumed to react by a k_s process. The α-*d* effect is said to be smaller in (**45**) and (**47**) than in (**44**) owing to bridging lagging ionization.[51] We are not certain that the last conclusion is justified since it is now known that the magnitude of the α-*d* effect is dependent on the leaving group (Shiner[50]), on the degree of nucleophilic solvation (Shiner[50]), and on substrate structure.[58,59,60] These special effects must be weighed against hybridization changes[61] and inductive effects[62] as factors in producing the observed isotope effect. Hence the interpretation of α-*d* effects requires considerable care, and the ambiguities devalue the importance of α-*d* effects in the assignment of participation.

More remote secondary isotope effects, such as γ-*d* effects, cannot be generalized, and thus each effect requires a specific explanation.[51] An example of the use of a combination of α-, β-, and γ-*d* effects in the study of participation by the neighboring aryl group is included below to represent the use of these effects in studies of neighboring groups. The interpretations presented are the reported ones[63] rather than our own.

Loukas *et al.*[63] measured α, β, and γ secondary deuterium isotope

Table 2. Secondary Isotope Effects in the Solvolysis of Methoxy-
Substituted Alkyl p-Bromobenzenesulfonates

Compound	Temperature (°C)	Type	k_H/k_D per atom D
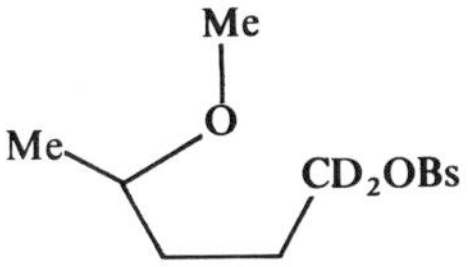 Me-C(Me)(OMe)-CH-CD₂OBs	39.7	α-d	1.00[a]
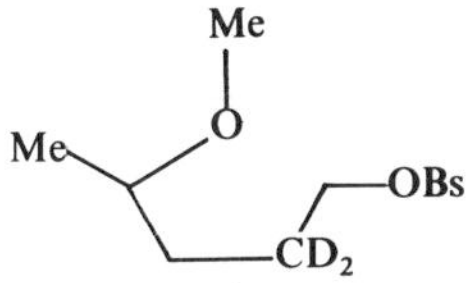 Me-C(Me)(OMe)-CH(-CD₂-OBs)	40.5	β-d	1.00[b]
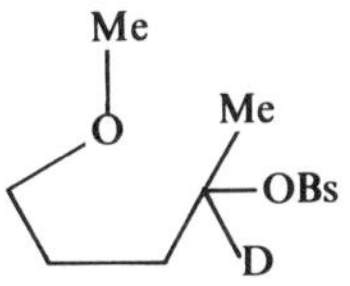 [tetrahydrofuran with Me, OBs, D]	40.3	α-d	1.08[b]
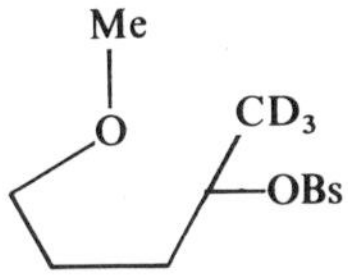 [tetrahydrofuran with Me, CD₃, OBs]	40.5	β'-d	1.03[b]

[a] Reference 55.
[b] Reference 56.

effects for solvolysis of threo-3-phenyl-2-butyl-p-toluenesulfonate (48), a substrate known to undergo competitive k_s and k_Δ processes[18,19] (see Figs. 1 and 2). The results of rate studies on (48a)–(48e) in acetic acid, formic acid, and trifluoroacetic acid are shown in Table 3. It was considered that ionization to the unbridged 3-phenyl-2-butyl cation through transition state (49) should be attended with an α-deuterium isotope effect of about 12–15% (i.e., $k_H/k_D \equiv 1.12$–1.15) (see Streitwieser[50]; actually, the "limiting" value of sulfonates in highly ionizing, nonnucleophilic solvents is much higher, i.e., 22–25%; cf. Refs. 64 and 65), a β-deuterium isotope effect of about 8–15%, and a γ-deuterium isotope effect of roughly 1%.[66] A near-symmetrical aryl bridged transition state (50), on the other hand, should

Table 3. Solvolysis of Deuterated Derivatives of threo-3-Phenyl-2-butyl-p-toluenesulfonate

Derivative	Solvent, temp. (°C)	Type	k_H/k_D
(48b)	AcOH, 75.10	α-d	1.104 ± 0.006^a
(48c)	AcOH, 75.10	β-d	1.094 ± 0.012^a
(48d)	AcOH, 75.10	β-d	1.073 ± 0.009^b
(48e)	AcOH, 75.10	γ-d	1.065 ± 0.009^b
(48b)	HCO_2H, 25.02	α-d	1.142 ± 0.010^a
(48c)	HCO_2H, 25.02	β-d	1.040 ± 0.010^a
(48d)	HCO_2H, 25.02	β-d	1.160 ± 0.009^b
(48e)	HCO_2H, 25.02	γ-d	1.015 ± 0.007^b
(48b)	TFA, −7.9	α-d	1.133 ± 0.014^a
(48c)	TFA, −7.9	β-d	1.009 ± 0.015^a
(48d)	TFA, −7.9	β-d	1.170 ± 0.026^b
(48e)	TFA, −7.9	γ-d	1.009 ± 0.019^b

[a] Corrected to 1.0 deuterium per molecule.
[b] Corrected to 3.0 deuteriums per molecule.

tend to equalize the electronic requirements on the C-2 and C-3 hydrogens and on the C-1 and C-4 hydrogens; thus this model would predict equaliza-

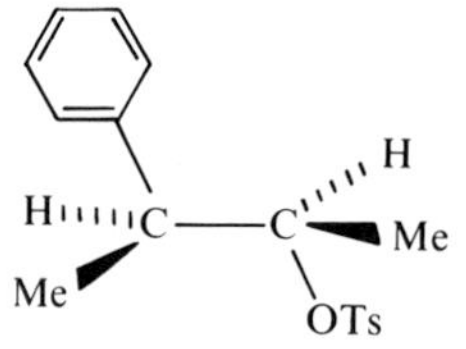

(48)

a: $CH_3CHPhCHOTsCH_3$
b: $CH_3CHPhCDOTsCH_3$
c: $CH_3CDPhCHOTsCH_3$
d: $CH_3CHPhCHOTsCD_3$
e: $CD_3CHPhCHOTsCH_3$

tion of the isotope effects expected (Streitwieser[50]).[65-67] The presence of the solvent-assisted route was expected to affect an interpretation of the results only to a minor extent, depending on how much charge at carbon is present in the rate-determining step [i.e., how much it resembles (49)]. The data for solvolysis of (48) in acetic acid were judged to be reasonably

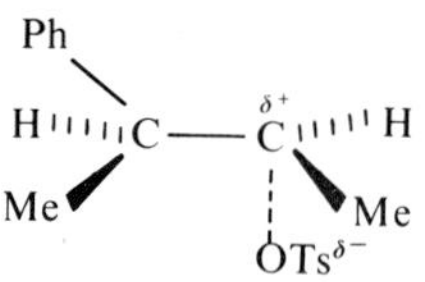

(49) 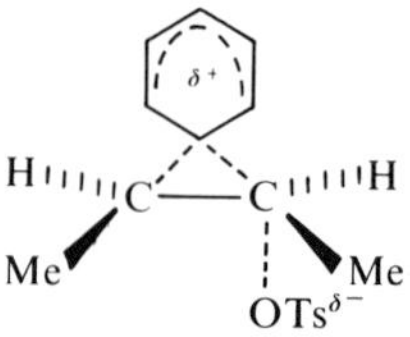   (50) 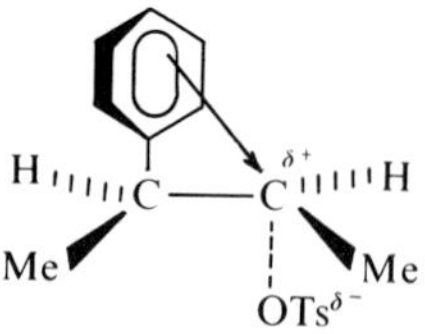   (51)

consistent with the bridged transition state. In the better ionizing solvents, formic acid and trifluoroacetic acid, the results are inconsistent with a model such as (**50**), and if bridging is important, the transition state was said to more resemble (**51**) for the fraction leading to a bridged intermediate.

3.2. Investigation of Solvent Effects

The ability of a neighboring group to participate in a reaction is obviously dependent on many factors. As already mentioned above, solvent effects are important in many cases in these processes; hence understanding the roles of solvent is important in defining the role of the neighboring group. We shall now discuss some ways in which the solvent affects neighboring group reactions and how changes in solvent may be used to probe the reaction mechanism.

Prototropic groups, such as the carboxyl and hydroxyl groups, may act either as general acids, as general bases, or as nucleophiles in intramolecular reactions. Such reactions will therefore be pH dependent since the species must be present in the proper form if it is to be effective. A sigmoidal rate constant versus pH curve that has its maximum at low pH indicates general acid catalysis by the un-ionized form of the neighboring group, while a sigmoidal curve that is maximum at high pH is indicative of general base or nucleophilic behavior by the ionized form. Bender[1] and Jencks[1] extensively treat pH dependence in intramolecular catalysis.

In addition to direct involvement as above, the carboxylate group in its ionized form may electrostatically stabilize a cationic transition state and resulting intermediate.[68,69] Such stabilization is solvent dependent since ionization is sensitive to solvent effects. The hydrolysis of (**52**) is an example where such effects may occur.[69] In 50% aqueous dioxane the *ortho/para* rate ratio, $k_{(52)}/k_{(53)}$, is 50. The rate ratio is only 12 in 50% aqueous dimethylsulfoxide. Although nucleophilic assistance was not excluded, Vitullo and Grossman[69] claim that electrostatic stabilization (ion pairing) can account for the change in the *ortho/para* rate ratio upon changing solvents since in a more polar solvent the need for electrostatic stabilization is lessened owing to effective solvation of the transition state and intermediate by the polar solvent. Subsequent studies, however, revealed that the *ortho* derivative (**52**) undergoes a mechanistic change when the solvent is changed from mixed water–dioxane to pure water. This is in-

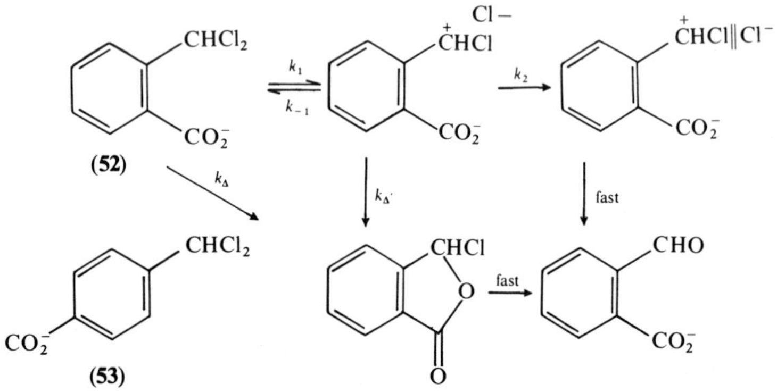

Scheme 2

dicated by a change in the α-deuterium isotope effect and by the variation in linearity of the Grunwald–Winstein plot (Fig. 8).[70] Other mechanistic probes[70] suggest that rate-determining formation of the solvent-separated ion pair occurs in water (rate-limiting k_2, Scheme 2) and that rate-determining ring closure from the tight ion pair occurs in less polar solvents (rate-limiting $k_{\Delta'}$, Scheme 2). The question of electrostatic stabilization was not resolved.

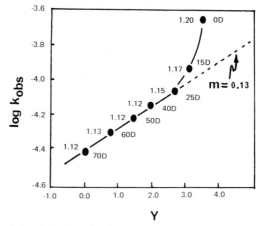

Fig. 8. Plot of log k for the hydrolysis of (5) versus Grunwald–Winstein values. Numbers to the left of each datum point refer to the α-d effect determined under these conditions. Numbers to the right of each datum point refer to the volume percent dioxane in the solvent; e.g., 70D = 70% dioxane.[70]

The solvent may also drastically affect the anchimerically assisted fraction in nucleophilically assisted ionization reactions. As the solvent becomes more nucleophilic, there is greater solvent participation, if possible, at the expense of neighboring group assistance. The data in Table 4 are illustrative of this effect. It is obvious that in both formic acid and acetic acid there is considerable MeO-5 and MeO-6 nucleophilic participation. In ethanol, however, the ratio k_Δ/k_s is diminished as solvent attack (k_s) becomes more effective. These effects may be quantitatively assessed. Winstein et al.[71] used a Taft treatment for the system RCH_2OBs in estimating the k_s portion of the solvolytic rate constant [Eq. (14)]. Values for k_t/k_s, where k_t is the titrimetrically observed rate constant, were determined,

$$\log\left(\frac{k_R}{k_{Me}}\right) = a + \rho^*\sigma^* \tag{14}$$

and, according to Eq. (15), k_Δ can also be estimated. Without correcting for ion-pair return, the k_t/k_s values obtained for the series in Table 4,

$$\frac{k_\Delta}{k_s} = \frac{k_t}{k_s} - 1 \tag{15}$$

where MeO-5 participation is occurring, are 22.2, 610, and 425 for EtOH, HCO_2H, and AcOH, respectively. These values show that the k_Δ process is dominant for the MeO-5 process in all solvents. For the MeO-6 process, however, k_t/k_s for EtOH is only 2.42, indicating that k_Δ and k_s are nearly equal.

Another effect of solvent in a solvolytic-type reaction is revealed by the plot of $\log k$ versus σ^* for a series of primary alkyl p-toluenesulfonates

Table 4. Relative Rates of Solvolysis of ω-Methoxyalkyl p-Bromobenzenesulfonates[71]

Compound	Relative rate		
	EtOH (75°C)	AcOH (25°C)	HCO_2H (75°C)
$Me(CH_2)_3OBs$	1.00	1.00	1.00
$MeO(CH_2)_2OBs$	0.25	0.28	0.10
$MeO(CH_2)_3OBs$	0.67	0.63	0.33
$MeO(CH_2)_4OBs$	20.4	657.0	461.00
$MeO(CH_2)_5OBs$	2.8	123.0	32.6
$MeO(CH_2)_6OBs$	1.19	1.16	1.13

(ROTs) in various solvents (Fig. 9).[72,73] The partitioning of k_s and k_Δ hydrogen or methyl participation) provides a dramatic illustration. Because of the varying requirements for nucleophilic solvent assistance and the varying abilities of a wide range of solvents to lend assistance, one can sometimes find what may appear to be anomalous trends in comparing results in more than one solvent.[73,74] It was found in studies of β-arylethyl derivatives, for example, that both the *p*-anisyl and π-phenyl chromium tricarbonyl groups seem to lose effectiveness as neighboring groups, when compared to the phenyl group, upon going from acetic to the more ionizing formic acid.[74] The same trend is present in the primary alkyl series when the results of acetolysis and formolysis are compared. The secondary series shows just the opposite effect: The better neighboring group gains effectiveness on going to formic acid as compared to acetic acid.[19] Figure 9

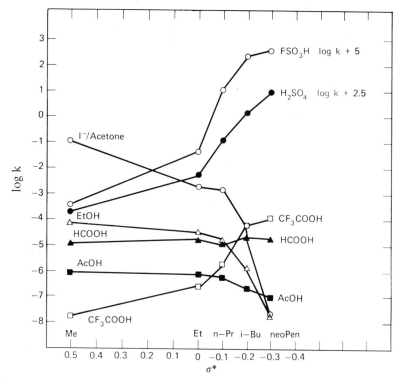

Fig. 9. Effect of solvent on rate as measured by the response of log k versus σ^* for a series of primary substrates in solvents of a wide range of ionizing abilities and nucleophilicities.[72,73] Reproduced by permission of John Wiley and Sons.

should suggest that an outright prediction among two closely related solvents is risky. It seems well established, however, that the removal of k_s as a competing process allows one to say with assurance that, unless a system is already 100% k_Δ, a change in solvent toward a more ionizing solvent will increase the fraction of k_Δ unless k_Δ is also difficult; in that instance, a k_c process is realized.

The 2-adamantyl system (**54**) has been proposed as a model k_c system.[75] Numerous studies have indicated that neither nucleophilic attack on (**54**)

(**54**)

nor rearrangement of the cation resulting from ionization of (**54**) are important. If the contribution from nucleophilic solvent assistance is assumed to be negligible in trifluoroacetic acid, the relationship in Eq. (16) allows one to calculate the nucleophilic solvent contribution for other substrates

$$\left(\frac{k_s}{k_c}\right)_{\text{ROTs,solvent}} = \frac{(k_{\text{ROTs}}/k_{\text{2-AdOTs}})_{\text{solvent}}}{(k_{\text{ROTs}}/k_{\text{2-AdOTs}})_{\text{TFA}}} \quad (16)$$

in any other solvent. Applying this relationship to the *exo-* and *endo-*2-norbornyl *p*-toluenesulfonates,[12] we find that the calculated ratio for k_s/k_c for the *exo-*OTs is 7.53 in acetic acid at 25°C, while the *endo* isomer has a calculated k_s/k_c of 30.0. Under the same conditions, 2-propyl *p*-toluenesulfonate shows a k_s/k_c ratio of 547. The differences in the assigned solvent assistance in the 2-propyl esters and that of the isomeric 2-norbornyl esters must reflect differences in the importance of electronic and steric effects in these systems.

Schleyer and his co-workers (see the first reference in Ref. 75) have defined criteria that, when used together, may aid in the determination of the importance of anchimeric assistance in solvolytic reactions. They have compiled data (Table 5) that allow one to assess reasonable values for a limiting (k_c) solvolysis and a solvent-assisted (k_s) solvolysis.

Grunwald–Winstein *m* values (see Ref. 76 and the last reference in Ref. 75) measure the susceptibility of a substrate to solvent ionizing power. The tertiary substrates and 2-adamantyl are all k_c substrates and all have

Table 5. Summary of Mechanistic Criteria for Solvolysis at $25°C^a$

	m (aq. alcohol)		$(k_{aq.alc.}/k_{AcOH})_Y$		k_{OTs}/k_{Br}	
	X = Br	X = OTs	X = Br	X = OTs	80% C_2H_5OH	CH_3COOH
Methyl	0.22	0.23 (75°)		97(75°)	11 (50°)	470 (100°)
Ethyl	0.34 (55°)	0.25 (50°)		70(50°)	10 (50°)	
Isopropyl	0.43	0.42	40 (50°)	7.8	40 (50°)	
Cyclohexyl		0.44		4.3		
2-Adamantyl	1.03	0.91	4.8	0.13	231	16,000
t-Butyl	0.94		3		4000	90,000
1-Bicyclo(2.2.2)octyl	1.03		1.6		5000	90,000
1-Adamantyl	1.20^b	0.99	4.2	0.16	9750	200,000

^{a}Literature references to the data can be found in the first reference in Ref. 75.
^{b}Taken from the second reference in Ref. 75.

m values of ca. 0.9–1.2 (see Table 5). The secondary and primary substrates listed in Table 5 are undergoing solvolysis in aqueous ethanol with nucleophilic solvent assistance. Unlike the k_c substrates, k_s substrates can be assumed to have m values between ca. 0.2 and 0.5 in aqueous ethanol.*

Substrates ionizing with neighboring group assistance should have m values much lower than k_c substrates, yet unlike k_s substrates their rate should be less dependent on solvent nucleophilicity. The amount of nucleophilic solvent assistance is indicated by $(k_{aq.alc.}/k_{AcOH})_Y$ [or by use of Eq. (16), vide supra]. One can see from Table 5 that primary substrates have large $(k_{aq.alc.}/k_{AcOH})_Y$ ratios, whereas k_c substrates have rather low values, as expected. Thus, a compound undergoing neighboring-group-assisted ionization will have a ratio similar to a limiting (k_c) substrate.

Although Schleyer (see the first reference in Ref. 75) has advocated the use of k_{OTs}/k_{Br} ratios, he has warned of the possible difficulties in interpreting the results since steric factors often dominate the ionization of tosylates but not bromides.[77] Nevertheless, that ratio is perhaps of use in the simpler systems. In other cases, k_{Br}/k_{Cl} ratios[78] have been used to indicate bond extension in the transition state of ring closures; normal values are, however, unpredictable (Table 6). The general purpose of k_{OTs}/k_{Br} ratios is to determine if nucleophilic assistance is occurring. If nucleophilic participation is occurring [indicated by a low m value and a low k_{OTs}/k_{Br} (see Table 5) or k_{Br}/k_{Cl} (Table 6) ratio], and this assistance is not solvent assistance [indicated by the use of $(k_{aq.alc.}/k_{AcOH})_Y$ or Eq. (16)], then neighboring

* m Values have now been calculated for a wider range of solvent ionizing ability; cf. the last two references in Ref. 75.

Table 6. k_{Br}/k_{Cl} Solvolytic Rate Ratios for Selected Alkyl Substrates

Substrate	Solvent, temp. °C	k_{Br}/k_{Cl}	Ref.
Isopropyl	80% aq. EtOH, 50°	76	17
t-Butyl	80% aq. EtOH, 25°	39.4	17
t-Amyl	80% aq. EtOH, 25°	42.1	17
1-Adamantyl	50% aq. EtOH, 25°	28.6^a	75
	80% aq. EtOH, 70°	29.6	77
3-Homoadamantyl	80% aq. EtOH, 70°	45.5	77

^{a}Chloride data from P. v. R. Schleyer and V. Buss, *J. Am. Chem. Soc.*, **91**, 5880 (1969).

group assistance is indicated. For further discussion of the use of these criteria in assigning a k_Δ mechanism, see Ref. 79.

3.3. Isolation or Trapping of Intermediates

Just as in other types of mechanistic studies the isolation of a suspected intermediate in a neighboring group process is proof of the involvement of that particular group. One must recognize that isolation of intermediates or products is not necessarily conclusive evidence of *anchimeric assistance* since the neighboring group could have captured the first-formed intermediate following the rate-determining step. When taken together with other knowledge about the reaction under study, intermediate isolation or trapping, or failure in an attempt to do so, may yield valuable information about the process.

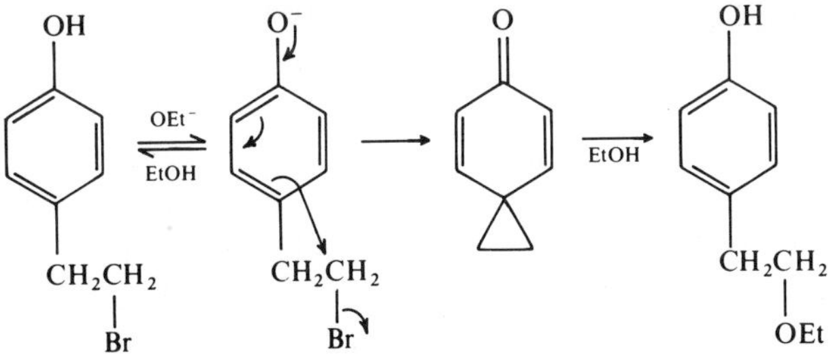

The first-order rate constant for the ethanolysis of the conjugate base of β-p-hydroxyphenylethyl bromide (**55**) is roughly 7 powers of 10 faster than the parent phenyl derivative.[80] This leaves little doubt that anchimeric assistance is involved, yet the isolation of the "phenonium" intermediate (**56**), at that time, was of some significance since it provided clear evidence of the potential intermediacy of other "phenonium" species in systems where kinetic evidence was less compelling. Winstein et al.[81] were later to trap an intermediate phenonium ion. By choosing an appropriate system and adjusting the reaction conditions, they obtained the kinetic product (**59**) from solvolysis of the β-9-anthrylethyl p-toluenesulfonate (**57**) through intermediacy of the cation (**58**). The thermodynamic product (**60**) was

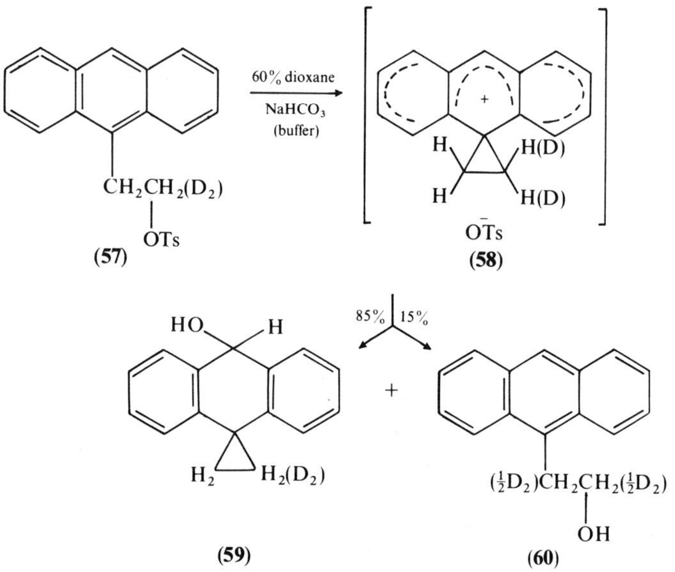

$$\text{(57)} \xrightarrow[\text{NaHCO}_3 \text{ (buffer)}]{60\% \text{ dioxane}} \text{(58)} \quad (18)$$

formed as the minor product under their carefully chosen reaction conditions. Deuterium labeling also served to implicate (**58**), as shown in Eq. (18).

Just as in the study above, it is often the case that the intermediates from neighboring group participation, being the kinetic but not the thermodynamic products, are destroyed under the normal reaction conditions; thus special conditions must be employed if the intermediate is to be isolated or trapped. Such was the case in the trapping[82] of the intermediate dioxolan-2-ylium cation (**61**) from the solvolysis of trans-2-acetoxycyclohexyl halides or arenesulfonates. The reaction is anchimerically assisted, but the assistance is somewhat obscured by the rate-retarding effect of the electron-withdrawing acetoxy group.[82] The isolation of the salt (**61**) assures the stability of the solvolytic intermediate.

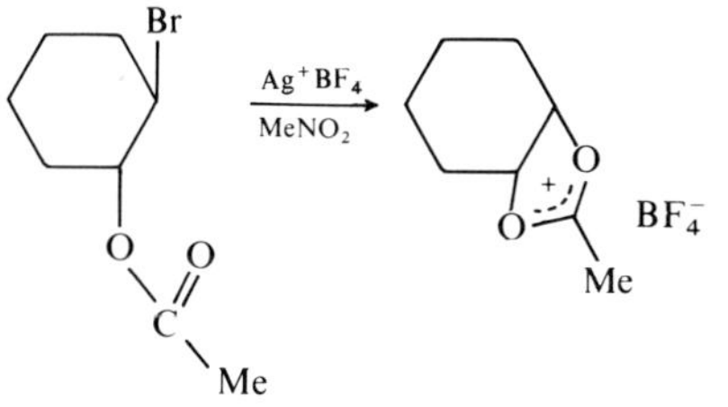

$$\text{(diagram 61)} \quad (19)$$

The elusive "tetrahedral intermediate," long implicated in acyl transfer reactions, has also been isolated.[83,84] Under the proper conditions, for example, the phthalimidium cation (62) gives the same intermediate (63) that one envisions from hydroxyl group attack on the carbonyl in (64).[84]

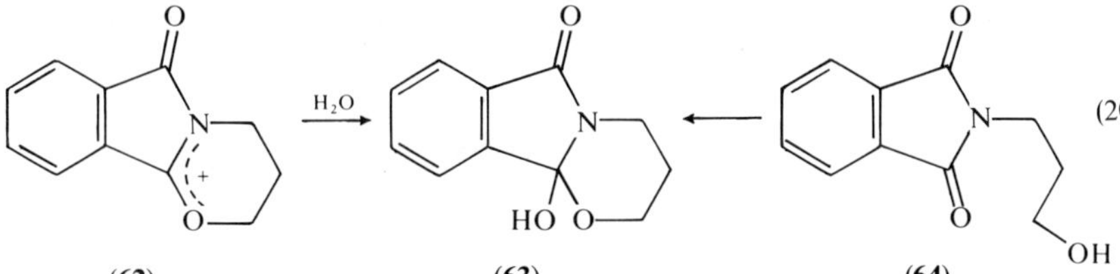

$$\text{(diagrams 62, 63, 64)} \quad (20)$$

Other suspected intermediates are not always amenable to isolation. One such example is the highly strained α-lactone from participation of the carboxylate group in the displacement of the halide ion in α-halocarboxylate salts. While much of the evidence points to the involvement of the lactone (66) in the hydrolysis of (65), many attempts at its isolation have failed.[85] (The alternative intermediate is an electrostatically stabilized species; cf. Ref. 68.) Based on the chemistry[86] expected of (66), however, one would not predict it to survive even the mildest conditions of hydrolysis reactions.

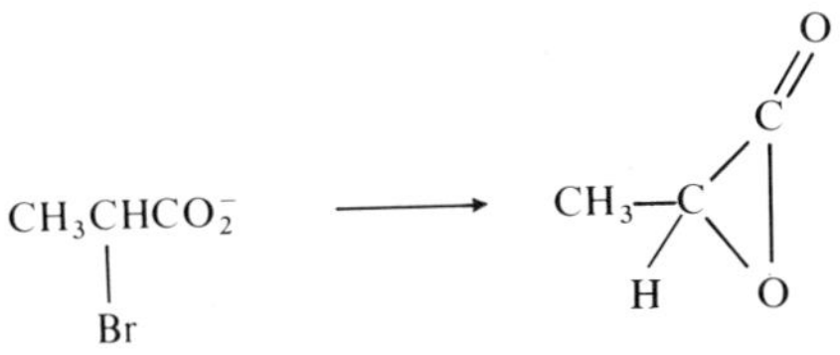

$$\text{CH}_3\text{CHCO}_2^- \ (\text{Br}) \quad \longrightarrow \quad \text{(diagram 66)} \quad (21)$$

(65)   (66)

3.4. Spectroscopic Observation of Intermediates

In recent years spectroscopic methods have yielded valuable information about species previously suspected as intermediates in neighboring group processes. It would be fruitless to try to recall here all or even a significant number of studies involving spectroscopic methods. Reviews have been provided in the series by Olah and Schleyer[87] and in other sources by Bethell and Gold[88] and McManus[1]; various other reviews and monographs are appearing periodically.[89] A summary of some of the useful methods is included in Table 7.

Direct observation of reaction intermediates by spectroscopic methods often requires the use of solvent media and temperatures unlike those where the intermediates are first suspected to occur. For this reason some skepticism may remain as to their existence as true intermediates under other conditions. This type of information is still of value, however, since

Table 7. Spectroscopic Methods Useful in Studying Ionic Organic Reactive Intermediates

Method	Information provided, applications, and limitations
NMR	
1. Proton	Most valuable for routine use in product studies. Temperature dependence gives information about static vs. equilibrating intermediates; chemical shifts give a measure of the electron density about the protons. Line-broadening analysis may give activation parameters for rearrangements. The proton spectra of many stable ions have been obtained and discussed. Time scale is slow enough that bridged ions cannot often be distinguished from equilibrating ions.
2. Carbon	With Fourier transform capability, it is possible to observe every carbon in the system; temperature dependence yields information on bridged vs. equilibrating nature of species. Difficulty may arise in calculating chemical shifts. $J_{^{13}C-H}$ allows calculation of hybridization.
ESCA	Gross atomic charge can be correlated with the electron binding energy chemical shift (normally 1s binding energy); offers a fast method: time scale $\sim 10^{-16}$. Difficulty arises in obtaining reproduceable spectra of ions in solid state (as salts).
Raman/IR	While direct interpretation is difficult, especially with large ions, this method offers an infrequently used complementary comparative method for identifying characteristic vibrations. Time scale for vibrations is $< 10^{-12}$.
UV/Vis	Useful in detecting small quantities of intermediates, starting materials, or products. This method is often used in following kinetics where it may be the detector in fast kinetic techniques.

one learns that such species can indeed exist albeit under controlled conditions. The 2-norbornyl cation falls into this category. It has now been studied in media of low nucleophilicity at low temperatures, and the combination of Raman, ^{1}H and ^{13}C NMR, and electron spectroscopy (ESCA) methods led Olah et al. to conclude, without any doubt, that **(67)** is the structure of the stable norbornyl cation.[90] This structure is essentially that proposed by Winstein and Trifan[10] to account for results obtained in their solvolytic studies of exo-2-norbornyl p-bromobenzenesulfonate **(68)**. The direct observation of **(67)** in solvents of low nucleophilicity at low tempera-

(22)

(67)   **(68)**

tures does not prove that **(67)** is an intermediate in the acetolysis of **(68)** and does not prove that participation is involved in the formation of any norbornyl ion pair from **(68)**. If, however, **(67)** is more stable than the equilibrating ions **(69)** and **(70)** in a nonnucleophilic media, it is reasonable to assume that **(67)** should also be more stable than the equilibrating ions in

(23)

(69)   **(70)**

acetic acid. Thus postulation of the involvement of **(67)** in the solvolytic process, and its likely formation via σ participation, in view of the other convincing data available,[24,26,31] seems justified if the recent challenges[91] to Olah's data and their interpretation are without adequate basis.

Halonium ions are another group of intermediates that have been characterized by spectral methods. Peterson et al.[92] have shown that five- and six-membered cyclic halonium ions are reasonably stable species capable of being isolated, stored, and subjected to reaction with a variety of nucleophiles. Three-membered cyclic halonium ions (ethylenehalonium ions) are not so stable, yet because of their long-postulated involvement in electrophilic addition reactions with alkenes and in the nucleophilic displacement of β-substituted leaving groups, they have attracted a significant amount of attention. Olah et al.[93] have recently reported spectroscopic studies on cyclic bromonium ions that reveal that, in media of low nucleophilicity, the bridged species may be static or in equilibrium with its isomeric carbocationic species, depending on structure. It is of interest that regardless of whether *meso-* or *d,l*-2,3-dibromobutane is used as the precursor the same mixture of the bridged bromonium ions **(71)** and **(72)** results.[94] This suggests

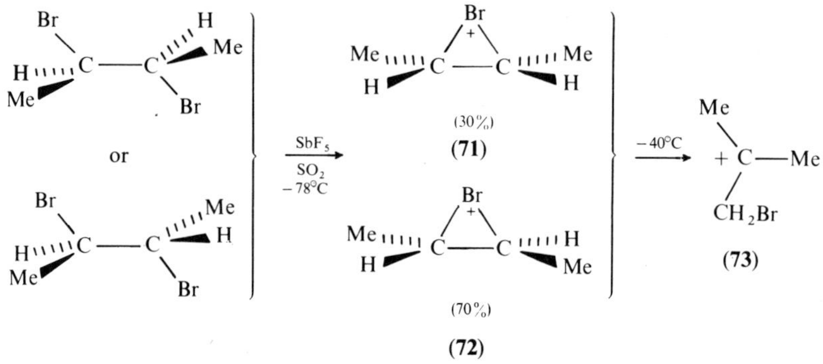

a common intermediate during their formation or that **(71)** and **(72)** are equilibrated by opening to the secondary carbocation, which is free to rotate. Upon warming to −40°C the mixture of **(71)** and **(72)** rearranges by sequential 1,2-H and 1,2-Me shifts to form the 1-bromo-2-methyl-2-propyl cation **(73)**, which probably exists in equilibrium with the bridged form.[93]

The halonium ion systems present quite a different problem in interpretation than did the 2-norbornyl system. In the 2-norbornyl system, the choice is between one carbocation versus another, although they admittedly present different bonding situations [i.e., **(67)** versus **(69)** ⇌ **(70)**]. In the case of halonium ions, the choice is between a charged carbon or a charged halogen (or perhaps a species with delocalized charge). There is evidence that as the solvent is changed from a polar highly solvating one to a nonpolar, poorly solvating one the amount of carbocationic character decreases and the halonium ion character increases.[95-97] This is attributed to the more polarizable halogen's ability to accommodate a positive charge with a lower requirement for solvation than is required by carbon.[95-97] Thus the observation that halonium ions exist in highly polar media of low nucleophilicity is evidence that they may exist under the conditions of neighboring group participation reactions in other media, but their intermediacy is likely dependent on the solvent as well as the substituents on the attached carbons.

3.5. Product Analysis

The analysis of reaction products is an obvious method applicable in discerning the presence of neighboring group participation. Frequently it is from a product analysis that the participation of a neighboring group is indicated, although simple analysis does not yield definitive information in all cases. In the case of the anchimerically assisted intramolecular catalysis by the carboxyl group in the hydrolysis of phthalamic acid (74), product analysis yields no information relative to the exact mode of participation. A double label tracer experiment[98] serves to provide the information necessary to decide between path a and path b in Eq. (24). The general acid

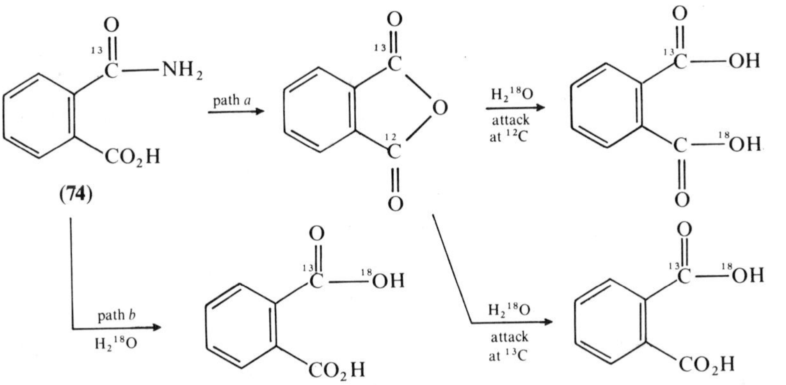

$$\text{(24)}$$

catalysis mechanism (path b) is ruled out in favor of direct nucleophilic involvement of the carboxyl group (path a) when mass spectrometry of the decarboxylated phthalic acid product reveals that the isotopically labeled carbon dioxides can arise only via path a.

In many examples of nucleophilic participation, rearrangement or cyclization is evidence of participation, but the extent of participation is not always easily established. There is no evidence for anchimeric assistance in the very fast reactions of N-methallylbenzamide (75) with sulfuric acid[99] or fluorine.[100] All evidence in both processes suggests electrophilic attack with subsequent capture of the carbocation intermediate by the amide group. Because of the presence of competing external nucleophiles (F$^-$ and MeOH), fluorination in methanol leads to only 21% of the product of neighboring group participation (76). When no good external nucleophile is present, such as in the reaction of (75) with sulfuric acid, ring closure is highly favored; in the case of (75), ring closure is quantitative.

In solvolytic reactions of primary or secondary carbon derivatives, the situation is different, and current theory predicts a correlation between

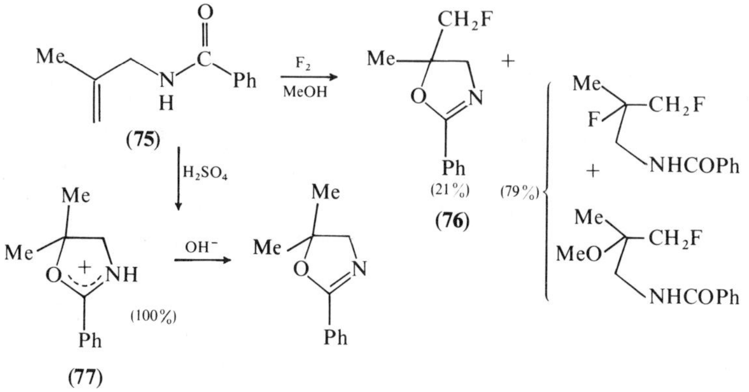

the fraction of rearranged or cyclized products and the degree of anchimeric assistance. Winstein et al.[71] (also cf. Refs. 19 and 81) first predicted such a correlation when they proposed that the observed rate constant (k_t) is the sum of the rate constants of the independently competing processes: solvent assistance (k_s), anchimeric assistance (k_Δ and k_R; see pp. 13–14), and no assistance (k_c).

$$k_t = k_s + k_\Delta + k_c + k_R \tag{26}$$

It was many years later when Schleyer et al.[19,75] conclusively demonstrated that solvent assistance was strong in the solvolysis of primary and secondary derivatives; thus the appearance of any rate enhancement in a solvolytic reaction is an indication that neighboring group assistance is yet stronger. It is now known that k_c is rarely significant in reactions where either solvent or neighboring group assistance occurs; the equation can then be modified to

$$k_t = k_s + k_\Delta \tag{27}$$

the form shown in Eq. (27). If internal return is occurring in the anchimerically assisted process, it is customary to designate as F the fraction of bridged ions that goes on to product; thus $1 - F$ is the fraction that returns to starting material (even rearranged starting material).[19,101] The equation is then rewritten as

$$k_t = k_s + Fk_\Delta \tag{28}$$

Internal return in a k_s process can also occur.[102] In such a case, the fraction Fk_Δ/k_t has been determined by extrapolating the scrambling observed in the substitution products back to time equals zero. This gives an Fk_Δ value that can be compared to the kinetic data for product and rate correlations (for a discussion of this topic, see Ref. 19 and the first reference in Ref. 75).

The use of Eq. (28) in verifying that the β-aryl- and solvent-assisted processes were indeed independent without crossover was of key importance in establishing an irrefutable experimentable basis for strong aryl participation in solvolysis reactions (see Ref. 19 and the first reference in Ref. 75). In the case of β-aryl sulfonate esters, the aryl group introduces a $-I$ effect that must be accounted for in determining the solvent-assisted or aryl-assisted rate. Schleyer et al.[19] first used Hammett or Hammett–Taft treatments to establish base-line reactivity assuming that deactivated aryls solvolyzed solely by a k_s process. It was later determined,[103] however, that even the deactivated compounds had a small Fk_Δ component. To account for this in dissecting k_t, a computer program was written that allows the calculation of a first set of approximate Fk_Δ values as deviations ($Fk_\Delta = k_t - k_s$) from the crude Hammett k_s correlation line defined by k_t for the deactivated substrates. A plot of σ^+ versus the approximate Fk_Δ values is constructed and is extended through the σ^+ constants for the deactivated substrates. This extrapolation yields Fk_Δ values of rather small magnitude for the deactivated substrates. Deducting these from their k_t values allows one to refine the k_s Hammett plot. From the deviations in this new k_s correlation line, better approximate Fk_Δ values are obtained. The cycle is repeated until ρ_s becomes constant by least-squares analysis.

The above approach was applied by Harris et al.[103] in dissecting k_t from rate data alone into k_s and Fk_Δ components for the solvolysis of

Table 8. Partitioning of Rate Constants for β-Arylethyl p-Toluenesulfonates at $115°C$[102,103]

Component	Method used for determination	Aryl substituent			
		p-Cl	H	p-Me	p-OMe
$k_s \times 10^7$ (sec^{-1})	Rate data	70	73	76	78
	Rate + prod.	64 ± 0.1	79 ± 11	78 ± 44	282 ± 250
$Fk_\Delta \times 10$ (sec^{-1})	Rate data	15	54	354	3922
	Rate + prod.	6.1 ± 0.08	48 ± 11	352 ± 43	480 ± 419
$Fk_\Delta/k_t \times 100$	Rate data	18	42	82	98
	Rate + prod.	8.7 ± 0.1	38 ± 8.5	82 ± 10	95
k_t/k_s	Rate data	1.21	1.73	5.63	51.5

β-arylalkyl systems. The results of their method and a comparison of their results with a combined rate and product (^{14}C scrambling) analysis by Coke et al.[102] are shown in Table 8. The agreement is quite satisfactory, especially at the high-k_Δ end of the scale.

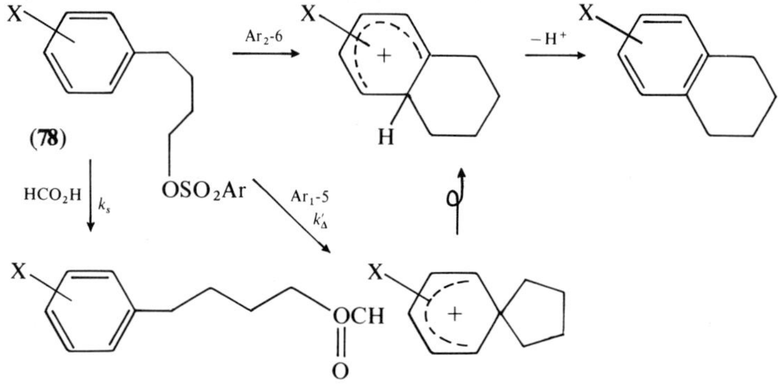

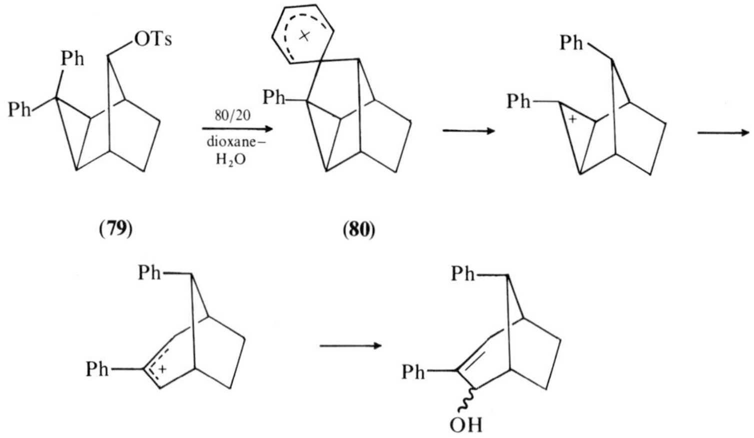

In applying Eq. (28) to neighboring group processes, one must recognize that the equation will fail when the system under study does not specifically lend itself to such a treatment, and, of course, there are many such examples. 4-Aryl-1-butyl arenesulfonates (**78**) undergo competitive Ar$_1$-5 and Ar$_2$-6 participation during formolysis.[104] The products of both k_Δ processes are tetralin derivatives, while the solvent-assisted route gives the noncyclic ester [Eq. (29)]. The process can readily be dissected into discrete k_Δ and k_s routes using Eq. (28) (see the first reference in Ref. 104). The tricyclic derivative (**79**) is also capable of undergoing Ar$_1$-5- and Ar$_2$-6-assisted ionization. It is doubtful that a solvent-assisted pathway can occur, owing to steric effects. Aryl assistance does occur with (**79**), but, unlike with (**78**), the cyclic intermediate is not stable.[105] A product analysis combined with rate comparison data allows one to invoke an aryl assistance mechanism [Eq. (30)]. Compound (**79**) is probably only capable of undergoing an Ar_1-5 process because of steric requirements. Since the rearrangement of (**80**) to a tetralin intermediate is likely a high-energy process while phenyl migration leads to a benzyl cation that can undergo an electrocyclic ring opening, the process shown in Eq. (30) seems reasonable.

Other examples of neighboring group cyclization that give no cyclic products are often results of kinetic versus thermodynamic control. The isolation of the bicyclic product (**7**) instead of a tricyclic acetate similar to (**6**) in the acetolysis of (**3**) is such an example. In summary, then, Eq. (28) cannot be used to predict products (i.e., there may be a rate/product discrepancy) when the routes of kinetic control and thermodynamic control are of similar energies, and the noncyclic product is thermodynamically favored.

References

1. S. P. McManus (ed.), *Organic Reactive Intermediates,* Academic Press, Inc., New York (1973); M. S. Newman, *Steric Effects in Organic Chemistry,* John Wiley & Sons, Inc. (Interscience Division), New York (1956); E. S. Huyser, *Methods in Free-Radical Chemistry,* Marcel Dekker, Inc., New York (1969); E. S. Gould, *Mechanism and Structure in Organic Chemistry,* Holt, Rinehart and Winston, Inc., New York (1959); A. R. Katrizky, *Physical Methods in Heterocyclic Chemistry,* Academic Press, Inc., New York (1963); R. S. Drago, *Physical Methods in Inorganic Chemistry,* Van Nostrand Reinhold Company, New York (1965); J. C. P. Schwartz (ed.), *Physical Methods in Organic Chemistry,* Holden-Day, Inc., San Francisco (1964); A. Weissberger (ed.), *Physical Methods of Organic Chemistry,* 3rd ed., John Wiley & Sons, Inc. (Interscience Division), New York (1959); L. P. Hammett, *Physical Organic Chemistry,* 2nd ed., McGraw-Hill Book Company, New York (1970); J. Hine, *Physical Organic Chemistry,* 2nd ed., McGraw-Hill Book Company, New York (1962); E. M. Kosower, *An Introduction to Physical Organic Chemistry,* John Wiley & Sons, Inc., New York (1968); F. L. J. Sixma and H. Wynberg, *A Manual of Physical Methods in Organic Chemistry,* John Wiley & Sons, Inc., New York (1964); K. B. Wiberg, *Physical Organic Chemistry,* John Wiley & Sons, Inc., New York (1964); J. E. Leffler and E. Grunwald, *Rates and Equilibria of Organic Reactions,* John Wiley & Sons, Inc., New York (1963); F. Basolo and R. G. Pearson, *Mechanisms of Inorganic Reactions,* 2nd ed., John Wiley & Sons, Inc., New York (1967; M. L. Bender, *Mechanisms of Homogeneous Catalysis from Protons to Proteins,* John Wiley & Sons, Inc., New York (1971); W. P. Jencks, *Catalysis in Chemistry and Enzymology,* McGraw-Hill Book Company, New York (1969).
2. E. N. Rengevich, V. I. Staninets, and E. A. Shilov, *Dokl. Akad. Nauk SSSR,* **146**, 111 (1962).

3. H. J. Lucas, R. R. Hepner, and S. Winstein, *J. Am. Chem. Soc.*, **61**, 3102 (1939); D. L. H. Williams, E. Bienvenue-Goetz, and J. E. Dubois, *J. Chem. Soc. B*, 517 (1969).
4. P. B. D. de la Mare and R. Bolton, *Electrophilic Additions to Unsaturated Systems*, Elsevier Publishing Company, Amsterdam (1966), p. 114.
5. S. Winstein, M. Shatavsky, C. Norton, and R. B. Woodward, *J. Am. Chem. Soc.*, **77**, 4183 (1955); S. Winstein and M. Shatavsky, *ibid.*, **78**, 592 (1956).
6. A. Diaz, M. Brookhart, and S. Winstein, *J. Am. Chem. Soc.*, **88**, 3133 (1966); M. Brookhart, A. Diaz, and S. Winstein, *ibid.*, **88**, 3135 (1966); H. Tanida, T. Tsuji, and T. Irie, *ibid.*, **88**, 864 (1966).
7. C. L. Liotta, W. F. Fisher, E. L. Slightom, and C. L. Harris, *J. Am. Chem. Soc.*, **94**, 2129 (1972).
8. E. L. Eliel, *Stereochemistry of Carbon Compounds*, McGraw-Hill Book Company, New York (1962).
9. J. B. Lambert and G. J. Putz, *J. Am. Chem. Soc.*, **95**, 6313 (1973); J. E. Nordlander and T. J. McCrary, Jr., *ibid.*, **94**, 5133 (1972); **96**, 4066 (1974), and references therein.
10. S. Winstein and D. Trifan, *J. Am. Chem. Soc.*, **71**, 2953 (1949); S. Winstein and D. Trifan, *ibid.*, **74**, 1154 (1952); S. Winstein and D. Trifan, *ibid.*, **74**, 1147 (1952).
11. H. C. Brown, *Acc. Chem. Res.*, **6**, 377 (1973).
12. J. E. Nordlander, R. R. Gruetzmacher, W. J. Kelly, and S. P. Jindal, *J. Am. Chem. Soc.*, **96**, 181 (1974).
13. M. L. Bender and M. S. Silver, *J. Am. Chem. Soc.*, **84**, 4589 (1962).
14. C. A. Grob, K. Seckinger, S. W. Tam, and R. Traber, *Tetrahedron Lett.*, 3051 (1973).
15. W. G. Bentrude and J. C. Martin, *J. Am. Chem. Soc.*, **84**, 1561 (1962).
16. R. W. Taft, in: *Steric Effects in Organic Chemistry* (M. S. Newman, ed.), p. 586, John Wiley & Sons, Inc. (Interscience Division), New York (1956).
17. A. Streitwieser, Jr., *Solvolytic Displacement Reactions*, McGraw-Hill Book Company, New York (1962).
18. C. J. Lancelot, D. J. Cram, and P. v. R. Schleyer, in: *Carbonium Ions* (G. A. Olah and P. v. R. Schleyer, eds.), Vol. III, Chap. 27, John Wiley & Sons, Inc. (Interscience Division), New York (1972).
19. C. J. Lancelot and P. v. R. Schleyer, *J. Am. Chem. Soc.*, **91**, 4291, 4296, 4297 (1969); C. J. Lancelot, J. J. Harper, and P. v. R. Schleyer, *ibid.*, **91**, 4294 (1969).
20. P. E. Peterson, C. Casey, E. V. P. Tao, A. Agtarap, and G. Thompson, *J. Am. Chem. Soc.*, **87**, 5163 (1965).
21. P. E. Peterson, R. E. Kelley, R. Belloli, and K. A. Sipp, *J. Am. Chem. Soc.*, **87**, 5169 (1965), and earlier papers in this series.
22. J. C. McGowan, *J. Appl. Chem.*, **10**, 312 (1960); H. Klootsterziel, *Rec. Trav. Chim. Pays-Bas*, **82**, 508 (1963).
23. D. L. H. Williams *et al.*; see Ref. 3.
24. J. M. Harris and S. P. McManus, *J. Am. Chem. Soc.*, **96**, 4693 (1974).
25. C. S. Foote, *J. Am. Chem. Soc.*, **86**, 1853 (1964).
26. P. v. R. Schleyer, *J. Am. Chem. Soc.*, **86**, 1854, 1855 (1964).
27. J. M. Harris and S. P. McManus, unpublished results.
28. P. v. R. Schleyer, personal communication.
29. H. C. Brown, I. Rothberg, P. v. R. Schleyer, M. M. Donaldson, and J. J. Harper, *Proc. Natl. Acad. Sci. U.S.A.*, **56**, 1653 (1966).
30. J. L. Fry, J. M. Harris, R. C. Bingham, and P. v. R. Schleyer, *J. Am. Chem. Soc.*, **92**, 2540 (1970).
31. P. D. Bartlett, *Nonclassical Ions*, W. A. Benjamin, Inc., Reading, Mass. (1965); H. C. Brown, *Boranes in Organic Chemistry*, Cornell University Press, Ithaca, N.Y. (1972), Chaps IX–XI; H. C. Brown, *Chem. Brit.*, **2**, 199 (1966); H. C. Brown, *Chem. Eng. News*, **45**, 87 (Feb. 13, 1967); H. C. Brown, *Chem. Soc. Spec. Publ.*, **No. 16**, 140 (1962); R. Bern-

hard, *Sci. Res.,* 26 (Aug. 18, 1969); H. C. Brown, *ibid.,* 5 (Dec. 22, 1969); G. D. Sargent, *Quart. Rev.* **20**, 301 (1966); G. D. Sargent, *in: Carbonium Ions* (G. A. Olah and P. v. R. Schleyer, eds.), Vol. III, Chap. 24, John Wiley & Sons, Inc. (Interscience Division), New York (1972); K. B. Wiberg, B. A. Hess, Jr., and A. J. Ashe, III, *ibid.,* Chap. 26; C. K. Ingold, *Structure and Mechanisms in Organic Chemistry,* Cornell University Press, Ithaca, N.Y. (1953), p. 514.

32. S. P. McManus and C. U. Pittman, Jr., *in: Organic Reactive Intermediates* (S. P. McManus, ed.), Chap. 4, Academic Press, Inc., New York (1973).
33. J. L. Fry, E. M. Engler, and P. v. R. Schleyer, *J. Am. Chem. Soc.* **94**, 4628 (1972); E. J. Jacob, H. B. Thompson, and L. S. Bartell, *J. Chem. Phys.,* **47**, 3736 (1967); R. H. Boyd, *ibid.,* **49**, 2574 (1968); N. L. Allinger, M. T. Tribble, M. A. Miller, and D. A. Wertz, *J. Am. Chem. Soc.,* **93**, 1637 (1971); N. L. Allinger, M. T. Tribble, and M. A. Miller, *Tetrahedron,* **28**, 1173 (1972); A. Warshel and M. Karplus, *J. Am. Chem. Soc.,* **94**, 5612 (1972); N. Bodor, M. J. S. Dewar, and D. H. Lo, *ibid.,* **94**, 5303 (1972); D. F. DeTar, *ibid.,* **96**, 1254, 1255 (1974); M. Simonetta, G. Favini, C. Mariani, and P. Grammaccioni, *ibid.,* **90**, 1280 (1968).
34. F. H. Westheimer, *in: Steric Effects in Organic Chemistry,* (M. S. Newman, ed.), Chap. 12, John Wiley & Sons, Inc., New York (1956).
35. P. G. Gassman, J. Zeller, and J. T. Lumb, *Chem. Commun.,* 69 (1968).
36. P. G. Gassman and A. F. Fentiman, Jr., *J. Am. Chem. Soc.,* **91**, 1545 (1969); **92**, 2549 (1970).
37. P. G. Gassman and A. F. Fentiman, Jr., *J. Am. Chem. Soc.,* **92**, 2551 (1970).
38. J. Haywood-Farmer, *Chem. Rev.,* **74**, 315 (1974) (review of participation by remote cyclopropyl groups).
39. H. Tanida, T. Tsuji, and T. Irie, *J. Am. Chem. Soc.,* **89**, 1953 (1967).
40. M. A. Battiste, C. L. Deyrup, R. E. Pincock, and J. Haywood-Farmer, *J. Am. Chem. Soc.,* **89**, 1954 (1967).
41. H. C. Brown and K.-T. Liu, *J. Am. Chem. Soc.,* **91**, 5909 (1969); H. C. Brown, S. Ikegami, K.-T. Liu, and G. L. Tritle, *ibid.,* **98**, 2531 (1976).
42. H. C. Brown, M. Ravindranathan, and E. N. Peters, *J. Am. Chem. Soc.,* **96**, 7351 (1974).
43. E. N. Peters and H. C. Brown *J. Am. Chem. Soc.,* **95**, 2398 (1973).
44. H. C. Brown, M. Ravindranathan, K. Takeuchi, and E. N. Peters, *J. Am. Chem. Soc.,* **97**, 2899 (1975); H. C. Brown and K. Takeuchi, *ibid.,* **90**, 2691, 2693 (1967); H. C. Brown, personal communication.
45. E. N. Peters and H. C. Brown, unpublished results; private communication.
46. H. C. Brown, E. N. Peters, and M. Ravindranathan, *J. Am. Chem. Soc.,* **97**, 2900 (1975).
47. R. Hoffman, P. D. Mollere, and E. Heilbronner, *J. Am. Chem. Soc.,* **95**, 4860 (1973).
48. M. A. Battiste and R. A. Fiato, *Tetrahedron Lett.,* 1255 (1975) (a brief critical analysis of some of the interpretive problems).
49. C. D. Ritchie and W. F. Sager, *Prog. Phys. Org. Chem.,* **2**, 323 (1964).
50. A. Streitwieser, Jr., *Solvolytic Displacement Mechanisms,* McGraw-Hill Book Company, New York (1962), pp. 98–103 and 171–175; E. A. Halevi, *Prog. Phys. Org. Chem.,* **1**, 109 (1963); K. B. Wiberg, *Chem. Rev.,* **55**, 713 (1955); R. P. Bell, *The Proton in Chemistry,* Cornell University Press, Ithaca, N.Y. (1959), Chap. XI; W. P. Jencks, *Catalysis in Chemistry and Enzymology,* McGraw-Hill Book Company, New York (1969), Chap. 4; L. Melander, *Isotope Effects on Reaction Rates,* The Ronald Press Company, New York (1960); V. J. Shiner, Jr., *in: Isotope Effects in Chemical Reactions* (C. J. Collins and N. S. Bowman, eds.), A.C.S. Monograph No. 167, Chap. 2, Van Nostrand Reinhold Company, New York, (1970); S. E. Scheppele, *Chem. Rev.,* **72**, 511 (1972).
51. D. E. Sunko and S. Borcic, *in: Isotope Effects in Chemical Reactions* (C. J. Collins and N. S. Bowman, eds.), A.C.S. Monograph No. 167, Chap. 3, Van Nostrand Reinhold Company, New York (1970).

52. K. B. Wiberg and E. L. Motell, *Tetrahedron*, **19**, 2009 (1963).
53. F. H. Westheimer, *Chem. Rev.*, **61**, 265 (1961).
54. V. J. Shiner, Jr., W. E. Buddenbaum, B. L. Murr, and G. Lamaty, *J. Am. Chem. Soc.*, **90**, 418 (1968).
55. R. Eliason, M. Tomic, S. Borčić, and D. E. Sunko, *Chem. Commun.*, 1490 (1968).
56. R. Eliason, unpublished results quoted in Ref. 51.
57. M. Tarle, unpublished results quoted in Ref. 51.
58. T. W. Bentley, S. H. Liggero, M. A. Imhoff, and P. v. R. Schleyer, *J. Am. Chem. Soc.*, **96**, 1970 (1974).
59. J. H. Ong and R. E. Robertson, *Can. J. Chem.*, **52**, 2660 (1974).
60. K. Mislow, R. Graeve, A. J. Gordon, and G. H. Wahl, Jr., *J. Am. Chem. Soc.*, **86**, 1733 (1964); R. E. K. Winter and M. L. Honig, *ibid.*, **93**, 4616 (1971).
61. A. Streitwieser, Jr., R. H. Jagow, R. C. Fahey, and S. Suzuki, *J. Am. Chem. Soc.*, **80**, 2326 (1958).
62. R. P. Bell and W. B. T. Miller, *Trans. Faraday Soc.*, **59**, 1147 (1963); R. P. Bell and J. E. Crooks, *ibid.*, **58**, 1409 (1962).
63. S. L. Loukas, M. R. Velkou, and G. A. Gregoriou, *Chem. Commun.*, 1199 (1969); 251 (1970).
64. J. M. Harris, R. E. Hall, and P. v. R. Schleyer, *J. Am. Chem. Soc.*, **93**, 2551 (1971); V. J. Shiner, Jr., and R. D. Fisher, *ibid.*, **93**, 2553 (1971); K. Humski, V. Sendijarvic, and V. J. Shiner, Jr., *ibid.*, **95**, 7722 (1973).
65. A. Streitwieser, Jr., and G. A. Dafforn, *Tetrahedron Lett.*, 1263 (1969).
66. W. H. Saunders, Jr., S. Asperger, and D. H. Edison, *J. Am. Chem. Soc.*, **80**, 2421 (1959).
67. M. Nikoletic, S. Borcic, and D. E. Sunko, *Tetrahedron*, **23**, 649 (1967); K. T. Leffek, J. A. Llewellyn, and R. E. Robertson, *J. Am. Chem. Soc.*, **82**, 6315 (1960); S. Borcic, *Croat. Chem. Acta*, **35**, 67 (1963); J. P. Schaefer, J. P. Foster, M. J. Dagnani, and L. M. Hornig, *J. Am. Chem. Soc.*, **90**, 4497 (1968).
68. F. G. Bordwell and A. C. Knipe, *J. Org. Chem.*, **35**, 2956, 2959 (1970).
69. V. P. Vitullo and N. R. Grossman, *J. Org. Chem.*, **38**, 179 (1973).
70. V. P. Vitullo and F. P. Wilgis, *J. Am. Chem. Soc.*, **97**, 5616 (1975).
71. S. Winstein, E. Allred, R. Heck, and R. Glick, *Tetrahedron*, **3**, 1 (1958).
72. A. Diaz, I. L. Reich, and S. Winstein, *J. Am. Chem. Soc.*, **91**, 5637 (1969).
73. J. M. Harris, *Prog. Phys. Org. Chem.*, **11**, 89 (1974).
74. R. S. Bly, R. A. Mateer, K. K. Tse, and R. L. Veazey, *J. Org. Chem.*, **38**, 1518 (1973).
75. J. L. Fry, C. J. Lancelot, L. K. M. Lam, J. M. Harris, R. C. Bingham, D. J. Raber. R. E. Hall, and P. v. R. Schleyer, *J. Am. Chem. Soc.*, **92**, 2538 (1970); P. v. R. Schleyer, J. L. Fry, L. K. M. Lam, and C. J. Lancelot, *ibid.*, **92**, 2542 (1970); S. H. Liggero, J. J. Harper, P. v. R. Schleyer, A. P. Krapcho, and D. E. Horn, *ibid.*, **92**, 3789 (1970); J. M. Harris, D. J. Raber, R. E. Hall, and P. v. R. Schleyer, *ibid.*, **92**, 5729 (1970); P. E. Peterson and F. J. Waller, *ibid.*, **94**, 991, 5024 (1972); T. W. Bentley, F. L. Schadt, and P. v. R. Schleyer, *ibid.*, **94**, 992 (1972); D. J. Raber, R. C. Bingham, J. M. Harris, J. L. Fry, and P. v. R. Schleyer, *ibid.*, **92**, 5977 (1970) (papers dealing with solvent assistance; see also Refs. 12 and 30).
76. E. Grunwald and S. Winstein, *J. Am. Chem. Soc.*, **70**, 846 (1948).
77. R. C. Bingham and P. v. R. Schleyer, *J. Am. Chem. Soc.*, **93**, 3189 (1971); J. Slutsky, R. C. Bingham, P. v. R. Schleyer, W. C. Dickason, and H. C. Brown, *ibid.*, **96**, 1969 (1974).
78. C. J. M. Stirling, *J. Chem. Educ.*, **50**, 844 (1973) (discusses the use of k_{Br}/k_{Cl} ratios in ring closure studies).
79. P. v. R. Schleyer, W. F. Sliwinski, G. W. Van Dine, U. Schollkopf, J. Paust, and K. Fellenberger, *J. Am. Chem. Soc.*, **94**, 125 (1972).

80. R. Baird and S. Winstein, *J. Am. Chem. Soc.*, **85**, 567 (1963).
81. L. Eberson, J. P. Petrovich, R. Baird, D. Dyckes, and S. Winstein, *J. Am. Chem. Soc.*, **87**, 3504 (1965).
82. C. B. Anderson, E. C. Friedrich, and S. Winstein, *Tetrahedron Lett.*, 2037 (1963).
83. G. Fodor, F. Letourneau, and N. Mandava, *Can. J. Chem.*, **48**, 1465 (1970); G. Lucente and A. Romeo, *Chem. Commun.*, 1605 (1971); S. Cerrini, W. Fedeli, and F. Mazza, *ibid.*, 1607 (1971); B. Borbránski and M. Sladowska, *Rocz. Chem.*, **46**, 451 (1972); G. A. Rogers and T. C. Bruice, *J. Am. Chem. Soc.*, **95**, 4452 (1973).
84. N. Gravitz and W. P. Jencks, *J. Am. Chem. Soc.*, **96**, 489 (1974).
85. S. Winstein and E. Grunwald, *J. Am. Chem. Soc.*, **70**, 841 (1948).
86. P. D. Bartlett and L. B. Gortler, *J. Am. Chem. Soc.*, **85**, 1864 (1963).
87. G. A. Olah and P. v. R. Schleyer (eds.), *Carbonium Ions*, Vols. I–IV, John Wiley & Sons, Inc., New York (1968–73).
88. D. Bethell and V. Gold, *Carbonium Ions, An Introduction*, Academic Press, Inc., New York (1967).
89. *Organic Reaction Mechanisms*, edited by B. Capon, M. J. Perkins, and C. W. Rees (1965–1967), B. Capon and C. W. Rees (1968–1972), and M. J. Perkins and A. R. Butler (1973–present), John Wiley & Sons, Inc., London, New York.
90. G. A. Olah, G. Liang, G. D. Mateescu, and J. L. Riemenschneider, *J. Am. Chem. Soc.*, **95**, 8698 (1973); G. A. Olah, *Acc. Chem. Res.*, **9**, 41 (1976); see also R. Haseltine, N. Wong, T. S. Sorensen, and A. J. Jones, *Can. J. Chem.*, **53**, 1891 (1975).
91. G. M. Kramer, *Adv. Phys. Org. Chem.*, **11**, 177 (1975).
92. P. E. Peterson, *Acc. Chem. Res.*, **4**, 407 (1971); P. E. Peterson, B. R. Bonazza, and P. M. Henrichs, *J. Am. Chem. Soc.*, **95**, 2222 (1973); P. M. Henrichs and P. E. Peterson, *ibid.*, **95**, 7449 (1973); P. E. Peterson, personal communication.
93. G. A. Olah, P. W. Westerman, E. G. Melby, and Y. K. Mo., *J. Am. Chem. Soc.*, **96**, 3565 (1974).
95. S. P. McManus and D. W. Ware, *Tetrahedron Lett.*, 4271 (1974).
96. R. D. Wieting, R. H. Staley, and J. L. Beauchamp, *J. Am. Chem. Soc.*, **96**, 7552 (1974).
97. S. P. McManus and P. E. Peterson, *Tetrahedron Lett.*, 2753 (1975).
98. M. L. Bender, Y.-L. Chow, and F. Chloupek, *J. Am. Chem. Soc.*, **80**, 5380 (1958).
99. S. P. McManus, J. T. Carroll, and C. U. Pittman, Jr., *J. Org. Chem.*, **35**, 3768 (1970).
100. R. F. Merritt, unpublished results; personal communication.
101. S. Winstein and A. F. Diaz, *J. Am. Chem. Soc.*, **91**, 4300 (1969).
102. J. L. Coke, F. E. McFarlane, M. C. Mourning, and M. G. Jones, *J. Am. Chem. Soc.*, **91**, 1154 (1969); M. G. Jones and J. L. Coke, *ibid.*, **91**, 4284 (1969).
103. J. M. Harris, F. L. Schadt, P. v. R. Schleyer, and C. J. Lancelot, *J. Am. Chem. Soc.*, **91**, 7508 (1969).
104. R. Heck and S. Winstein, *J. Am. Chem. Soc.*, **79**, 3105, 3114 (1957); S. Winstein and R. F. Heck, *J. Org. Chem.*, **37**, 825 (1972); M. Gates, D. L. Frank, and W. C. von Felton, *J. Am. Chem. Soc.*, **96**, 5138 (1974); L. M. Jackman and V. R. Haddon, *ibid.*, **96**, 5130 (1974).
105. J. W. Wilt and T. P. Malloy, *J. Am. Chem. Soc.*, **92**, 4747 (1970).
106. W. E. McEwen, J. E. Fountaine, D. N. Shulz, and W.-I. Shiau, *J. Org. Chem.*, **41**, 1684 (1976).

Part 2
Participation by Simple Oxygen, Sulfur, and Nitrogen Groups

In this part we shall focus our attention on participation by specific neighboring groups. The literature making reference to involvement of the common neighboring groups is immense, and hence an encyclopedic coverage is not practical. Thus we shall treat our topics subjectively, hoping that the reader will gain insight into (1) the driving force for participation by specific groups in specific reactions, (2) the relative driving forces of neighboring groups,* and (3) the scope and limitations of participation by certain neighboring groups; we shall also provide a useful summary of some of the more important studies of neighboring group effects.

As stated in Part 1, neighboring group effects may take many forms. One of the most common forms is neighboring group participation in nucleophilic displacement reactions. Participation by groups containing n, π, or σ electrons may occur. One can see from the geometries depicted in the accompanying figure that there should be differences in the driving forces for participation by different groups. In this part we shall begin our treatment of some common groups that bear n electrons. Their mode of participation in nucleophilic displacement processes should be conceptually similar to the common $S_N 2$ reaction. We shall review studies of these processes as well as participation by these groups in other reactions. We are deferring to future parts discussion of ambient neighboring groups con-

---

* For a general review of nucleophilic reactivity, the reader is referred to the treatment by R. F. Hudson in *Chemical Reactivity and Reaction Paths* (G. Klopman, ed.), Chap. 5, John Wiley & Sons, Inc. (Interscience Division), New York (1974).

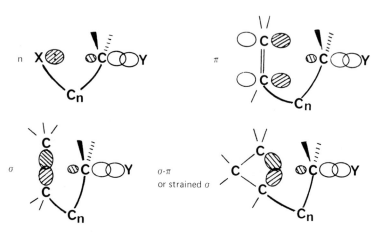

Representation of the different geometries for participation by neighboring groups in nucleophilic displacement reactions. The shaded lobes are those that will overlap to stabilize the transition state.

taining oxygen, sulfur, and nitrogen (e.g., esters and thioamides). We are also deferring to future parts detailed discussion of participation by various groups in certain types of reactions (e.g., hydrolysis reactions and reduction reactions).

4
Participation by Oxygen Groups

4.1. Ether Groups

4.1.1. In Solvolytic Displacement Reactions

4.1.1.1. The Methoxy Group

The absence of a steady trend in the rates of solvolysis of a series of ω-methoxyalkyl *p*-bromobenzenesulfonates (see Table 1) indicates the importance of an effect by the methoxy groups other than the inductive effect. In particular the high rates measured for the 4-methoxybutyl and 5-methoxypentyl compounds [(4) and (5)] suggest nucleophilic participation. Winstein and his co-workers[1] formulated these reactions as proceeding through the cyclic oxonium ions (7) (MeO-5) and (8) (MeO-6 participation). Hence the primary methoxyl group provides assistance when a five- or six-membered cyclic oxonium ion is possible, but methoxy group assistance is not readily apparent when this intermediate would have seven, three, or four members.

The increase in rate owing to MeO-5 participation in the acetolysis of secondary *p*-bromobenzenesulfonates [cf. compounds (11) and (15) in

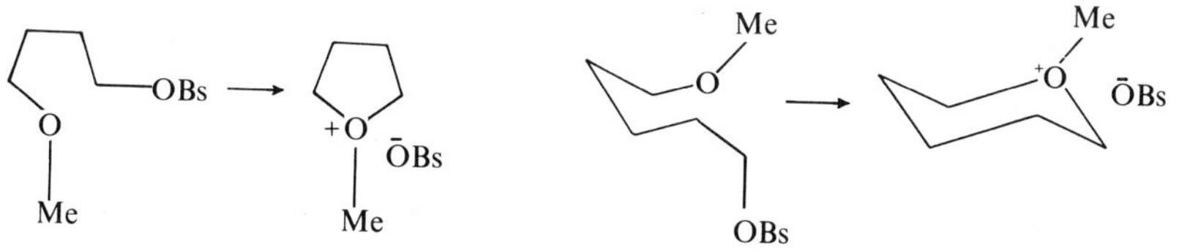

(1)

Table 2] is about 20-fold less than for the analogous primary compounds [cf. compounds (9) and (10)]. The α-methyl group in the secondary compound

Table 1. Relative Rates of Solvolyses of ω-Methoxyalkyl p-Bromobenzenesulfonates[1]

Compound	Relative rates			Calc. k_Δ/k_s (75°C)		
	EtOH (75°C)	AcOH (25°C)	HCO_2H (75°C)	EtOH	AcOH	HCO_2H
Me-$(CH_2)_3$-OBs (1)	1.00	1.00	1.00	0.93	0.81	1.07
MeO-$(CH_2)_2$-OBs (2)	0.25	0.28	0.10	1.33	1.51	0.85
MeO-$(CH_2)_3$-OBs (3)	0.67	0.63	0.33	1.14	1.20	0.84
MeO-$(CH_2)_4$-OBs (4)	20.4	657.0	461.0	22.0	425.0	610.0
MeO-$(CH_2)_5$-OBs (5)	2.8	123.0	32.6	2.42	47.2	30.6
MeO-$(CH_2)_6$-OBs (6)	1.19	1.16	1.13	0.87	0.71	0.85

helps to disperse the positive charge in the transition state, and hence the need for assistance from the neighboring methoxy group is less. Compound (12), with a secondary methoxy group, reacts faster than the corresponding compound (10) with a primary methoxy group. This accelerating effect of alkyl substituents on ring closure reactions is frequently encountered, and its origin is discussed in Chapter 2.

The slightly greater rate of acetolysis of *threo*-5-methoxy-2-hexyl p-bromobenzenesulfonate (14) relative to its *erythro* isomer (13) probably results from steric retardation in formation of the cyclic oxonium ion from

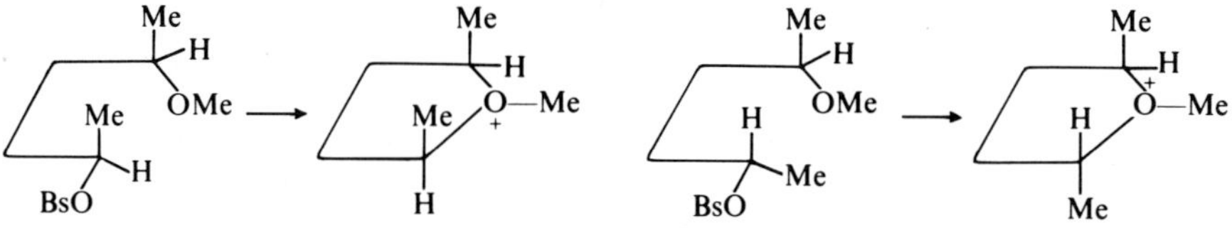

(16)   (17)   (2)

Table 2. Relative Rates of Acetolysis of Some Methyl-Substituted Butyl p-Bromobenzenesulfonates at 25°C

Compound	Relative rates			
Me-$(CH_2)_4$-OBs (9)	1.00	—	—	
MeO-$(CH_2)_4$-OBs (10)	659	1.00	—	
MeO-$(CH_2)_3$-CHMe-OBs (11)	4,135	6.29	30	
MeO-CHMe-$(CH_2)_3$-OBs (12)	4,111	6.26	—	
MeO-CHMe-$(CH_2)_2$-CHMe-OBs				
erythro (13)	9,447	14.38	67	
threo (14)	22,800	34.70	163	
Me-CH_2-CHMe-OBs (15)	140	—		1.00

Participation by Oxygen Groups

the *erythro* isomer. In the ion (**16**) from the *threo* isomer, the three methyl substituents may all be staggered, but in ion (**17**) from the *erythro* isomer, two of the methyl groups must be eclipsed.

According to the mechanism postulated, (**11**) and (**12**) should yield the same oxonium ion (**18**), and if counterion effects are unimportant, their products also should be identical. Allred and Winstein[2] have conducted detailed studies showing that both the rate constants (Table 2) and the product ratios are nearly identical for the acetolyses of (**11**) and (**12**) at 25°C.

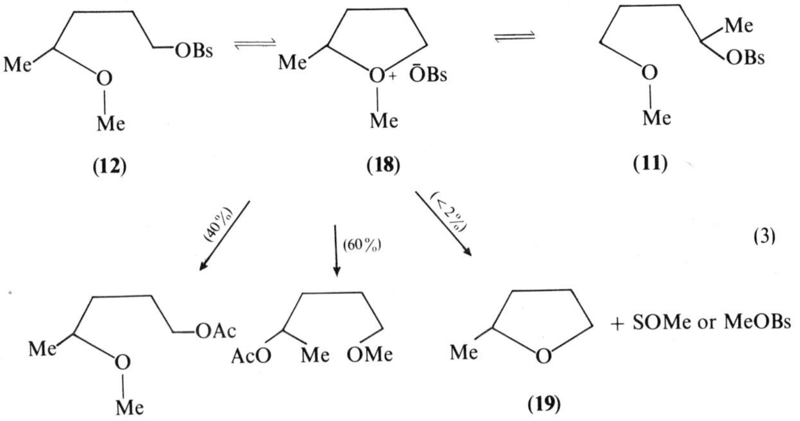

It is somewhat surprising that the rate constants for acetolysis and ethanolysis are so similar for (**11**) and (**12**). These similarities have been rationalized in terms of the blending of k_Δ, k_s and substituent effects.[2] Since (**11**) and (**12**) are interconverted during acetolysis, attack on (**18**) may be partly rate limiting.

Allred and Winstein[3] found that acetolyses of (**11**) and (**12**) are accompanied by extensive isomerization, and after 70% reaction the mixture of *p*-bromobenzenesulfonates is composed of 69% of (**11**) and 31% of (**12**) regardless of the starting material. This is the result of return from free ions as shown by the observation of a small common-ion rate depression and of the conversion of the *p*-bromobenzenesulfonates into toluene-*p*-sulfonates in the presence of lithium toluene-*p*-sulfonate. The reaction also shows a special salt effect in the presence of lithium perchlorate, yet even at high salt concentrations some isomerization is still evident. Since return from solvent-separated ion pairs such as (**23**) is excluded in the presence of lithium perchlorate, and since the geometry of the initially formed ionic species does not allow direct rearrangement to occur, a common, symmetrical intimate ion pair, such as (**21**), is implicated in the isomerization. The solvolysis

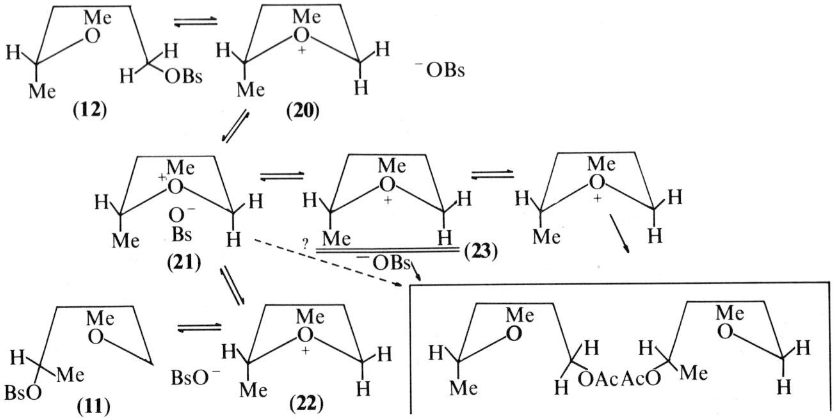

Scheme 1

was formulated to occur as shown in Scheme 1.[3] Product was not thought to arise from the intimate ion pairs (20) and (22) owing to the shielding by the anion that would be expected to affect the product ratios.

Tarbell et al.[4] have determined the stereochemistry of the reaction of optically active 5-methoxy-2-pentyl p-bromobenzenesulfonate (24) with lithium chloride in pyridine. Their results are consistent with the formation

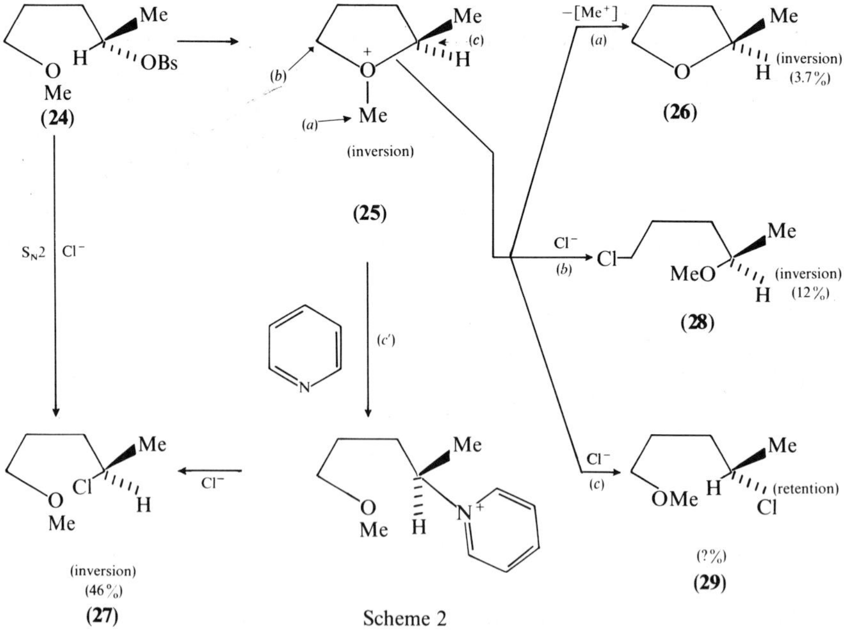

Scheme 2

of the cyclic oxonium ion (**25**) as an intermediate since the products included 3.7% of (**26**), formed with 100% inversion of configuration, and the primary chloride (**28**), which was found to be 93% inverted. The chlorides (**27**)–(**29**) were isolated as an optically active mixture in a total yield of 58%. By the correlation of absolute configurations it was found that 80% of the mixture of chlorides was (**27**). This leaves a nil amount of (**29**) that would be formed by attack of Cl⁻ on (**25**) by path (*c*). It is somewhat surprising that the major amount of the secondary chloride is (**27**) and not (**29**). Tarbell *et al.*[4] assumed that (**27**) was derived from the direct S_N2 displacement of the *p*-bromobenzenesulfonate by chloride ion as shown in Scheme 2. This is inconsistent with Allred and Winstein's product ratios from acetolysis of the inactive *p*-bromobenzenesulfonate (**11**)[2] and for reactions of other oxonium[5] and halonium ions[6] where the major product is derived from ring opening at the more substituted carbon (e.g., see Scheme 1). Since a 12% yield of the primary chloride (**28**) is obtained from the reaction of (**24**) with pyridine and chloride ion, one expects > 15% of the secondary chloride (**29**) if the reactivity of (**25**) toward chloride is the same as the reactivity of the free ion and ion pairs in Scheme 1 toward acetic acid.

An alternative explanation for the above results is available. In view of the apparent absence of path (*c*), the large amount of (**27**) (Scheme 2), and the high k_Δ/k_s measured for acetolysis of (**11**),[2] it seems reasonable to postulate oxonium ion formation as the predominant reaction. To account for the secondary chloride (**27**) involving a single inversion when compared to (**24**), two subsequent inversions must occur after (**25**) is formed. This could occur by reaction of the oxonium ion with pyridine [path (*c′*), Scheme 2], yielding the pyridinium adduct, which subsequently reacts by chloride attack to yield (**27**). Notice that reaction of (**25**) with pyridine at the secondary position [path (*b′*)] would proceed undetected since the stereochemistry is established upon ring opening and the subsequent displacement of pyridine by chloride to give (**28**) would be of no consequence in a stereochemical analysis.

(**25**) $\xrightarrow[\text{pyridine}]{\text{path }(b')}$ [pyridinium intermediate] $\xrightarrow{\text{Cl}^-}$ (**28**)   (4)

Acetolysis of 5-methoxypentyl *p*-bromobenzenesulfonate (**5**) is accompanied by extensive conversion into tetrahydropyran and methyl *p*-bromobenzenesulfonate.[7] None of the latter ester is formed in ethanolysis or formolysis. No ester results from acetolysis with the addition of lithium perchlorate, suggesting that, like five-membered cyclic oxonium ions (see Scheme 1), a spectrum of ion pairs is involved. Unlike the tetrahydrofuranium ions, displacement of the O-methyl group is a significant reaction of tetrahydropyranium ions in acetic acid in the absence of added lithium perchlorate.

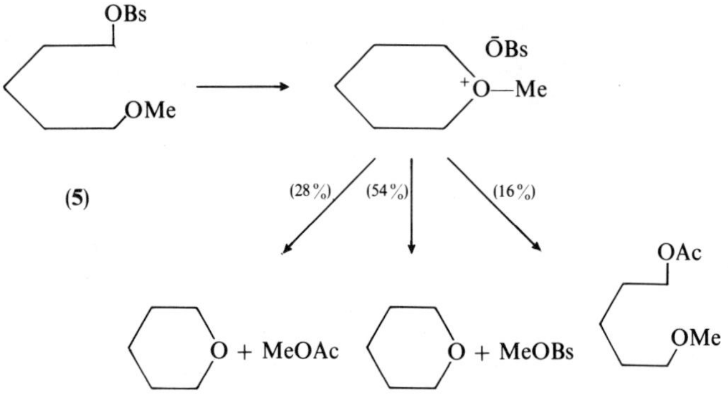

The reasons for the different reaction patterns of the five- and six-membered ring can be rationalized even without relative rate data for Me—O and methylene C—O cleavage for the two systems. The results imply either that attack at the methyl group of (**30**) is much faster than at the methyl group of (**31**) or that attack at the ring carbon of (**31**) is much

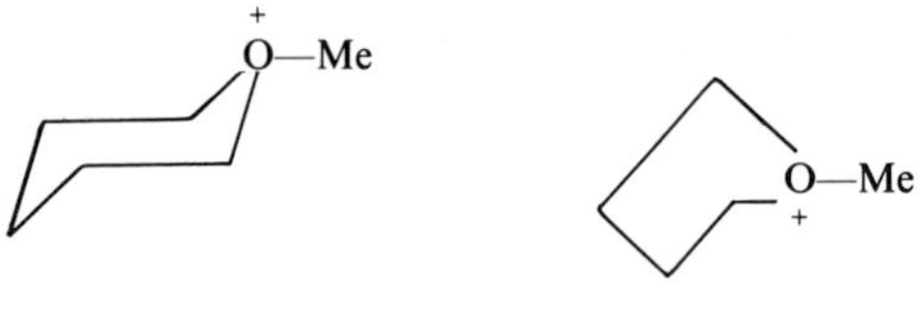

faster than at the ring carbon of (**30**). It is reasonable to assume that nucleophilic attack on the methyl groups of (**30**) and (**31**) is very similar. Therefore, the difference in observed reactivities of (**30**) and (**31**) must arise from ring opening of (**31**) being faster than ring opening of (**30**). Allred and Winstein[7] attributed the differences in ring opening characteristics to a combination of ring strain and steric hindrance at the methylene carbons of the two ring systems. Cation (**30**) can assume a relative strain-free chair form [i.e., (**32**)]. but (**31**) cannot adopt a conformation free of strain. The preferred form for (**31**) is probably (**33**). Conformation (**32**) provides a somewhat hindered approach (by the neighboring equitorial H) for the nucleophile. Conversely, a less hindered approach to the methylene carbon is available in conforma-

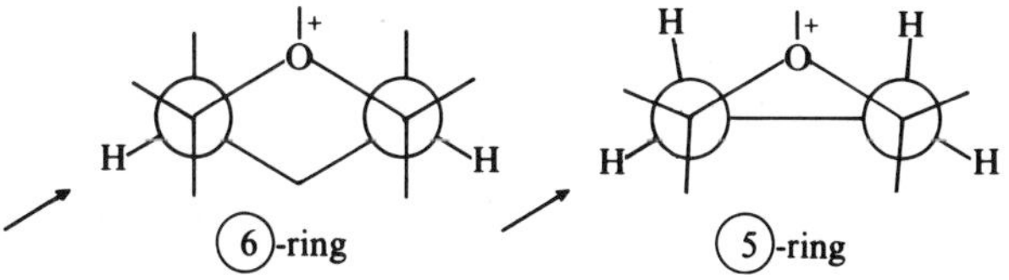

(32) ⑥-ring   (33) ⑤-ring

tion (33) of the five-membered ring. Hence ring opening of the five-membered ring is more rapid than Me—O cleavage.

The differences in ring opening for the five- and six-membered rings may also be viewed in terms of the free-energy differences for the opening and closing reactions.[8] Ring closure to form (31) is much faster than ring closure to form (30). Since the free energies of the acyclic species should be similar [with only $\delta\Delta S$ (internal rotation) needing to be taken into consideration], ring opening of (31) should be faster than (30) unless (31) is much more stable than (30). Of course (30) is actually more stable owing to the strain considerations mentioned above. Hence the free-energy–reaction coordinate diagrams are

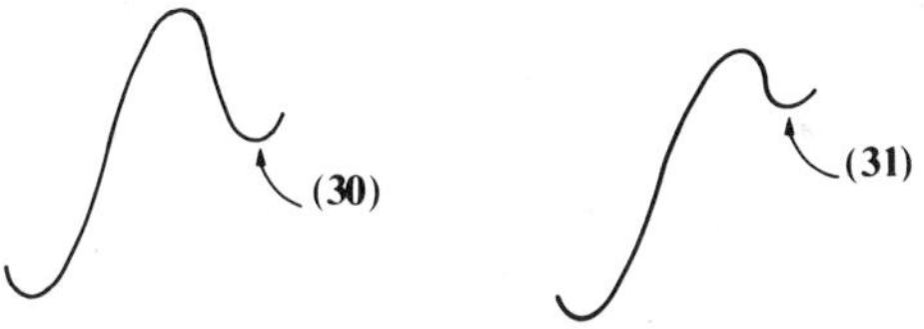

Therefore it may be argued that the observation that (30) appears to be demethylated more easily than (31) is related to the fact that (31) is formed more rapidly than (30).

The trifluoroethanolysis of 5-methoxypentyl toluene-p-sulfonate (34) is accompanied by 29% conversion to methyl toluene-p-sulfonate. This is reduced to only 19% by the addition of lithium perchlorate, suggesting that only two-thirds of the methyl toluene-p-sulfonate is formed from intimate ion pairs and that intimate ion pair (35) is converted into intimate ion pair (36) without passing through a solvent-separated ion pair.[9]

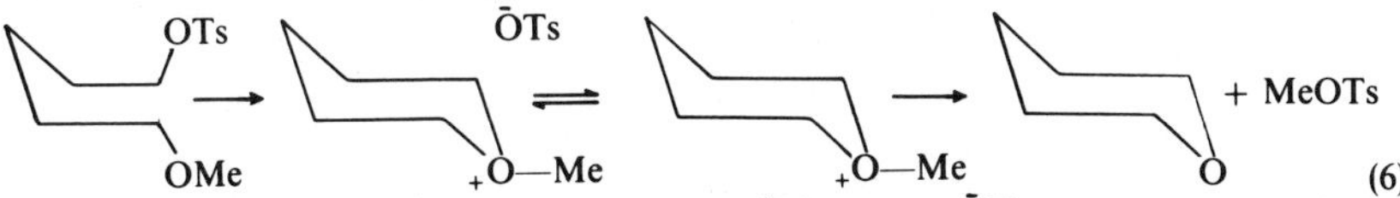

(34) OTs, OMe → (35) OTs, $^+$O—Me ⇌ (36) $^+$O—Me, OTs → O + MeOTs   (6)

Noyce and co-workers[10] have made a thorough study of the acetolysis of the interesting 4-methoxycyclohexyl toluene-*p*-sulfonate system for which MeO-5 participation is possible in the *trans* isomer (37). By comparing the reaction rate of (37) with that of the *cis* isomer and making a small allowance for the amount of *cis* isomer in a conformation with the toluene-

(37)   (38)   (39)

sulfonyl group in a more reactive axial position, they estimate that k_Δ/k_s for the *trans* compound is about 5.6. If this is true, a large amount of the product probably results from reaction involving participation by the methoxy group. The products are 4-methoxycyclohexene (67.7%), *cis*-4-methoxycyclohexyl acetate (8.7%), and *trans*-4-methoxycyclohexyl acetate (23.6%). This suggests that much of the 4-methoxycyclohexene results from a reaction involving participation. The products of acetolysis of *trans*-4-methoxy[1-^{3}H]cyclohexyl toluene-*p*-sulfonate are given in Table 3. Any products derived from the bicyclic oxonium ion (38) should have the tritium scrambled equally between the 1 and the 4 position, or possibly have a slight excess in the 4 position owing to a secondary isotope effect. The *trans*-4-methoxycyclohexyl acetate has 35% more tritium in the 1 position than in the 4 position, and hence some of it cannot have been formed by way of the oxonium ion (38). Although the 4-methoxycyclohexene contains all its tritium in the 1 position, it is possible, but not probable, that it came from an assisted reaction. Consistent with these observations, Noyce and Bastian[10] suggested that participation by the methoxyl group first involves partial bonding to the reaction center to give an internally solvated ion-pair intermediate (39), the anchimeric assistance being associated with the formation of this species. This ion pair may then react with acetic acid to

Table 3. Products of Acetolysis of trans-4-Methoxy-[1-^{3}H]cyclohexyl Toluene-p-sulfonate[10]

4-Methoxy[1-^{3}H]cyclohexene	66.4%
cis-4-Methoxy[1-^{3}H]cyclohexyl acetate	9.6%
trans-4-Methoxy[1-^{3}H]cyclohexyl acetate	13.8%
trans-4-Methoxy[4-^{3}H]cyclohexyl acetate	10.2%

give *trans*-acetate, collapse to give a bicyclic oxonium ion, or undergo elimination, the complete reaction scheme being as shown in Scheme 3. The

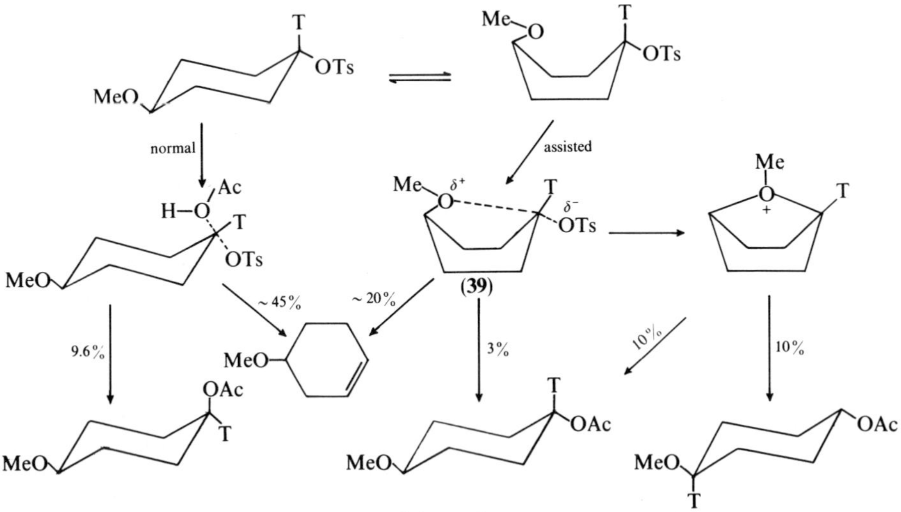

Scheme 3

driving force for participation in the acetolysis of 4-methoxycyclohexyl toluene-*p*-sulfonate is much less than in the acetolysis of a similar acyclic compound, e.g., (**14**), for which $k_\Delta/k_s = 162$ (see Table 2), because much of the energy gained from participation in the 4-methoxycyclohexyl compound is lost in nonbonded interactions on going to the unfavorable boat conformation. Because of the greater strain in the oxonium ion, relative to acyclic models, internal return may also account for a lower overall fraction proceeding to product through the oxonium ion. The effect of added lithium perchlorate would be interesting to determine in view of the results of this study.

As shown from product studies, MeO-5 participation also results upon treating (**40**) with boiling 50% aqueous methanol containing sodium acetate.[11] The formation of other products in the reaction may also be accounted for through invoking (**41**) as an intermediate. The high yield of the

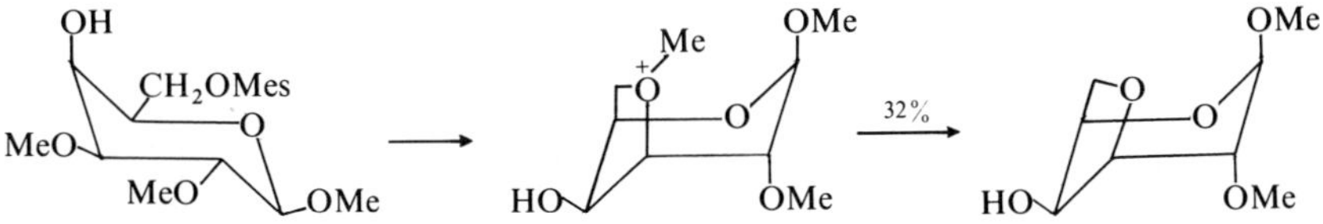

Mes = methanesulfonate

tetrahydrofuran derivative (relative to acetolysis studies involving MeO-5 participation) is an interesting example of structure and solvent variation affecting product formation. In another study, MeO-5 participation was found to compete with HO-3 participation in the hydrolysis of methyl 4-O-*p*-nitrobenzenesulfonyl-β-D-xylopyranoside.[12]

When a double bond is added to an acyclic ω-methoxy arenesulfonate competition between π and methoxy participation is possible. The acetolysis of *cis*-5-methoxypent-3-en-1-yl *p*-bromobenzenesulfonate (**42a**) is anchimerically assisted and yields compounds (**43**)–(**46**) as well as the corresponding methyl sulfonate ester unless lithium perchlorate is present.[13] It was shown that lithium perchlorate also prevented (by eliminating return from solvent separated ion pairs) formation of the allylic toluene-*p*-sulfonate (**47**), thus products (**45**) and (**46**) must be formed from allylic cation (**48**).

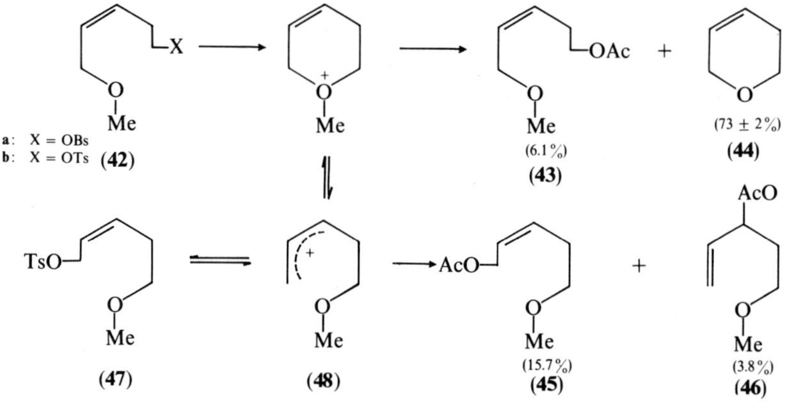

Methoxy participation in solvolytic studies involving leaving groups other than sulfonates has not received much study. The action of Lewis acids on ω-methoxy alkylhalides has been studied by Kirrmann *et al.*[14,15] and Meerwein.[16] A concerted reaction is proposed. In these studies the

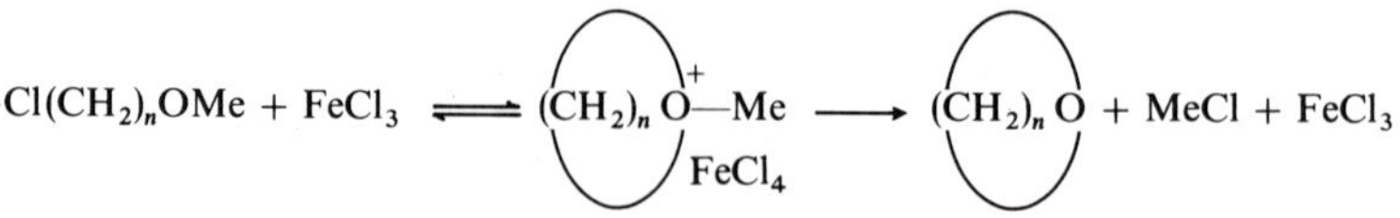

$$Cl(CH_2)_n OMe + FeCl_3 \rightleftharpoons \underset{FeCl_4}{(CH_2)_n \overset{+}{O}-Me} \longrightarrow (CH_2)_n O + MeCl + FeCl_3 \qquad (9)$$

products obtained are quite different from those from solvolytic reactions [e.g. Eq. (9)], usually consisting of only cyclic ether. Meerwein[16] suggests that the high yields of the cyclic ether in both the MeO-5 and MeO-6 participation reactions result from the rapid reaction to form the cyclic oxonium ion. Also, ring opening, if it occurs in the presence of a Lewis acid, will likely result in cyclization back to the oxonium ion. If halide ion attacks the methyl group, the high volatility of the methyl halides makes this pathway essentially irreversible.

The intermediacy of (**50**) was not proposed in the studies of the hydrolysis of 4-bromobutyl methyl ether (**49**), although its rate of hydrolysis at 50°C is 61.9 times that of *n*-propyl bromide.[17] A symmetrical transition state or intermediate is indicated, however, since the hydrolysis of (**49**) with ^{13}C at C-4 resulted in complete scrambling of the label between C-1 and C-4 in the sole product 4-methoxy-1-butanol. We believe that the data are more readily and simply accounted for by invoking the intermediacy of (**50**) rather than the suggested transition state (**51**).[17]

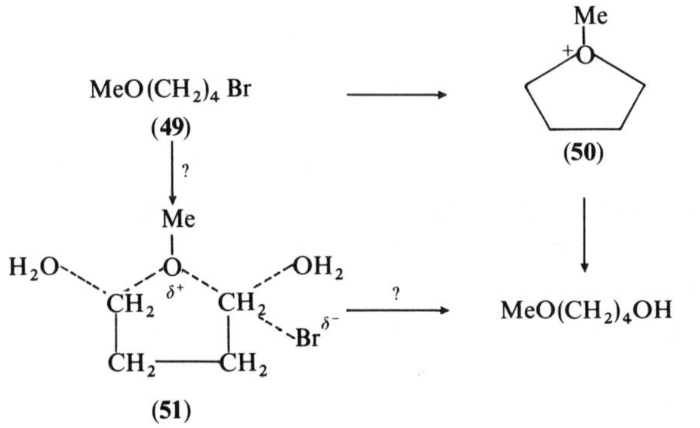

(10)

Although the primary methoxyl group in 2-methoxyethyl *p*-bromobenzenesulfonate provides no anchimeric assistance,[1] the absence of cyclization in a step after the rate-determining one cannot be discounted. MeO-3 assistance by secondary and tertiary methoxyl groups, however, quite definitely occurs. Examples of participation by secondary methoxyl groups are found in the reactions of compounds (**52**), (**53**), and (**54**) with silver acetate in acetic acid. The products are methoxyacetates of retained configuration.[18] Anchimeric assistance is said to occur to a limited degree in the solvolysis of *trans*-2-methoxycyclopentyl and *trans*-2-methoxycyclohexyl tosylates, although the rates are much slower than the respective unsubstituted cycloalkyl tosylates.[19] The relative rates are solvent dependent. MeO-3 participation is also noticeable from the products formed in solvolytic studies of *cis*- and *trans*-2-methoxycyclooctyl toluene-*p*-sulfonates[20] and in the reaction of (−)-inositol-2,3,4,5,6-pentamethyl ether with PCl_5.[21] The tertiary methoxyl group in 2-bromo-3-methoxy-3-methylbutane (**55**) participates in hydrolysis and migrates to give 3-methoxy-2-methylbutan-2-ol (**56**).[5] MeO-3 participation by tertiary methoxyl groups can provide

(**52**) = *threo*
(**53**) = *erythro*

(**54**)

(**55**) (**56**) (11)

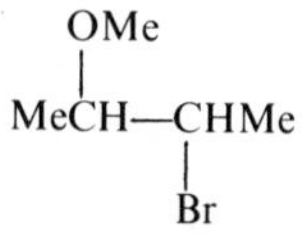

considerable anchimeric assistance; k_{Δ}/k_s for the acetolysis of 2-methoxy-2-methylpropyl *p*-bromobenzenesulfonate (**57**) is about 1500, and when the solvent is aqueous dioxane the major product is isobutyraldehyde (**58**).[22] There is no evidence for anchimeric assistance in the hydrolysis studies of 2-chloro-2-methylpropyl methyl ether (**60**), which theoretically could form the same oxonium ion (**59**) as is derived from (**57**).[17] The relative rate for the hydrolysis of (**60**) is 0.0049 relative to *t*-butyl chloride. This rate depression

(**57**) (**59**) (**58**) (12)

(**60**) (**59**) (13)

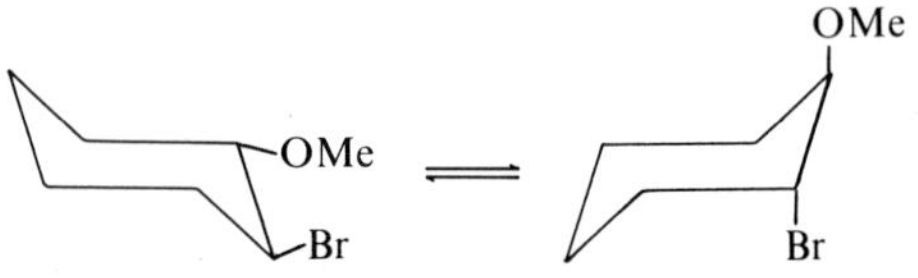

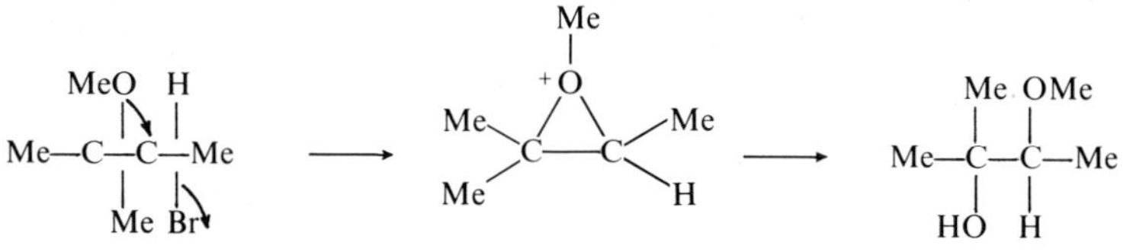

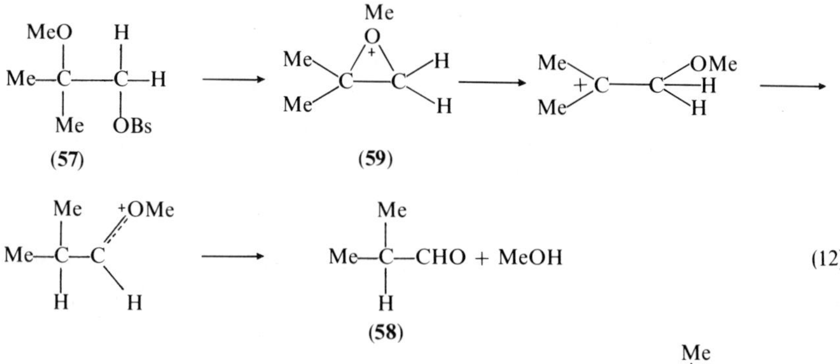

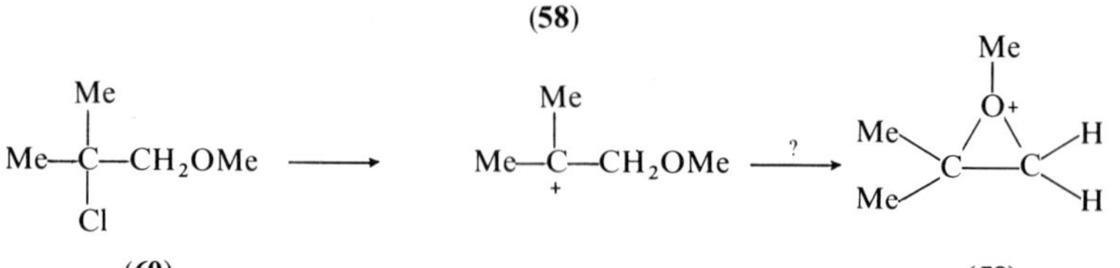

Table 4. Relative Rates of Solvolysis of Methoxyphenyl-alkyl Toluene-p-sulfonates[23]

Substrate, ROTs, R =	HCO$_2$H (25°C)	AcOH (75°C)
2-Methyl-2-phenylpropyl	1.0	1.0
2-(p-Methoxyphenyl)-2-methylpropyl (**63**)	72.0	88.0
2-(o-Methoxyphenyl)-2-methylpropyl (**64**)	6.47	5.5
2-(o-Methoxyphenyl)ethyl (**65**)	0.66	0.99
2-(p-Methoxyphenyl)ethyl (**66**)	0.67	1.3

results from the inductive effects of the neighboring oxygen. The lack of product studies for the reaction[17] prohibits a conclusion relative to the formation of (**59**) after the transition state.

Participation by methoxyl groups attached to benzene rings, as shown in (**61**), sometimes occurs. The nucleophilicity of such groups is reduced by the resonance interaction with the benzene ring, but this is at least partly counterbalanced by the increased rigidity of the system. In compounds

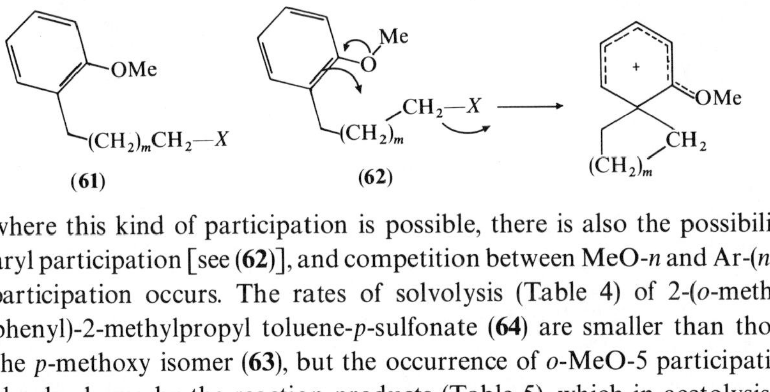

$$\text{(61)} \quad \text{(62)} \longrightarrow \text{(14)}$$

where this kind of participation is possible, there is also the possibility of aryl participation [see (**62**)], and competition between MeO-n and Ar-(n − 2) participation occurs. The rates of solvolysis (Table 4) of 2-(o-methoxyphenyl)-2-methylpropyl toluene-p-sulfonate (**64**) are smaller than those of the p-methoxy isomer (**63**), but the occurrence of o-MeO-5 participation is clearly shown by the reaction products (Table 5), which in acetolysis, for-

Table 5. Products of Solvolysis of 2-(o-Methoxyphenyl)-2-methylpropyl Toluene-p-sulfonate[23]

Solvent	Temp. (C°)	Added base	Total yield	Olefin (**67**) + (**68**)	Alcohol Tert. (**69**)	Alcohol Prim. (**70**)	Furan (**71**)
AcOH	25	NaOAc, 0.301M	96	8	15	4	73
HCO$_2$H	25	NaOCHO, 0.0515M	92	32	42	—	26

molysis, and ethanolysis all contain some of the benzofuran (71). The solvolyses therefore proceed as shown in Scheme 4. The ethanolysis and formolysis reactions obey the first-order rate law and give the theoretical infinity titers, but the acetolysis reaction gives infinity titers (presumably after 10 half-lives) corresponding to 63% reaction. This is due to the formation of methyl toluene-p-sulfonate, which undergoes acetolysis 500 times more slowly. In the presence of sodium acetate or lithium perchlorate, fairly good infinity titers are obtained owing to the exclusion of internal return.[1]

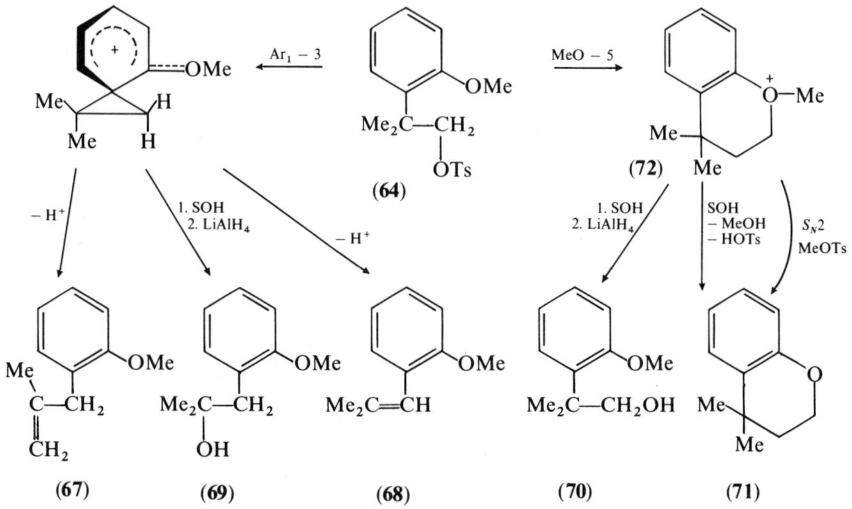

Scheme 4. SOH = solvent.

The rates of solvolysis in ethanol, formic acid, and acetic acid of 2-(o-methoxyphenyl)ethyl toluene-p-sulfonate (65) are approximately the same as those of its *para* isomer (66). Because of the lack of product analyses and therefore quantitative knowledge of the importance of aryl participation[24] and the k_s pathway,[25] it is not possible to determine the extent of o-MeO-5 participation. However, o-MeO-6 participation occurs extensively in the acetolysis and formolysis of the analogous 3-(o-methoxyphenyl)-1-propyl p-bromobenzenesulfonate (73) and 3-(o-methoxyphenyl)-3-methyl-1-butyl toluene-p-sulfonate (74) (see Table 6).[1] Any competing aryl participation would be Ar$_1$-4 participation and, as shown by the reaction rates for the *para* isomers, is unimportant. The reaction products (Table 6) contain large amounts of the pyran derivatives (75) and (76) and of the methyl arenesulfonates, and the reactions may therefore be formulated as shown in

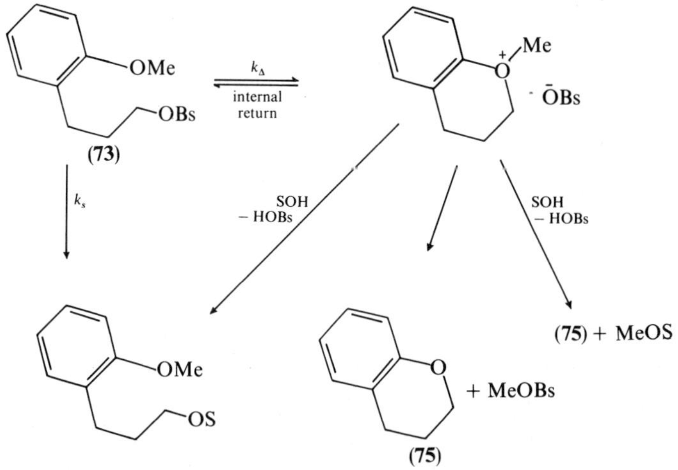

Participation by Oxygen Groups

Scheme 5

Schemes 5 and 6. The infinity titers for the acetolysis of (**74**) correspond to 34% reaction at 75°C, and this value is decreased by the addition of lithium toluene-*p*-sulfonate, a result that indicates that formation of methyl toluene-*p*-sulfonate involves more than just ion-pair return.

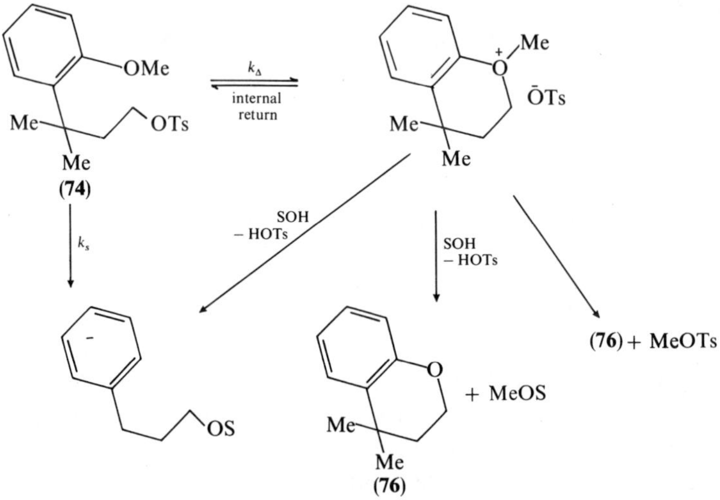

Scheme 6

Table 6. Rates and Products for the Formolyses of 3-(o-Methoxyphenyl)-1-propyl p-Bromobenzenesulfonate (**73**) and 3-(o-Methoxyphenyl)-methyl-1-butyl Toluene-p-sulfonate (**74**) at 25°C[1]

Compound	$10^5 k$	$10^5 k$ for para isomer	Products	
			Alcohol	Pyran
(**73**)	0.2	~0.01	46%	49%
(**74**)	22.0	~0	5%	95%

MeO-5 participation also occurs in the reaction of (**77**) with potassium mercaptoacetate in anhydrous methanol.[26] The products obtained were the benzofuran derivative (**78**), S-methylmercaptoacetic acid, and a small amount of (**79**).[27] No kinetic studies were performed to measure anchimeric assistance, but based on studies of similar systems,[1,23,24] assistance would

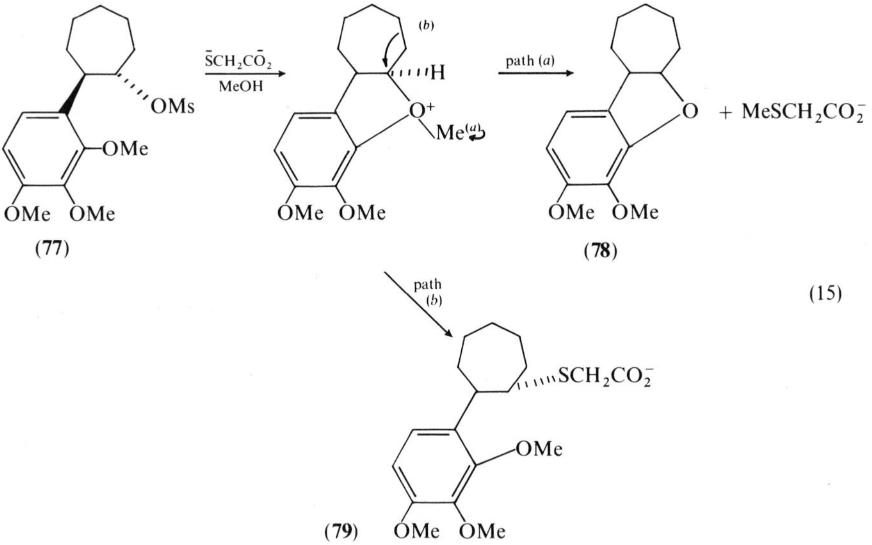

(15)

be expected. A more thorough study of the cyclohexyl derivatives has appeared.[27] In addition to the products of nucleophilic displacement on the intermediate oxonium ion, (**82**) and (**83**) [and (**86**)?], products of competing direct displacement, (**81**) and (**86**), and elimination, (**84**) and (**85**), were isolated in varying amounts as the number of methoxy groups and the ratio of nucleophile sulfonate were changed (see Table 7). It is interesting that a 3-methoxy group nearly doubles the yield of the benzofuran derivative (**83**)

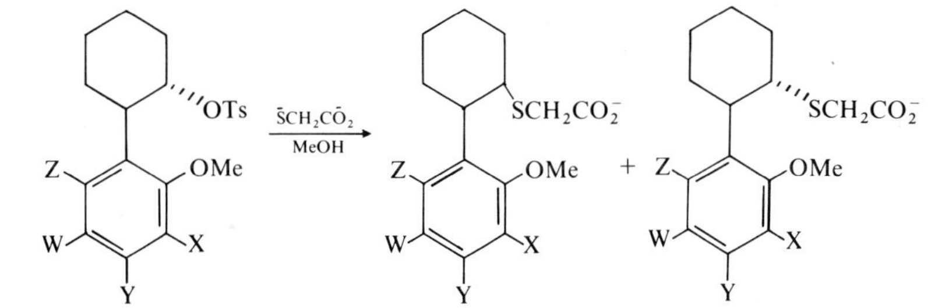

(80) (81) (82)

W, X, Y, Z = H or OMe

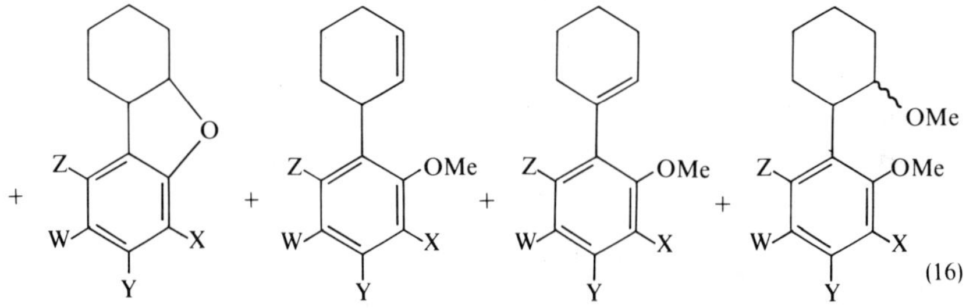

(16)

(83) (84) (85) (86)

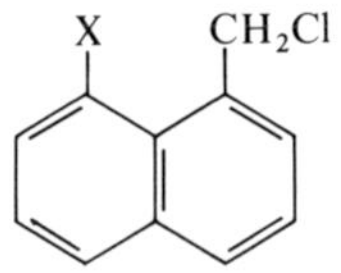

(87)

X = H, OMe, Br, Cl, Me, NO_2

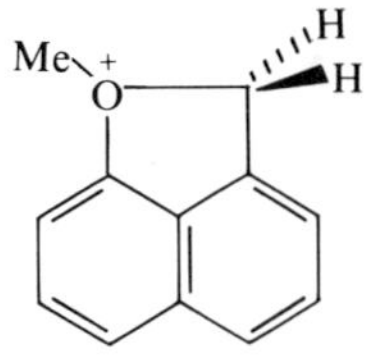

(88)

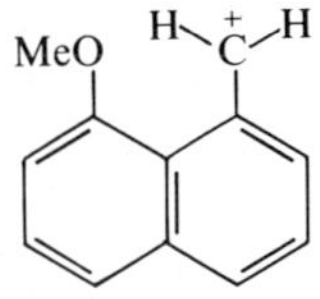

(89)

probably through increased MeO-5 participation. A 4-methoxy group, however, lowers the amount of (**83**), probably through increased Ar_1-3 participation.

Anchimeric assistance is not detected in the solvolysis of a series of 8-substituted chloromethylnaphthalenes (**87**) in 80% acetone.[28] The lack of MeO-5 assistance and/or subsequent formation of the cyclic oxonium ion (**88**) may be the result of the greater stability derived from benzylic-type resonance interactions in (**89**) relative to (**88**) despite the favorable stereochemistry for MeO-5 participation and the steric strain in the *peri*-substituted carbonium ion (**89**).

Neither *o*-MeO-6, O-3, Ar_1-4, nor Ar_2-5 anchimeric assistance was detected in the reaction of (**90**) with dimethylamine in 95% aqueous ethanol at 100°C.[29]

Table 7. Yields of the Important Products from the Displacement Reactions of Methoxy-Substituted Phenylcyclohexyl p-Toluene-sulfonates[22]

Ring substituent	% yield	% (81) + (82)	% (83)	% (84)	% (85)	% (86)	% recovered (80)	Ratio (81)/(82)	Ratio of anion/tosylate
2-OCH$_3$	102	11.77	47.30	15.66	1	24.99	9	27:73	2:1
2-OCH$_3$	100	42.42	21.03	32.60	0	3.97	0	78:22	50:1
2,3-OCH$_3$	97	6.17	87.50	4.44	2.22	0	0	58:42	2:1
2,3-OCH$_3$	101	25.06	45.70	26.28	0	3.13	0	100:0	50:1
2,3,4-OCH$_3$	107	31.10	48.70	5.69	1.70	12.84	0	0:100	2:1
2,3,4-OCH$_3$	96	45.27	27.49	24.00	1	3.06	0	0:100	50:1
2,5-OCH$_3$	98	6.07	67.42	8.06	4.92	13.37	0	a	2:1
2,5-OCH$_3$	100	40.78	30.51	27.29	1	1.29	0		50:1
2,6-OCH$_3$	98	7.30	82.60	7.30	2.79	0	0	b	2:1
2,6-OCH$_3$	98	0	83.30	12.36	4.34	0	0		50:1
2,4-OCH$_3$	85	64.52	8.57	0	3.71	23.00	0	0:100	2:1
2,4-OCH$_3$	104	78.20	5.45	15.19	0	1.32	0	0:100	50:1
4-OCH$_3$	94	23.18	0	0	0	33.03	43.68	18:82	2:1
4-OCH$_3$	100	63.94	0	35.09	0	1	0	85:15	50:1
3,4-OCH$_3$	107	22.30	0	0	0	16.02	61.67	31:69	2:1
3,4-OCH$_3$	101	64.86	0	33.59	1	1.24	0	97:3	50:1
3-OCH$_3$	85	19.26	0	0	0	0	80.74	100:0	2:1
3-OCH$_3$	106	45.58	0	41.84	4.52	8.16	0	100:0	50:1

^{a}Isomers could not be separated; NMR showed acid at low concentration to be primarily the *trans* isomer, and at high concentration, the *cis* isomer.
^{b}Isomers could not be identified due to low yields.

Participation by Oxygen Groups

Methoxy participation has also been encountered in reduction reactions with lithium aluminum hydride[30] and with lithium in liquid ammonia[31] and in the addition of Grignard-type reagents to ketones.[32]

(90) — o-(2-bromoethoxy)anisole derivative with OCH$_2$CH$_2$Br and OMe substituents

(91) — methyl 2,3-O-dibutyl-stannylene-α-D-glucopyranoside with Sn---OMe coordination

Methoxy coordination to tin in methyl 2,3-O-dibutyl-stannylene-α-D-glucopyranoside (91) and similar compounds has been proposed to account for selective formation of esters at C-2.[33]

A synthetically useful neighboring group effect by oxygen is observed in the metallation of certain aromatic substrates.[34] When anisole is treated

$$\text{anisole (OMe)} + C_4H_9Li \longrightarrow \text{2-lithioanisole (OMe, Li)} \quad (17)$$

with n-butyl lithium metal–oxygen coordination directs formation of the 2-lithiated anisole [Eq. (17)] (Gilman and Morton[34]). Metallation reactions are dependent on time, solvent, and substituents.[35] Meyers et al.[36] have also demonstrated methoxy group participation in alkylation reactions with alkyl lithium derivatives.

It is clear from the above discussion that the methoxy group may be an effective neighboring group in solvolytic reactions and, when properly positioned, may lend significant anchimeric assistance. Despite the vast amount of research, however, there are only few examples where the extent of methoxy group participation can be quantified.

4.1.1.2. Benzyloxy and Aryloxy Groups

Other than the methoxy group, the benzyloxy group is the only other neighboring ether group frequently encountered in displacement reactions. Owing to its use as an easily removed hydroxyl blocking group, participation by this group has been observed often in reactions of carbohydrate derivatives (for reviews of neighboring group interactions in carbohydrate reactions, see Ref. 37). Barker et al.[38] have studied some simple acyclic models. 4-Benzyloxypentyl toluene-p-sulfonate gives 2-methyltetrahydrofuran, toluene-*p*-sulfonic acid, and benzyl ethyl ether when reacted with ethanol at 75°C. Although anchimeric assistance was clearly indicated, the kinetics, determined by the rate of formation of acid, were complex. It was thought the reaction involved the ion pair (**92**), which yielded the furan and benzyl-*p*-

$$\text{Me} \underset{}{\overset{\text{CH}_2\text{Ph}}{\diagup\!\!\!\diagdown\!\!\!\diagdown}}\text{OTs} \longrightarrow \text{Me}\underset{(\mathbf{92})}{\overset{\text{CH}_2\text{Ph}}{\diagup\!\!\!\diagdown}}\bar{\text{O}}\text{Ts} \longrightarrow \text{Me}\underset{}{\diagup\!\!\!\diagdown} + \text{PhCH}_2\text{OTs} \xrightarrow{\text{EtOH}} \text{PhCH}_2\text{OEt} \qquad (18)$$

toluenesulfonate. The latter ester apparently undergoes a slightly faster solvolysis. The rates for the corresponding reactions of 2,3,4-tri-0-benzyl-1,5-di-0-toluene-*p*-sulfonyl-ribitol, -arabitol, and -xylitol were measured. After symmetry corrections are made, the rate constant for the ribitol compound is about 10 times smaller than those for the other two, apparently owing to conformational effects. Brimacombe and Ching[39] have reported several examples of neighboring benzyloxy participation in reactions of carbohydrate derivatives.

As mentioned above, the nucleophilicity of an atom attached to an aromatic ring is decreased owing to resonance interactions with the ring. Thus examples of phenoxy group participation are rare, yet some may be found. Attempts to recrystallize the diazoketone (**93**) from methanol resulted in its complete destruction and the formation of four products. Three products were ordinary products that required no special explanation. The fourth product (**94**) was said to result from ArO-4 participation, and the mechanism shown in Eq. (19) was proposed.[40] Since N_2 is such a good

$$\begin{array}{c}\text{Me}\\|\\\text{ArOCCOCHN}_2\\|\\\text{Me}\\(\mathbf{93})\end{array} \xrightarrow{\text{H}^+} \cdots \qquad (19)$$

Ar = Biphenylyl

$$\begin{array}{c}\text{Me}\\|\\\text{ArOCH}_2\text{COC}=\text{CH}_2\\(\mathbf{94})\end{array}$$

Participation by Oxygen Groups

leaving group, the concerted loss of nitrogen is by no means the only reasonable mechanism, and hence kinetic substantiation of ArO-4 assistance out to be sought.

4.1.1.3. Cyclic Systems

A number of studies of participation by a ring oxygen have now appeared. Closson et al.[41] found that the tetrahydrofuranyl group was two to three times better than the methoxy group in participation (see Table 8). The

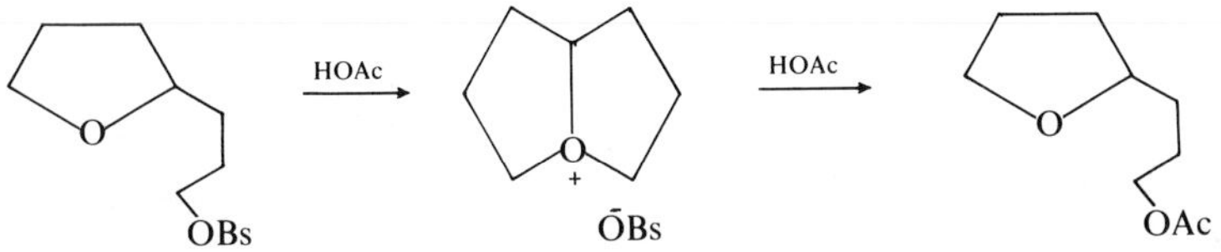

(20)

(95)

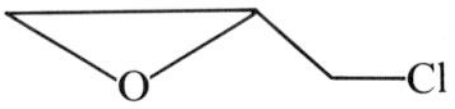

(96)

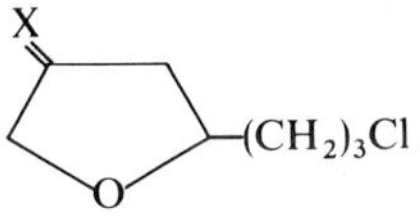

(97) X = H$_2$ (Gilman and Hewlett[41])
X = O (Paquette et al. [43])

bicyclic oxonium ion (95) was proposed as an intermediate in the solvolysis involving O-5 participation. We suspect that a similar mechanism accounts for the behavior of the chlorides (96)[42] and (97)[43] under solvolytic conditions. Epichlorohydrin, for example, undergoes acetolysis 3 times faster than

Table 8. Rates of Solvolysis of p-Bromobenzene-sulfonates in Acetic Acid[41]

Compound	Temp. (°C)	$10^5 k$ (sec^{-1})
MeO(CH$_2$)$_4$OBs	50	4.58
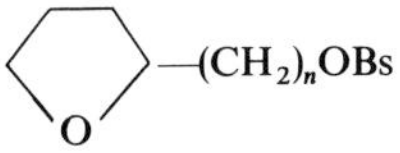 ⟨O⟩—(CH$_2$)$_n$OBs		
n = 1	80	0.094
n = 2	80	0.193
n = 3	50	12
n = 4	80	188
n = 5	80	21.5

cyclopropylmethyl chloride and 100 times faster than allyl chloride. Both of the model compounds are, of course, unusually reactive for primary derivatives. Additional evidence for O-3 participation in (**96**) is the isolation of 3-acetoxyoxetane, which could be formed as shown in Eq. (21).

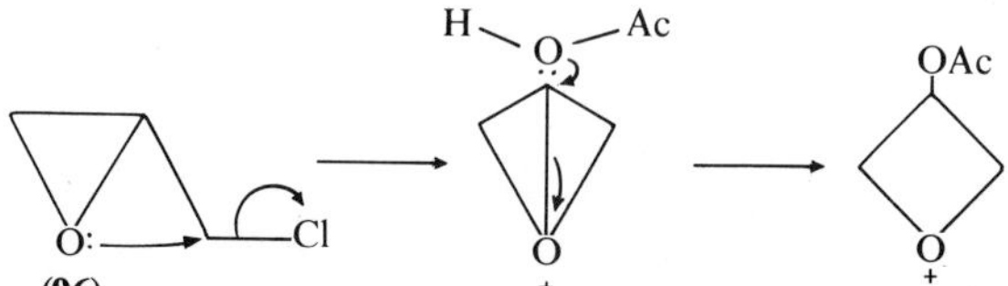

(**96**)

(21)

Tarbell and Hazen[44] measured the rates of acetolysis of arenesulfonates (**98**)–(**103**). The heterocyclic compounds showed substantial rate reductions when compared to (**101**) or (**102**) or to the appropriate cycloalkyl derivative.

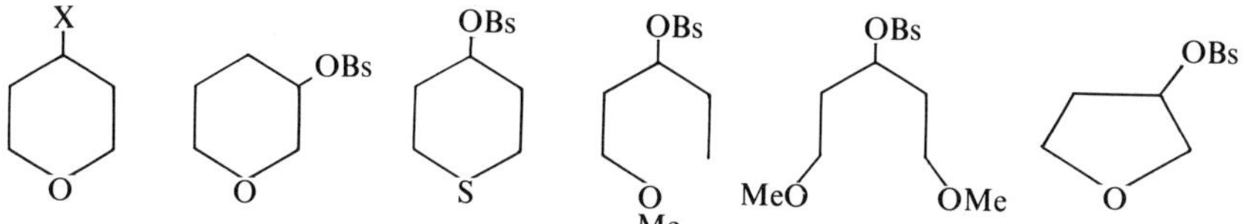

(**98**) (**99**) (**100**) (**101**) (**102**) (**103**)

a: X = OBs
b: X = OTs

This retardation was attributed to a transannular dipolar field effect exerted by the heterocyclic atom. In spite of the dipolar field effect, (**99**) solvolyzed faster than predicted. Since no tetrahydrofuranyl acetate (**105**) was formed, as would be expected from (**104**), a partially bonded structure (**106**) was

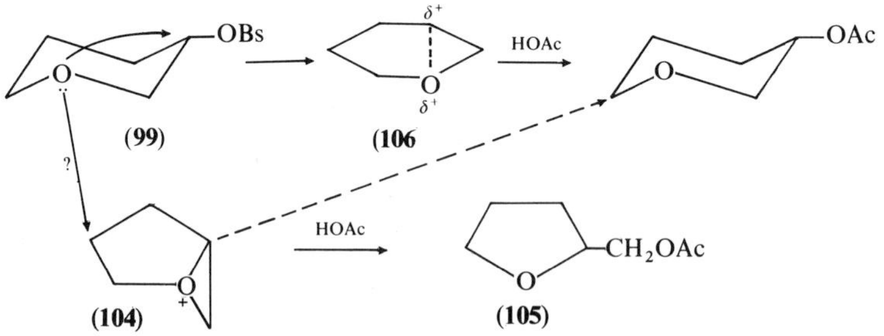

(**99**) (**106**) (**104**) (**105**)

(22)

preferred over the structure (**104**) as the intermediate [see Eq. (22)]. The intermediacy of (**104**) is not necessarily ruled out, however (cf. p. 234).

Paquette and his co-workers[43,45-52] have now published a number of papers dealing with neighboring ether oxygen in medium ring systems and bicyclic systems (for a study involving bridged androstanes, see Ref. 53). Virtually complete transannular reaction was observed in the acid-promoted rearrangement of (**107**) (Paquette et al.[43]). The experimental findings

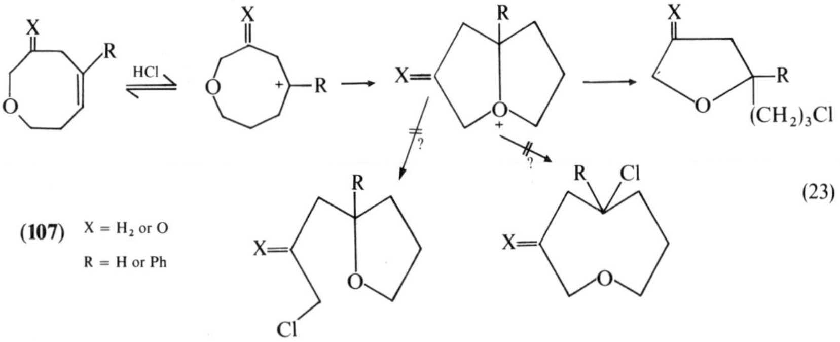

(23)

(**107**) X = H_2 or O
R = H or Ph

show a dependence of the propensity for rearrangement upon the intensity of electron deficiency at the developing carbocation center.

Kinetic and product studies were reported to support the conclusion the homoallylic participation (see Volume 2) overwhelms R_2O-3 participation in a medium-sized ring. This differs from the results in the acyclic system (**42**).[13] While a weak anchimeric assistance was noted in the acetolysis of (**108**), the products obtained are explicable on the basis of the derived homoallylic cation (**109**). The products from (**110**),[45] however, may arise exclusively from (**111**).

The kinetics of acetolysis of the *p*-bromobenzenesulfonate (**112**) show a sevenfold rate retardation relative to cyclooctyl *p*-bromobenzenesulfonate.

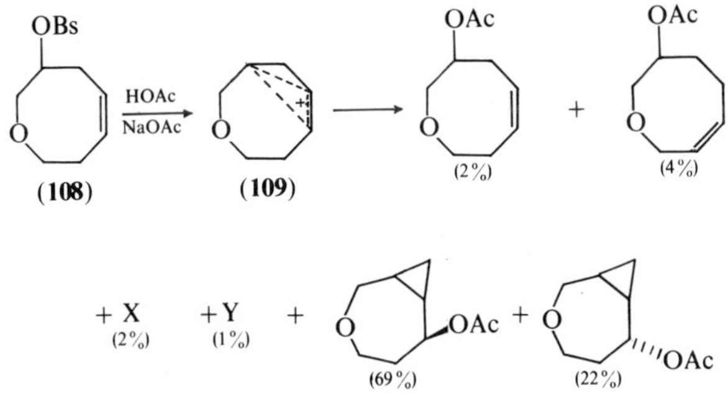

(**108**) (**109**)   (2%)   (4%)   (24)

+ X   + Y   +
(2%)   (1%)
              (69%)   (22%)

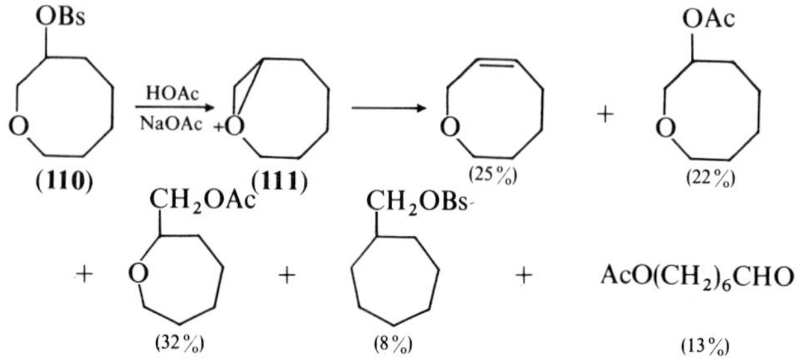

The 3,5-dinitrobenzoate (**113**), however, hydrolyzes at a rate 48,500 times that of cyclooctyl *p*-bromobenzenesulfonate, after appropriate leaving group conversions.[50] These results dramatically demonstrate transannular anchimeric R_2O-5 assistance in this medium ring system. Paquette and Scott[50] discussed inductive, steric, dipolar field, and anchimeric effects that are operative in the medium ring systems.

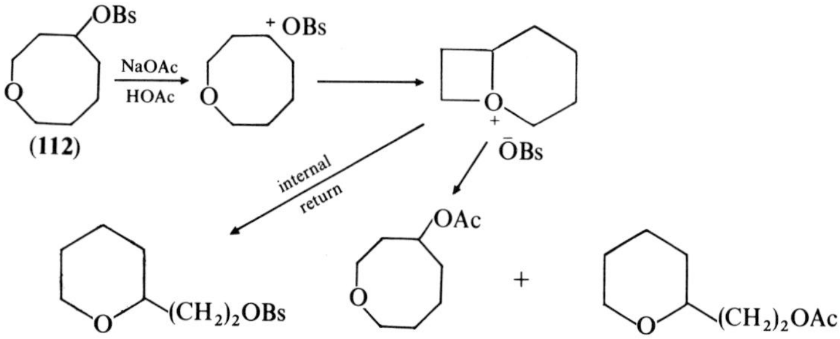

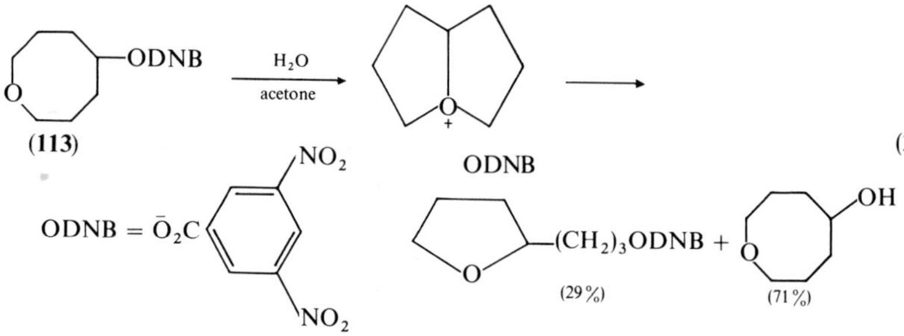

In a related study, the absence of products expected from oxonium ions (**115**) and (**116**) was taken as evidence against R_2O-4 participation in the formolysis of (**114**).[49] The isomeric derivative (**117**) was also found to undergo reaction without evidence of transannular oxygen participation. However, the nontransannular involvement of oxygen was proposed as shown in Scheme 7. Oxygen participation, which is probably a competitive process in all these formolysis studies, allows cleavage to form the cyclopropyl carbinyl cation (**118**) (see Volume 2). The homologous ether (**119**) reacts quite in contrast to (**114**) and (**117**). The formolysis of (**119**) leads ex-

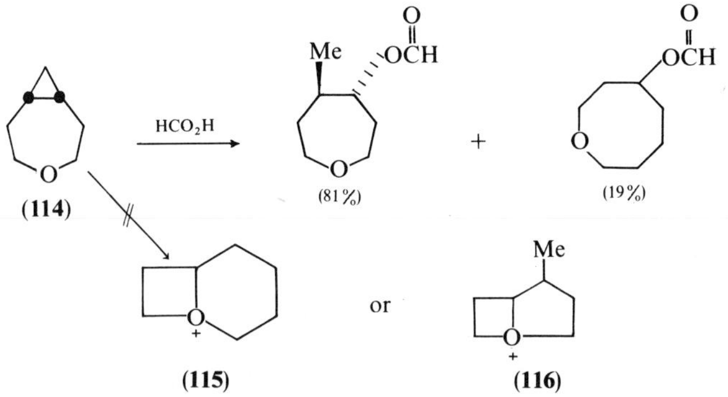

(28)

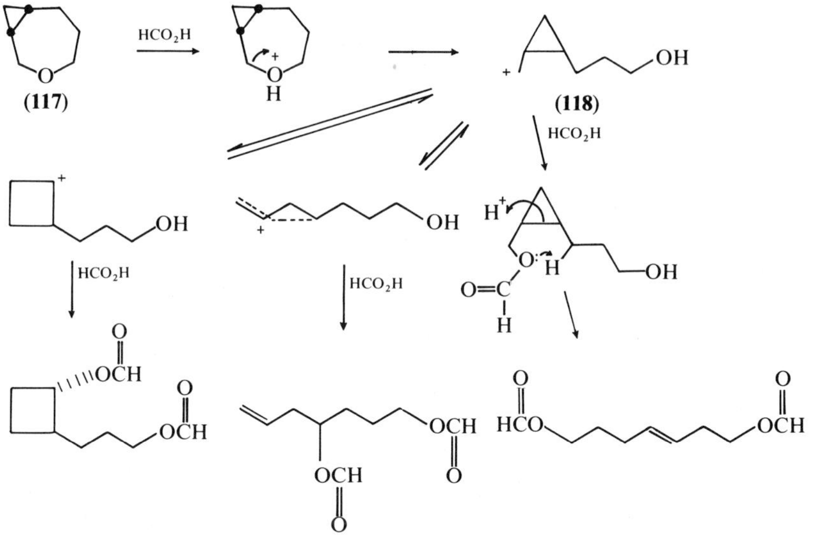

Scheme 7

clusively to (**120**) by either intermediate formation of carbocation (**121**) followed by oxygen attack to give the oxonium ion (**122**) or by direct RO-5 participation during heterolysis to give (**122**). Because of the comparatively different behavior of (**119**) (only internal bond cleavage) and its parent hydrocarbon *cis*-bicyclo[6.1.0]nonane (only external bond cleavage), the concerted mechanism involving R_2O-5 participation was indicated.[50]

Martin and Bartlett[54] found little evidence for oxygen participation in the solvolysis of the isomeric 2-chloro-7-oxabicyclo[2.2.1]heptanes [endo-

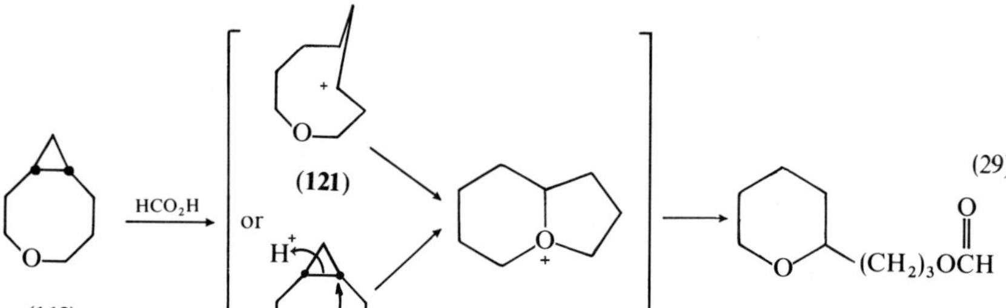

and exo-oxa-2-norbornyl chloride (123)]. A rate retardation, as estimated by rate comparisons with the respective 2-norbornyl models, accompanied the solvolysis of each isomer, and the product in each case was 3-formyl-cyclopentanol [Eq. (30)]. Paquette and Dunkin[52] observed a similar

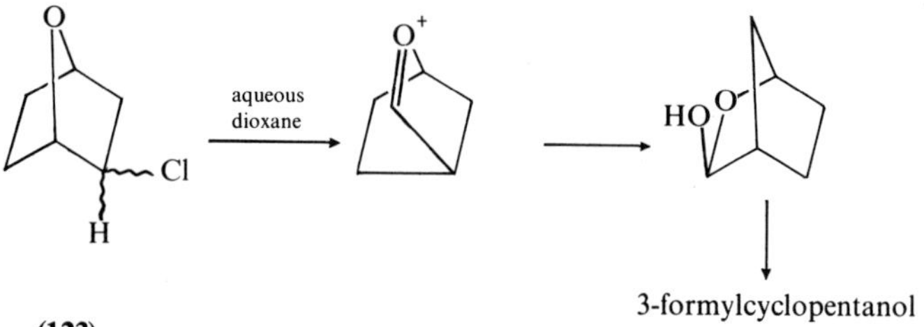

rate decelerative effect upon acetolysis of the epimeric benzo-7-oxabicyclo-[2.2.1]heptan-2-yl-p-bromobenzenesulfonates.

When the ether oxygen is substituted for C-6 of a bicyclo[2.2.1]heptyl ester, a dramatically different result is observed. Thus the solvolysis of exo-p-bromobenzenesulfonate (124) in buffered acetic acid[55] offers an extraordinary demonstration of anchimeric assistance. While the endo isomer (125) undergoes solvolysis at the rate predicted by the Foote–Schleyer method, the exo isomer (124) reacts some 7×10^7 times faster. This suggests that the ring oxygen is unusually capable of interacting with the back-side lobe of the sp^3 orbital of the exo substituent. R_2O-3 participation after ionization was suggested for the endo isomer from product analysis of solvolysis of the deuterated derivative (126). The endo derivative produced 85%

exo-acetate, 1.66% *endo*-acetate, and some unidentified products. The *exo-p*-bromobenzenesulfonate, however, produced 99.64% *exo*-acetate and 0.36% *endo*-acetate. It was suggested that the latter probably was formed

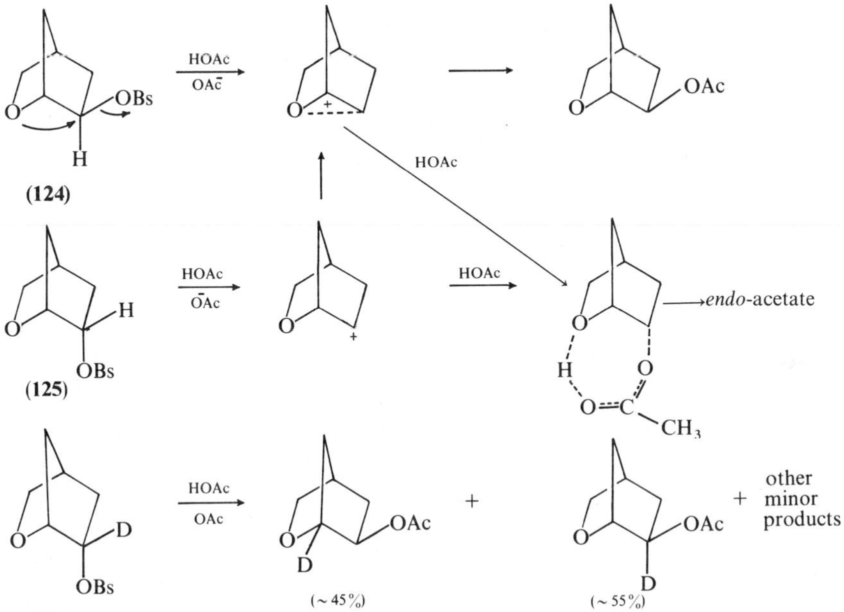

(124) **(125)** **(126)**

Scheme 8

by solvent transfer from the ring oxygen. These reactions are depicted in Scheme 8.

Unlike the results with *endo* (**123**), the homologous systems, (**127**) and (**131**), have been shown to solvolyze with marked anchimeric assistance. The *p*-bromobenzenesulfonate (**127**) was found to directly produce the oxonium ion (**128**) upon acetolysis, while its *exo* isomer (**129**) appears to react via the carbocation (**130**), which rearranges to (**128**). Kinetic and product studies support these conclusions.[46] Some elegant studies by Paquette

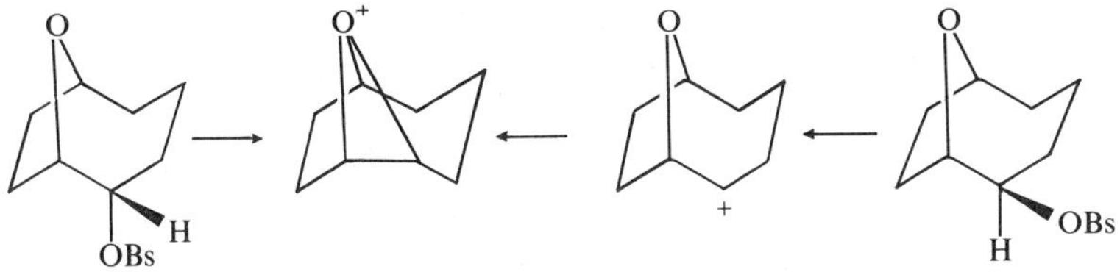

(127) **(128)** **(130)** **(129)** (31)

(131) → HOAc/NaOAc → (132) → (133) racemic (78%) OBs + (134) racemic (22%) OAc   (32)

et al.[51] on the optically active ester (131) have provided further insight into the conformational requirements for ether oxygen participation. The solvolysis in buffered acetic acid of optically active *endo*-8-oxabicyclo[3.2.1]-octan-2-yl *p*-bromobenzenesulfonate (131) gives only the corresponding racemic *endo*-acetate (134). Titrimetric and polarimetric rate measurements indicated that R_2O-3 participation occurs to give the symmetrical bridged oxonium ion (132), which partitions itself between solvent attack to give racemic *endo*-acetate (134) and internal return to give racemic *p*-bromobenzenesulfonate (133). The 4:1 ratio between internal return and solvent capture measured in this study demonstrates that the titrimetric rate constant is not a realistic measure of R_2O-3 (or possibly R_2O-n) participation. Thus, previous studies using only titrimetric kinetic methods must now be questioned. The polarimetric method, which is realistic in the case of (131), reflects significant rate-retarding inductive and field effects by the oxygen, although the ring oxygen in (131) unquestionably offers anchimeric assistance to ionization under the conditions of kinetic control. Acetolysis of the optically active *exo*-2-deuterio-*endo*-8-oxabicyclo[3.2.1]octan-2-yl *p*-bromobenzenesulfonate allowed the measure of the α-deuterium isotope effect in the ionization process. The observed effect ($k_H/k_D = 1.08$) is compatible with a high degree of R_2O-3 participation.[51]

Because of the demonstrated kinetically important participation by oxygen in the ionization of the esters of (124), (127), and (131) and by sulfur in the analogous sulfur derivative of *endo* (123) (see p. 201), it is somewhat surprising to find no participation in the solvolysis of *endo* (123). Martin and Bartlett[54] preferred to classify the solvolysis of *exo* (123) as a weak k_Δ process, but it is more likely a k_R process with any assistance weakened because of poor overlap by the *n* electrons of oxygen in the transition state. In addition to a k_R mechanism in the hydrolysis of *endo* (123), the k_Δ mechanism cannot be excluded (Scheme 9). The k_Δ process could lead to a weakly bridged oxonium ion (135), which rearranges to the oxocarbocation (136) before product formation. The formation of the oxonium ion (135) could be favor-

Participation by Oxygen Groups 153

able enough to exclude a k_s process, yet, owing to the strain attending tricyclic ring formation, no rate enhancement is discernable from the use of *exo/endo* rate ratios. In light of the results from the other bicyclic systems discussed in this section, the 7-oxanorbornyl system could bear reexamination. The results of acetolysis of optically active sulfonate esters would be especially interesting to have for comparison purposes.

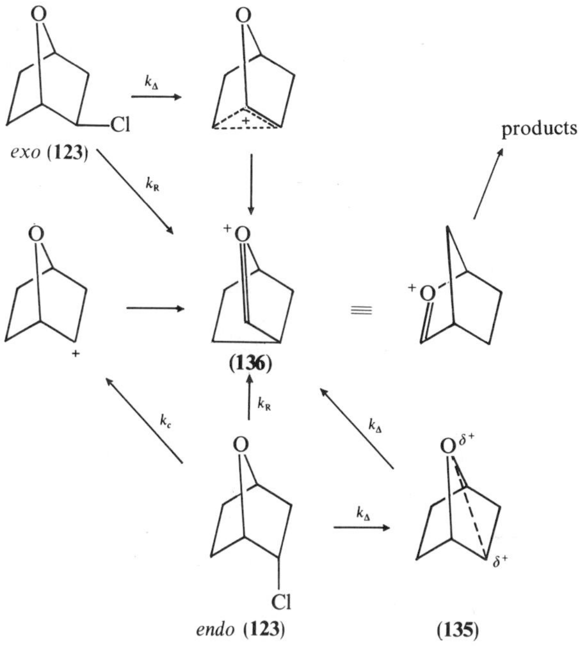

Scheme 9

In summary, the results discussed above reveal that an ether oxygen in a medium or a bicyclic ring system may offer significant anchimeric assistance in reactions in solvolytic displacement reactions. Because of inductive effects, steric effects, field effects, and internal return, anchimeric assistance, when present, is often difficult to demonstrate kinetically. Products formed from bicyclic and tricyclic oxonium ions probably more often result from participation after carbocation formation. Participation by oxygen apparently occurs to the exclusion of transannular hydride shifts in medium ring systems, but other types of participation such as homoallylic or formation of cyclopropyl carbinyl cations will most likely occur to the exclusion of ether oxygen participation.

4.1.1.4. Acetals and Ketals

Acetals or ketals, like simple ethers, participate in many displacement reactions. Although a smaller amount of data is available upon which to draw conclusions about the facility of the acetal participation, the results seem comparable to those of simple ethers. For example, 4-chloro-1,1-diethoxybutane (137) is said to liberate chloride ion in the presence of potassium hydroxide much faster than 5-chloro-1,1-diethoxypentane.[56] The rate difference was attributed to participation by the acetal ethoxy

EtO OEt   Cl $\xrightarrow{\text{KOH, diethylene glycol}}$ EtO—(Et,O+)—   ⟶   EtO OEt OH   (33)

(137)   (138)   (90%)

group in (137) to form the oxonium ion (138). Although the product could be accounted for by invoking S_N2 chloride displacement by hydroxide ion, the rate studies are consistent with the greater propensity for O-5 versus O-6 participation (*vide supra*).

Hazen[57] has compared the reactivity of acetals to simple ω-methoxy alkyl sulfonates. Some of his results are shown in Table 9. As expected, all the methoxy-containing compounds solvolyze much faster than the model substrate (139). Because of differences in the nucleophilicity of the solvents, only the solvolyses in trifluoroethanol occur with $Fk_\Delta = 100\%$ (calculated from kinetic data). The kinetic results show that a simple methoxy group is about 40% better than the acetal group in MeO-6 participation [e.g., compare (141) versus (34), Table 9]. Consistent with the results of Winstein *et al.*[1,2] (*vide supra*), MeO-5 participation in acetals is more effective than MeO-6 participation. While solvolysis of (34) in buffered

Table 9. Solvolysis of Some Toluene-*p*-sulfonates

Compound	In methanol at 59.86°C		In trifluoroethanol at 69.90°C		
	Rel. rate	Fk_Δ	$CF_3CH_2ONa(M)$	Rel. rate	Fk_Δ
Me(CH$_2$)$_6$OTs (139)	1.0			1	
(MeO)$_2$CH(CH$_2$)$_3$OTs (140)	7.49	87	0.0505	3.9 × 10^3	100
(MeO)$_2$CH(CH$_2$)$_4$OTs (141)	1.95	49	0.0560	2.45 × 10^2	>99
MeO(CH$_2$)$_5$OTs (34)	2.72	63	0.0560	6.80 × 10^2	>99

trifluoroethanol is accompanied by internal return and formation of methyl toluene-*p*-sulfonate (p. 131), no methyl toluene-*p*-sulfonate is formed from ion (**142**). This is apparently indicative of a rapid equilibrium between the cyclic oxonium ion (**142**) and the resonance-stabilized oxocarbonium ion (**143**) with solvent capture occurring through (**143**). The homolog of (**142**) [i.e. the oxonium ion from (**140**)] is reported to behave similarly.[57]

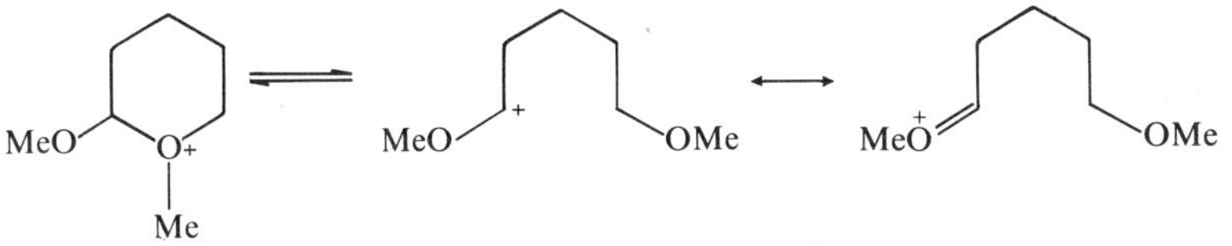

(34)

(**142**)   (**143**)

In attempting to determine the effect of electron-withdrawing substituents on the solvolysis of 2-norbornyl derivatives, Gassman *et al.*[58,59] found that the *endo*-ketal (**144**) solvolyzed with MeO-4 participation to produce (**148**) through methoxy group migration. The other products, (**146**) and (**147**), were also thought to arise from MeO-4 participation, since ketonic

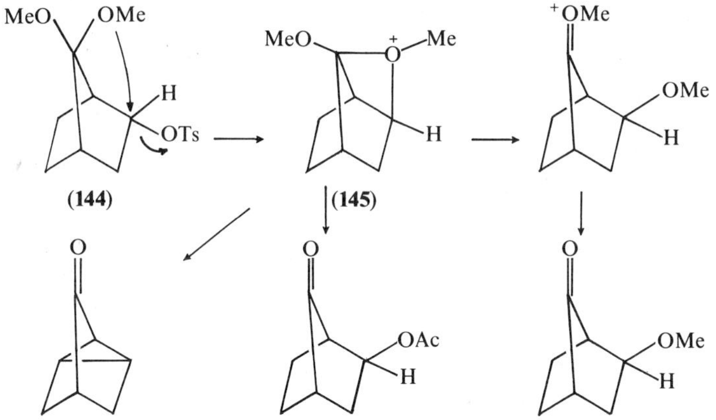

(35)

(**144**)   (**145**)

(**146**)   (**147**)   (**148**)

products are not produced from the corresponding ketals under the reaction conditions. The *exo* isomer of (**144**) acetolyzed more slowly by a factor of 38 at 25°C. Since the products were different (mostly the *exo*-acetal) from those obtained on acetolysis of (**144**), no MeO-4 participation was postulated. The ethylene ketal (**149**), unlike (**144**), cannot lend anchimeric assistance owing to transition-state strain. The acetolysis of (**149**) gave a 57% yield of ester (**151**) accompanied by other products normally expected from the reactions of the substituted norbornyl cation.[60] The ester was thought to be derived through fragmentation of (**149**), as shown, to the cation (**150**) which suffers attack by acetate ion to yield (**151**). The results with (**144**) and (**149**) suggest that, since (**144**) failed to undergo fragmentation, $k_\Delta > k_f$ in (**144**) (see Chapter 6).

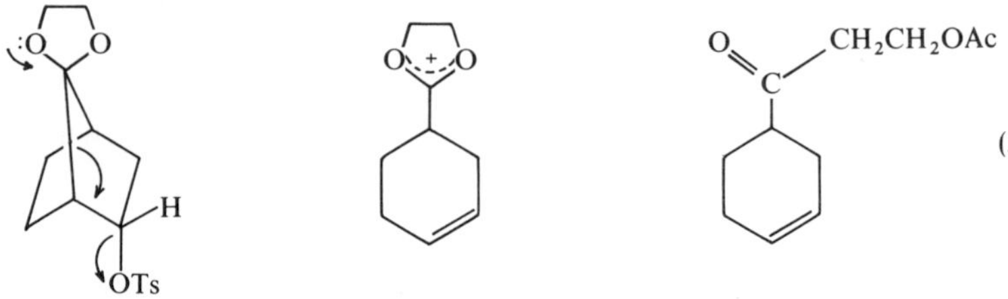

(36)

(**149**)   (**150**)   (**151**)

Participation by the acetal function is common in displacement reactions of carbohydrate derivatives.[37] An example of participation by an acetal group in an open-chain carbohydrate has been provided by Hughes and Speakman.[61] Treatment of ribose derivative (**152**) with tetrabutylammonium benzoate in 1-methylpyrrolidone gave 2,3,5-tri-O-benzyl-4-O-methyl-L-lyxose methyl hemiacetal 1-benzoate (**154**) instead of the expected 4-O-benzoyl-L-lyxose derivative. The oxonium ion (**153**) was thought to be an intermediate.

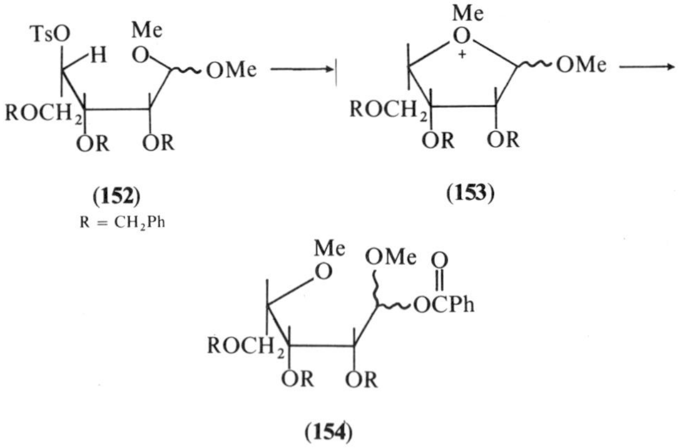

(**152**)
R = CH$_2$Ph

(**153**)

(37)

(**154**)

In slight contrast with solvolytic studies of simple cyclic ethers [e.g. (**99**)], Stevens et al. obtained evidence for R$_2$O-3 acetal participation in displacement reactions of pyranosides and furanosides.[62] On treatment with sodium acetate in dimethylformamide, (**155**) yields the ring-contracted products (**156a**) and (**157a**) in the ratio of 7:1. When heated in 90% aqueous dioxane in the presence of sodium bicarbonate (**155**) yielded (**156b**), (**157b**), and (**158**) in the ratio 1:2:6. The formation of (**158**) was rationalized on the basis of rearrangement of the intermediate oxonium ion (**159**) to (**160**) by ion-pair return and subsequent methoxy participation (see the first reference in Ref. 62).

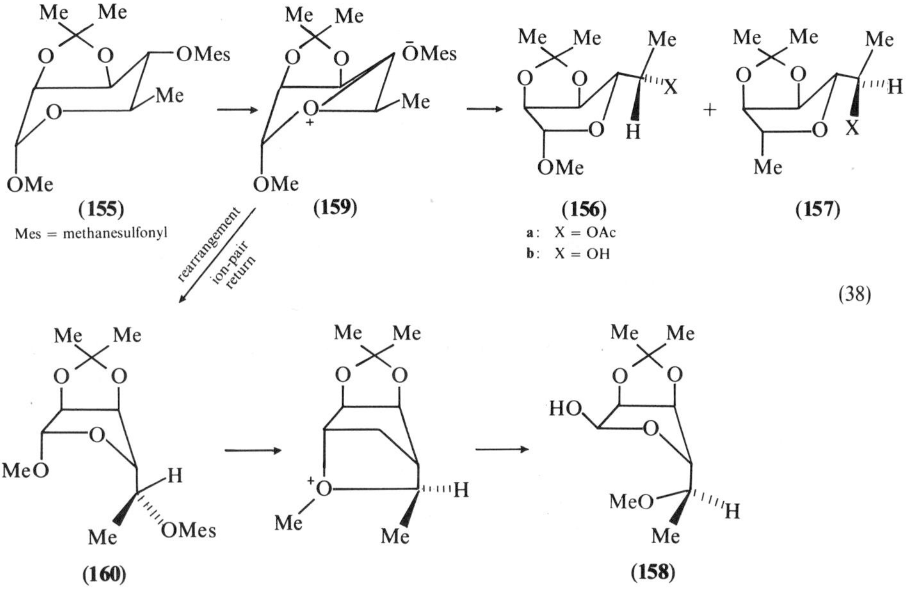

(38)

The formation of the resonance-stabilized oxocarbonium ion (**162**) probably accounts for the formation of a mixture of methyl 2,5-anhydro-3,4-isopropylidene-α- and β-talopyranoside (**163**) from (**161**).[63] It is probable

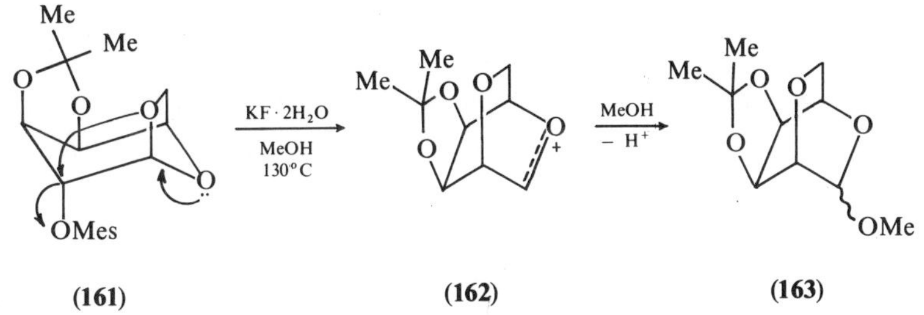

(39)

that (**162**) is formed by concerted bond migration with displacement of the methanesulfonate ion rather than by ring opening after initial formation of the tetracyclic oxonium expected from R_2O-3 participation.

The solvolysis of the glucosides (**164**) leads to products that apparently result from both R_2O-3 participation by ring oxygen and O-4 participation by the acetoxy alkyl oxygen.[64] The concerted nature of the proposed second step [Eq. (40)] is doubtful owing to the strain involved. A prior rearrangement, as previously suggested, may occur [e.g., see (**97**) and (**155**)].

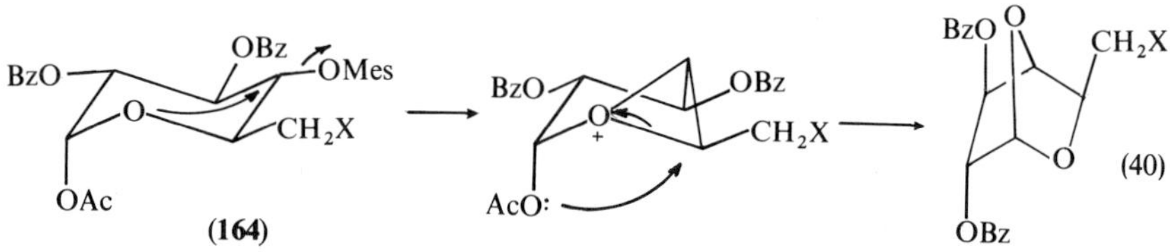

(40)

X = N_3 or OAc

The brominolysis of the mannoside (**165**) in the presence of silver acetate gives about 20% of the glucose derivative (**167**) through the postulated oxonium ion (**166**) by methoxy participation.[65]

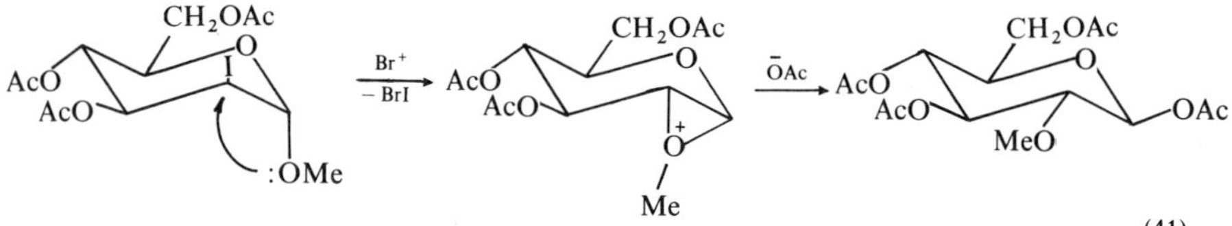

(41)

The reported studies of ether oxygen participation in carbohydrate derivatives demonstrate that O-3, O-4, O-5, and O-6 participation can occur. In all the cases illustrated, the participation occurs with the oxygen interacting from an antiperiplanar stereochemistry. Since this is the normal stereochemistry required for neighboring group participation, ether oxygen anchimeric assistance would not be expected unless such a configuration is attainable. Participation, of course, may result after ionization has produced a carbocation intermediate. The latter pathway, however, seems to be unimportant for product formation in most cases discussed here.

4.1.2. In Reactions at Carbonyl Carbon

When heated to 100–150°C, γ- and δ- (but not ε-) alkoxyacyl chlorides (**168**) rearrange to give the alkyl γ- or δ-chloro esters (**170**), respectively.[66,67] Support for the proposed intermediate oxonium ion (**169**) is derived from

Participation by Oxygen Groups

the observations that *cis*-3- and 4-methoxycyclohexane carbonyl chloride undergo this rearrangement, while the respective *trans* isomers do not,[68] a result that excludes an intramolecular reaction between the acyl group of one molecule and the ether group of another. Inversion of configuration also results. Thus the *cis*-3- and 4-cyclohexanecarbonyl chlorides[68] and (+)-4-methoxyvaleryl chloride[69] yield chloro esters with inverted configuration. If it is assumed that internal return to acyl chloride occurs to the same extent in EtO-5 and EtO-6 participation in these systems, EtO-5 is faster than EtO-6 participation, since 4-ethoxybutyryl chloride rearranges about 40 times faster than 5-ethoxyvaleryl chloride.[67]

$$(168) \longrightarrow (169) \longrightarrow (170) \quad (42)$$

(168)   (169)   (170)

It is interesting that no products resulting from alkoxyl cleavage of the oxonium ion (169) have been reported. This probably results from the conditions of the reaction, which raises an interesting point. Meerwein et al.[70] found that the stable dialkoxycarbocation (173), and not the oxonium ion (172), formed when (171) was treated with antimony pentachloride. While both cations (172) and (173) can produce ortho ester, lactone, or the ring-opened chloro ester,[71] cations such as (173) prefer to form lactones when formed by rearrangement directly from the chloro ester by carbonyl oxygen participation.[72] Thus, it appears to us that cation (172) and not (173) is involved in the rearrangements of γ- and δ- alkoxy alkyl carbonyl chlorides and that the rearrangements of (172) to the more stable (173)

$$(171) \xrightarrow{SbCl_5} (172) \longrightarrow (173) \quad (43)$$

(171)   (172)   (173)

may occur if a Lewis acid (or possibly a poor nucleophile) is present. Obviously more research is needed before a detailed understanding of the rearrangement of (172) to (173) is known.

Finally, the presence of a β-ether or amine group in certain bicyclic β, γ-unsaturated ketones alters the course of rearrangement of the latter group. This has been discussed by Cargill et al.[73]

4.1.3. In Electrophilic Addition Reactions

Peterson and his co-workers[74] have included the methoxy group in a study of substituent effects on the rate of addition of trifluoroacetic acid to a homologous series of ω-substituted 1-alkenes (**174**). They found that certain substituents [e.g., X = cyano, acetoxy, or trifluoroacetoxy in (**174**)] give a completely linear plot of log Δ log k versus n [the number of carbon atoms in (**174**)]. The finding of a constant slope for these substituents was

$$CH_2\!\!=\!\!CH(CH_2)_{n-2}X + CF_3CO_2H \longrightarrow CH_3CH(CH_2)_{n-2}X \quad (44)$$
$$\underset{\text{(174)}}{} \qquad\qquad\qquad\qquad\qquad \underset{O\!\!-\!\!CCF_3}{\overset{|}{}}$$
$$\qquad\qquad\qquad\qquad\qquad\qquad\qquad \overset{\|}{O}$$

$n = 4\text{-}11$
X = H, O_2CCF_3, CN, O_2CMe, OPNB, Br, Cl, or OMe

interpreted as evidence for a constant attenuation (ε) of the inductive effect per methylene group. The attenuation factor ε was found to be 0.65 for X = CN, O_2CCF_3, and O_2CMe (p. 85). Based on this same attenuation factor, some of the other groups studied showed deviant behavior. Figure 1 illustrates the behavior of the chloro and methoxy groups. When $n = 5$ the rates were significantly faster than predicted (by the solid line); thus anchimeric assistance was suggested (see Volume 2 for a more complete discussion of the other substituents). The rate data were corroborated by the observation of rearrangement products [e.g., Eq. (45)]. Since the amount of

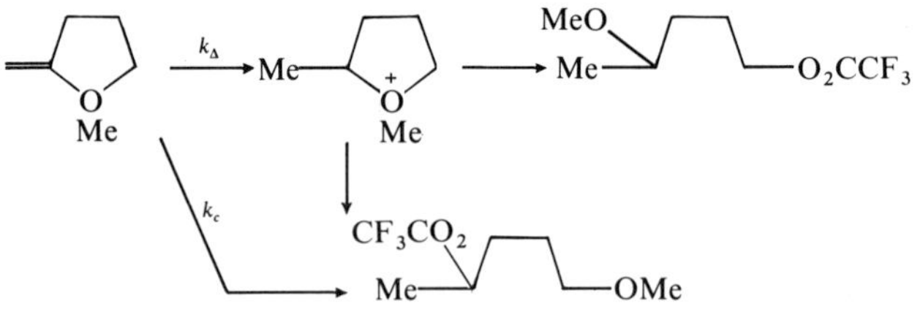

$$\text{(45)}$$

methoxy group migration (30%) is less than that formed in acetolysis of 5-methoxy-2-pentyl p-bromobenzenesulfonate (40%), it is reasonable to suggest dual pathways for the formation of the secondary trifluoroacetate [Eq. (45)].

Jernow et al.[75] have obtained evidence for the participation of ether groups in oxymercuration reactions, although rate studies (Table 10) show that ether groups provide less anchimeric assistance than hydroxyl groups

Table 10. Relative Rates of
Oxymercuration in Tetrahydrofuran at $25°C$[75]

Compound		$k_{rel}{}^a$
 (cyclooctene)		0.005
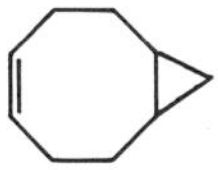 (bicyclic)		0.02
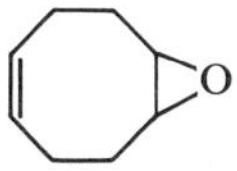 (epoxide)	(175)	1
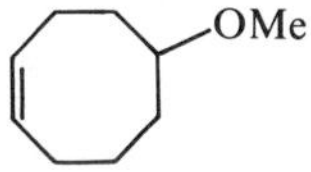 (OMe)	(176)	0.003
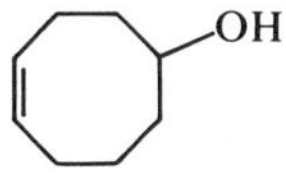 (OH)	(177)	2.5^b
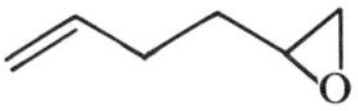 (alkenyl epoxide)	(178)	1.1^b

[a]Determined from competition reactions unless otherwise noted.
[b]Estimated from the time required for decolorization of mercuric complex in THF-water.

(also see Section 4.2.2). The monoepoxide (175) of cis,cis-1,5-cyclooctene presumably reacts via the tricyclic oxonium ion (179) to yield, after demercuration, a 1:3 ratio of the O-5,6 products, (180) and (181), respectively. The methoxy derivative is considerably less accelerated, perhaps owing to conformational differences, yet it too leads to a mixture (30% of total products) of the bicyclic ethers (180) and (181). By the same procedure, (178) gave, as the only isolable products, a 3:2 mixture of 2,5 dimethyltetrahydrofuran (cis and trans) and 2-methyltetrahydropyran.[75] Transannular participation by the ether oxygen in (182) has also been described.[76]

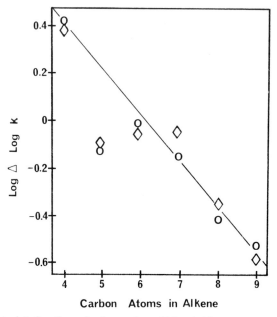

Fig. 1. Plot of rate data for alkenes having methoxy (◊) and chloro (○) substituents. The solid line has a slope corresponding to an attenuation factor ε of 0.65.

4.2. Hydroxyl Groups

Participation by hydroxyl groups occurs in many reactions, often yielding cyclic ether or lactone products. Owing to the acidic hydrogen, the hydroxyl group is more complex in its behavior than ether groups, and three distinct types of effects are possible: (1) direct oxygen ($-$**OH**) participation, (2) alkoxide ($-O^-$) participation, and (3) acidic hydrogen ($-OH$) participation. These effects are shown by appropriately substituted alcohols in reactions involving solvolytic displacement, elimination, addition, deamination, reduction, and hydrolysis. Two extensive reviews[37] reflect the importance of hydroxyl group participation in carbohydrate reactions.

4.2.1. In Solvolytic Displacement Reactions

The most thorough mechanistic studies of hydroxyl group participation have been in solvolytic studies of ω-hydroxy alkyl halides. The rates of hydrolysis in water of a series of alkyl halides (Tables 11 and 12) indicate that anchimeric assistance is occurring where HO-5 and HO-6 assistance is possible[17,77-79]; neighboring group participation in the Gabriel synthesis has been reviewed.[81] Anchimeric assistance is also indicated in some cases where three-membered rings are formed.[17] The increase in rate associated with participation is not very great in these reactions because the solvent (water) is itself relatively basic and nucleophilic, and the rate in the absence of participation by the neighboring group is already quite high.

Since the loss of a proton in aqueous media is fast under these conditions, one expects to find ethers as products when hydroxyl group partici-

$$\text{(OH)} \xrightarrow[-X^-]{\text{slow}} \text{(}^+\text{OH)} \xrightarrow[:B:]{\text{fast}} \text{(O)} + BH \qquad (47)$$

pation produces a cyclic oxonium ion as an intermediate [e.g., Eq. (47)]. Tetrahydrofuran is the sole product from the hydrolysis of 4-chloro-1-butanol.[17] Quite surprisingly, this result is not explicable as simply as depicted in Eq. (47). Instead, this cyclization is claimed to be the sole example of general base catalysis during displacement at saturated carbon.[82]*

General base catalysis[83] was first proposed for the cyclization of 4-chloro-1-butanol on the basis of a Brønsted plot involving the reactions catalyzed by water, hydroxide ion, borate ion, and water.[84] The three-point Brønsted plot of the 4-chloro-1-butanol studies yielded a slope of 0.25, which was interpreted as implying that the transfer of the alcoholic proton is about 25% complete at the transition state.[84] Solvent isotope effects of $k_{H_2O}/k_{D_2O} = 1.28$ and $k_{OD^-}/k_{OH^-} = 1.07$ were measured and interpreted as

---

*A second example, again involving the HO-5 model, has been reported; cf. ref. 156.

Table 11. Relative Rates of Reaction of Some ω-Halohydrins in Water[77-79] at 70.5°C and in Aqueous Sodium Hydroxide[77,80]

Halohydrin	G-n	k_{rel}^a (H_2O)	k_{rel}^b (OH^-)
$Cl(CH_2)_2OH$	O-3	0.026	185
$Cl(CH_2)_3OH$	O-4	0.111	0.05
$Cl(CH_2)_4OH$	O-5	24.4	245
$Cl(CH_2)_5OH$	O-6	(1.00)	(1.00)

[a] n-Propyl chloride = 0.21 relative to $Cl(CH_2)_5OH$; cf. R. E. Robertson, Prog. Phys. Org. Chem., **4**, 213 (1967).
[b] Uncorrected for polar effects; cf. Ref. 80.

demonstrating that alcoholic hydrogen is excluded from the reaction coordinate so that the asymmetric stretch of the OHO is a real vibration with zero-point energy and not a translation. Thus the asymmetric stretch of the OCCl system alone, as in (**189**), leads to cyclization. The transition state was visualized as (**190**) with a long, loose bond between carbon and oxygen.

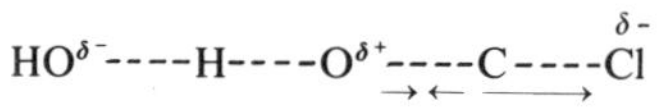

$$HO^{\delta-}\text{----}H\text{----}O^{\delta+}\text{----}C\text{----}Cl \qquad\qquad HO^-\text{----}H\text{----}O\text{----}CH_2^+\text{----}Cl^+$$

(values: 0.75, 0.25, 0.25, 0.25 above second structure, with CH_2–CH_2–CH_2 ring)

(**189**) 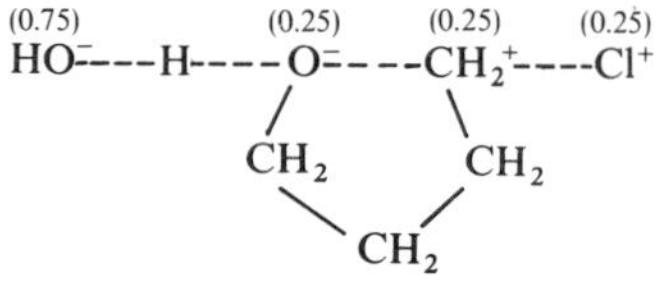   (**190**)

Jencks et al.[85] verified the general base catalysis mechanism for 4-chloro-1-butanol and showed that the carbonate and phenolate ions were

Table 12. Relative Rates of Hydrolysis of Some Bromides in Water at 50°C[17]

Halide	Compound	Rel. rate
Ethyl bromide	(**183**)	(1.0)
2-Bromoethanol	(**184**)	0.0289
n-Propyl bromide		(1.00)
3-Bromo-1-propanol		0.276
Cyclohexyl bromide	(**185**)	(1.00)
Isopropyl bromide	(**186**)	1.30
trans-2-Bromocyclohexanol	(**187**)	0.0347
3-Bromo-2-methyl-2-butanol	(**188**)	2.61

Table 13. Rates of Hydrolysis of
4-Chloro-1-butanol at 50°C in
Aqueous Media[85]

Catalysts[a]	k (mole sec^{-1})
Water	0.067 × 10^b
Phenolate	8.6 × 10^{-5}
Carbonate	11 × 10^{-5}
Borate	46 × 10^{-5}
Hydroxide	1,300 × 10^{-5}

[a]Except in the phenol experiment, ionic strengths were brought to $\mu = 0.6$ with NaClO$_4$.
[b]Corrected for 55.5 M water.

catalysts for the cyclization, yet the statistically uncorrected constants for the latter two catalysts were said to be well below the Brønsted plot for water, hydroxide, and borate (Jencks[85]). The measured rate constants are given in Table 13.

To ascertain whether or not nucleophilic displacement reactions were occurring in some of the catalyzed reactions, Jencks et al.[85] analyzed the reaction products for the presence of alcohols. With water, hydroxide, and borate, the amounts of alcohols found were less than 3%, and with carbonate less than 8%; the reaction products of the phenolate experiment were not analyzed. The results show, in agreement with the earlier work,[84] that at least a large part of the catalysis represents catalysis of cyclization.

Cromartie and Swain[82] have now redetermined the Brønsted slope with a series of oxygen bases (Fig. 2). A slope of 0.36 was determined, excluding the datum for water.[82] Notably, carbonate and borate, known to occasionally show "abnormal" behavior (Jencks[83]), were excluded from the latest treatment.

Determination[82] of the chlorine isotope effects for the base-catalyzed cyclization of 2-chloroethanol and 4-chloro-1-butanol (see Table 13) have provided support for the conclusions drawn from solvent isotope effects; that is, carbon–chlorine bond rupture in 2-chloroethanol is greater than that in 4-chloro-1-butanol at the transition state (i.e., compare[86] entries in the first and second columns of Table 14). Following a similar line of reasoning in the analysis of the chlorine isotope data of the hydroxide- and water-catalyzed cyclizations (i.e., compare entries in the second and third columns of Table 14) requires the conclusion that there is more carbon–chlorine

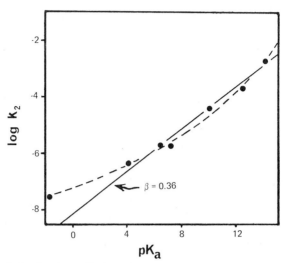

Fig. 2. Brønsted plot for the cyclization of 4-chlorobutanol in water at 25°C. Data points (left to right) correspond to water, 2,4-dinitrophenoxide, 2,4,6-trichlorophenoxide, 4-nitrophenoxide, phenoxide, 2,2,2-trifluoroethoxide, and hydroxide.

bond breaking in the water-catalyzed reaction than in the hydroxide-ion-catalyzed reaction.

Two possible explanations were offered[82] for the apparent curvature of the Brønsted plot (Fig. 2). It was suggested that the cyclization in water may represent the rather abrupt onset of an uncatalyzed reaction [i.e., similar to Eq. (47)]. We believe that the curvature is too gradual to be called abrupt. Thus, the alternative explanation[82] of a gradually changing response in rate to the difference in pK between the base and alcohol appears to be more reasonable, yet it awaits verification.

Table 14. Chlorine Isotope Effects for the Cyclization of ω-Chloroalcohols in Water at $25°C$[82]

Compound	Base	Isotope effect[a] (k_{35}/k_{37})
$Cl(CH_2)_2OH$	$Me_4N^+OH^-$	1.00815 ± 0.00011
$Cl(CH_2)_4OH$	$Me_4N^+OH^-$ [b]	1.00757 ± 0.00015
$Cl(CH_2)_4OH$	H_2O	1.00796 ± 0.00013

[a] 4-Chloro-1-butanol and hydroxide each initially 0.03–0.08M.
[b] Average of five to seven runs; R > 99%.

Measurements of the temperature coefficient of activation,[17] the volume of activation,[87] and the entropy of activation[78,88] have also been reported for the cyclization of 4-chloro-1-butanol. While there are some differences of opinion as to the interpretation of these values,[17] support of the kinetic evidence of a weakly anchimerically assisted process is indicated.

Like the methoxy group, the hydroxy group shows weak HO-3 anchimeric assistance when the neighboring group is on a primary carbon [cf. Table 12; compare (183) to (184) and (186) to (188)]. The enhanced assistance by the hydroxyl group in (188) relative to (184), $k_{(188)}/k_{(184)} = 698$ is said to be a result of steric interactions of the adjacent methyl groups.[17,89]

There are conflicting conclusions drawn from the reports on the hydrolysis of *trans*-2-bromocyclohexanol (187). In one report [17] (see Table 12), (187) was shown to hydrolyze considerably slower than cyclohexyl bromide (185) at 50°C. The observed result was termed "not surprising" on the basis that (187) was not in the necessary diaxial conformation. In dilute aqueous perchloric acid the rates of solvolysis of (185), *cis*-2-bromocyclohexanol (191), and (187) were found to be 1:2.12:0.64, suggesting that the *trans*-2-hydroxyl group does provide anchimeric assistance.[90] The *cis* isomer (191) is surprisingly reactive relative to (185). Anchimeric assistance by the adjacent axial hydrogen was suggested to account for the observed effects since cyclohexanone was isolated as the product. It is also possible to account for the high reaction rate of (191) by invoking an elimination mechanism [Eq. (48)]. Certainly the stereochemistry of (191) allows a facile elimination

pathway. Hence, a final conclusion on the mechanism of formation of cyclohexanone awaits evidence that will decide between the elimination or participation options.

Weak anchimeric assistance by the hydroxyl group has been suggested on the basis of kinetic evidence in the solvolysis of five-,[91,92] six-,[91] seven-,[91] and eight-membered[93] *trans*-2-hydroxyl cycloalkyl toluene-*p*-sulfonates. The data of Roberts[91] suggested an S_N1 mechanism and revealed that the influence of the hydroxyl group is sensitive to medium and *I*-strain effects but is insensitive to conformational effects. Owing to the sole formation of *trans* products[94] in *trans*-2-hydroxy cycloalkyl halide and sulfonate solvolyses, participation is indicated. The conflicting results noted in the kinetic studies, however, suggest the need for additional scrutiny of the cyclopentyl and cyclohexyl systems in view of the importance of such reactions in carbohydrate chemistry.

A majority of the remaining solvolytic studies involve base catalysis, and O^- is probably the reactive oxygen species present. O^--5 participation was suggested to account for the formation of the bicyclic ether **(192)** from the alkaline hydrolysis of *trans*-4-chlorocyclohexanol.[95] The rate of reaction is 1100 times less than that of 4-chloro-1-butanol under the same conditions, but, as discussed above, O^- attack was not suggested in the hydrolysis of the latter.

(49)

(192)

The most studied example of O^--3 participation is probably the base-catalyzed hydrolysis of 2-chloroethanol to produce ethylene oxide. As mentioned above, the reaction is catalyzed by hydroxide but not by water.[84] Studies of the solvent isotope effect[96,97] along with spectroscopic[97] and conductivity measurements[96] have confirmed the postulated two-step process.[98] The reaction is often used as a stereospecific route to the more hindered epoxide derived from an olefin (via initial halohydrin addition). The carbohydrate field serves as a rich field for the application of epoxide opening and closing reactions.[37,99] In this respect, Černý and co-workers[100] have observed that **(193)**, **(194)**, and **(195)** undergo hydroxide-catalyzed epoxide formation with relative rates of 1:23.3:180, respectively.

An interesting example where the alkoxy anion apparently does not displace the leaving group directly is the reaction of **(196)** in the presence of

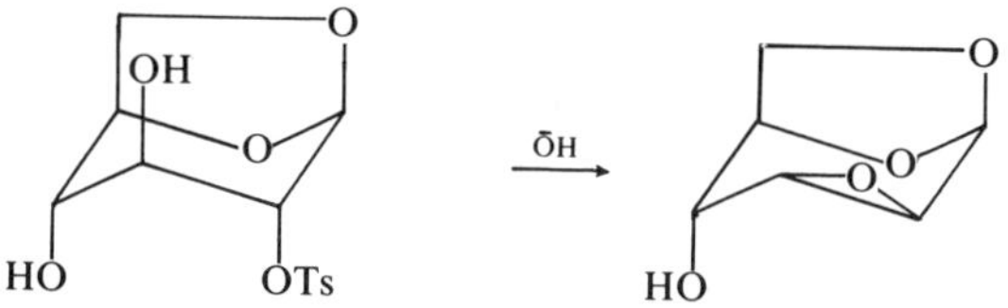

(50)

(193)

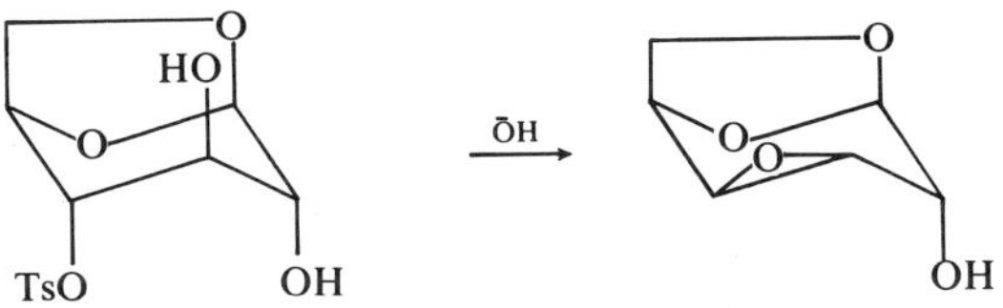

Participation by Oxygen Groups

(51)

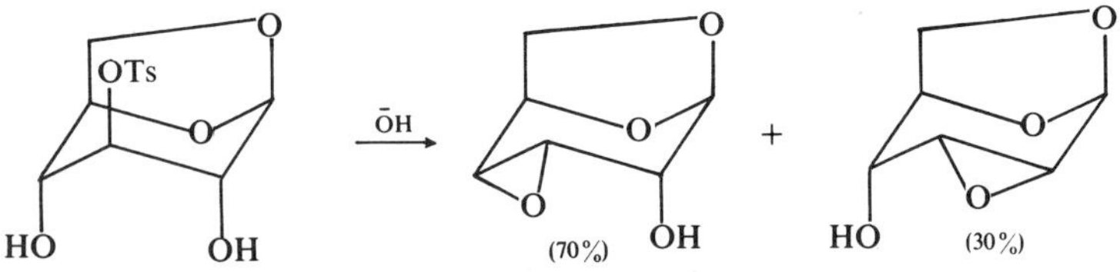

(194)

(52)

(195)

t-butoxide/*t*-butanol.[101] The isomeric *exo*-hydroxy-*exo*-*p*-toluenesulfoxy compound gives the *exo*-epoxide. The epoxides in these cases are said to be formed by capture of the intermediate ion pair after solvolysis.

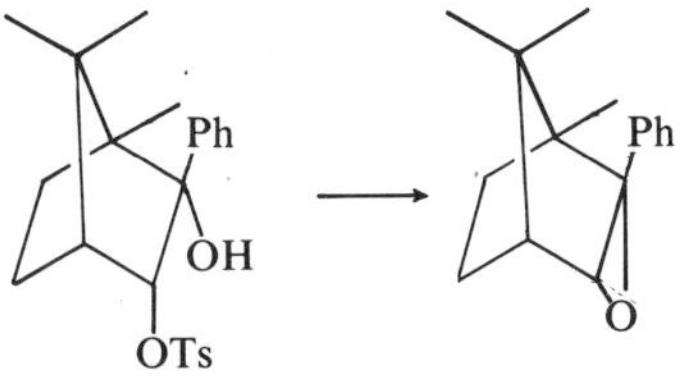

(53)

(196)

While participation resulting in a four-membered ring is generally rare, O-4 participation is common in base-catalyzed reactions of appropriately substituted alcohols. For example, the epoxide (**197**) undergoes oxide migration to form the oxetane (**198**).[102]

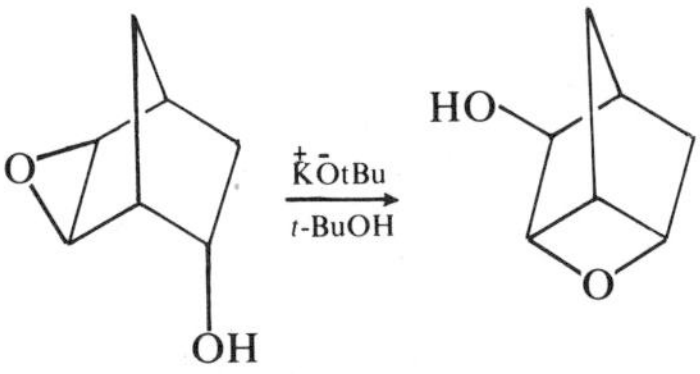

(54)

(197)   (198)

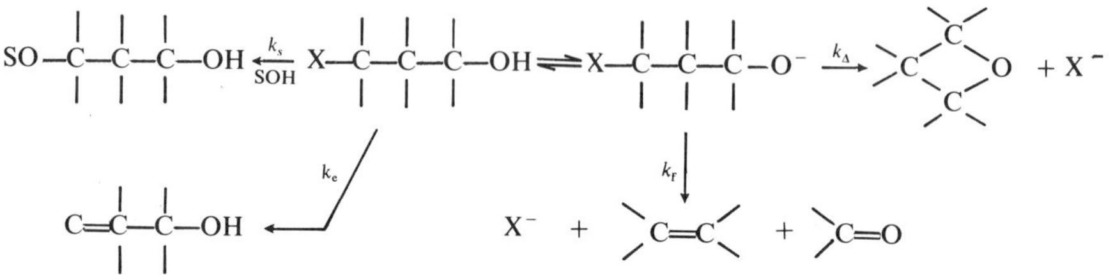

$$\text{SO}-\overset{|}{\underset{|}{C}}-\overset{|}{\underset{|}{C}}-\overset{|}{\underset{|}{C}}-\text{OH} \underset{\text{SOH}}{\overset{k_s}{\rightleftharpoons}} X-\overset{|}{\underset{|}{C}}-\overset{|}{\underset{|}{C}}-\overset{|}{\underset{|}{C}}-\text{OH} \rightleftharpoons X-\overset{|}{\underset{|}{C}}-\overset{|}{\underset{|}{C}}-\overset{|}{\underset{|}{C}}-\text{O}^- \overset{k_\Delta}{\longrightarrow} \overset{\diagdown}{\diagup}C\overset{C\diagdown}{\underset{C\diagup}{}}O + X^-$$

with k_e branch leading to $C=\overset{|}{\underset{|}{C}}-\overset{|}{\underset{|}{C}}-\text{OH}$ and k_f branch to $X^- + \diagup C=C \diagdown + \diagup C=O$

Scheme 10

This type of oxide migration also occurs when O-3, O-5, and O-6 participation is possible. These processes have been treated in an excellent review by Buchanan and Sable.[99] Simple displacement reactions involving O^--4 participation are also common with compounds of suitable configuration. Fragmentation, elimination, and solvent-assisted displacement (Scheme 10) are competing processes. 3-Chloro-1-propanol, for example, undergoes reaction (see Table 11 for relative rates) in 40% aqueous methanolic sodium hydroxide solution to yield a mixture of products in the ratios shown in Eq. (55).[80] The addition of alkyl or aryl groups at C-2 favor the fragmenta-

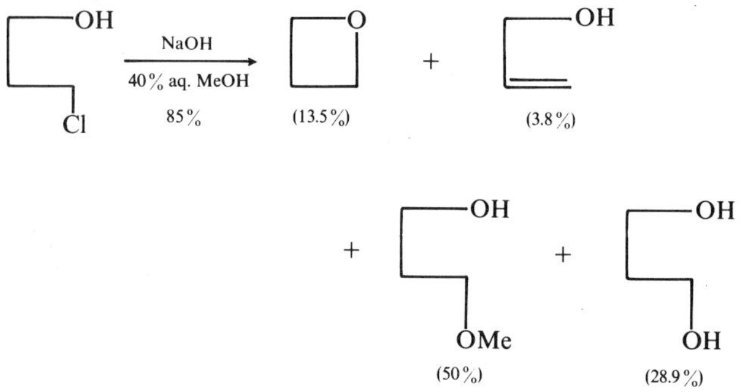

$$\text{HOCH}_2\text{CH}_2\text{CH}_2\text{Cl} \xrightarrow[40\% \text{ aq. MeOH}]{\text{NaOH}} \text{oxetane} (13.5\%) + \text{cyclopropanol-like} (3.8\%) \tag{55}$$
$$+ \text{HOCH}_2\text{CH}_2\text{CH}_2\text{OMe} (50\%) + \text{HOCH}_2\text{CH}_2\text{CH}_2\text{OH} (28.9\%)$$

(85%)

tion process, which is absent in Eq. (55). Searles et al.[103] have provided data on the relative driving forces for fragmentation versus cyclization in a series of 3-bromo-1-propanol derivatives (see Table 15). It is seen that substituents at C-2 favor fragmentation (1,4 elimination) over cyclization in the order

Ph ≫ alkyl > H

Gaylord et al.[104] found that both 2-chloro-4-hexanol and 1-phenyl-3-chloro-1-butanol, neither of which contain β-substituents, undergo fragmentation. Hence substitution in other than the β-position seems to make fragmentation more energetically favorable when compared to cyclization.

Table 15. Products from 3-Bromo-1-propanols and
15% Aqueous Potassium Hydroxide

	Yields (%)		Ratio oxetane/ olefin
3-Bromo-1-propanol	Oxetane	Olefin[a]	
2,2-Dimethyl	12	60	0.2
2,2-Diethyl	27	54	0.15
2-Ethyl-2-isopropyl	14	50	0.3
2-Ethyl	23	19[b]	1.2
2-Butyl	21	24	0.9
2-Phenyl	0	26[c]	0.1
2-Methyl-2-phenyl	0	15	0.0
2-Ethyl-2-phenyl	0	47	0.0
2,2-Tetramethylene	8	63	0.13
2,2-Pentamethylene	33	50	0.66

[a] Fragmentation product.
[b] 2-Ethylallyl alcohol, the product of ordinary 1,2 elimination, was isolated in 19% yield.
[c] Forty percent of 2-phenyl alcohol was also obtained.

5α-Hydroxycholestan-3-β-yl toluene-*p*-sulfonate (**199**) reacts rapidly at 50° with potassium *t*-butoxide in *t*-butanol to give a mixture of (**200**) and (**201**) as shown. The isomeric 5β,3α-ester (**202**) reacts at approximately the same rate as (**203**) but yields only the *seco*-ketone (**201**).[105] While many simple sulfonate esters [e.g., (**203**)–(**205**)] show only oxetane formation,[106] Corey et al.[107] showed that the fragmentation is predictable enough to incorporate into complex syntheses [e.g., Eq. (57)].

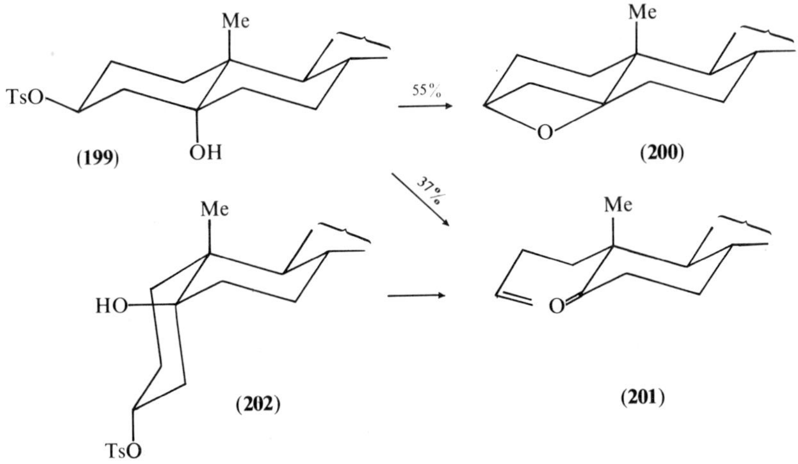

(57)

(203)

(204)

(205)

(57)

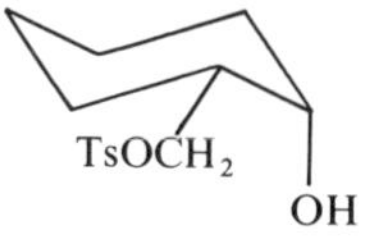

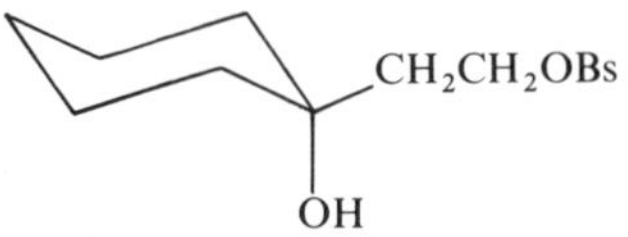

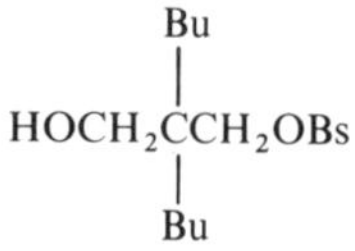

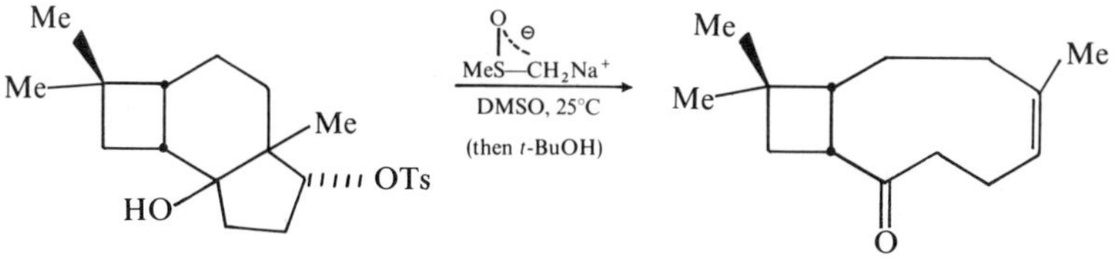

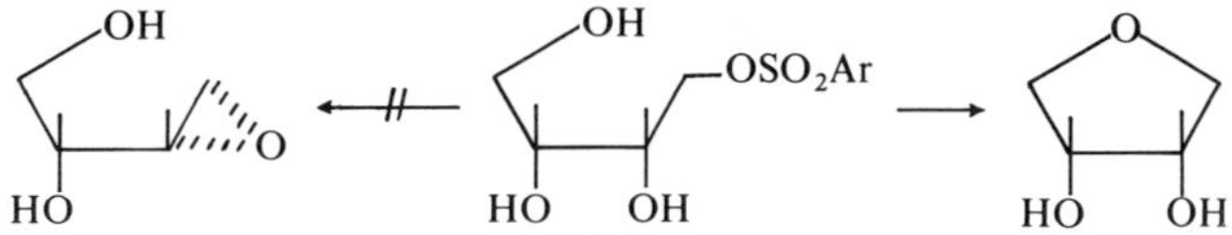

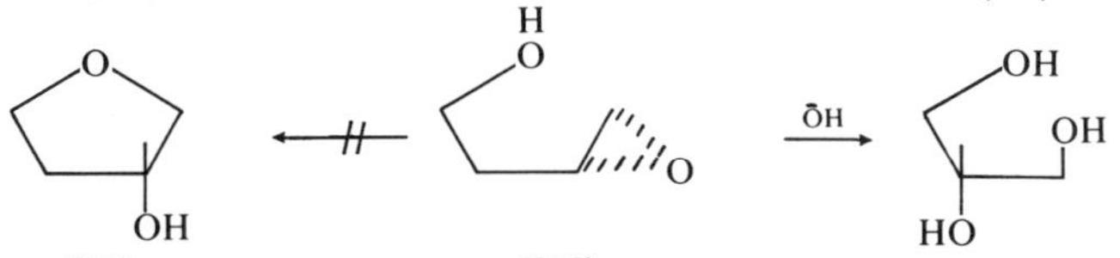

A rather careful study of the competition between O$^-$-5, O$^-$-6, and O$^-$-3 participation in the reactions of a number of O-toluene-*p*-sulfonyl-alditols and -deoxyalditols has appeared.[108] While the order of anchimeric assistance in the alkaline hydrolysis of ω-chloro alcohols is O$^-$-3 > O$^-$-5 > O$^-$-6, Hartman and Barker (second reference in Ref. 108) found the order primary O$^-$-5 > secondary O$^-$-5 = secondary O$^-$-3 > primary O$^-$-6 from their studies of primary O-toluene-*p*-sulfonyldeoxyalditol displacements. Thus, alkaline treatment of (**206**) yields (**207**) directly and not via the epoxide (**208**), since epoxide (**209**) yielded only glycol and none of the tetra-

(208) (206) (207) (58)

(210) (209) (59)

hydrofuran (**210**) on treatment with alkali. From the oxide migrations discussed earlier in this section, one might expect (**210**) to result from treatment of (**209**) with a stronger base. The monoester (**211**) shows a slight preference for O$^-$-5 over O$^-$-3 participation, while the monoester (**212**) shows a definite preference for O$^-$-3 over O$^-$-6.[108] In both cases control studies or configuration analyses ruled out optional pathways for product formation. Capon has reviewed these reactions elsewhere[37] and has discussed the relative order of these cyclizations versus those of ω-alcohols. The differences noted in the studies may be primarily a result of electronic effects operative in the systems. For example, in (**206**), O⁻-3, O⁻-4, and O⁻-5 participation are possible, but only O⁻-5 participation is observed. In the ω-chloro alcohols previously discussed, only 2-chloroethanol has an adjacent

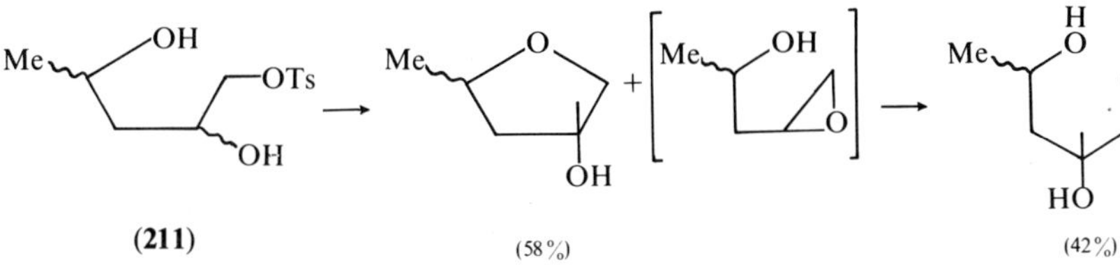

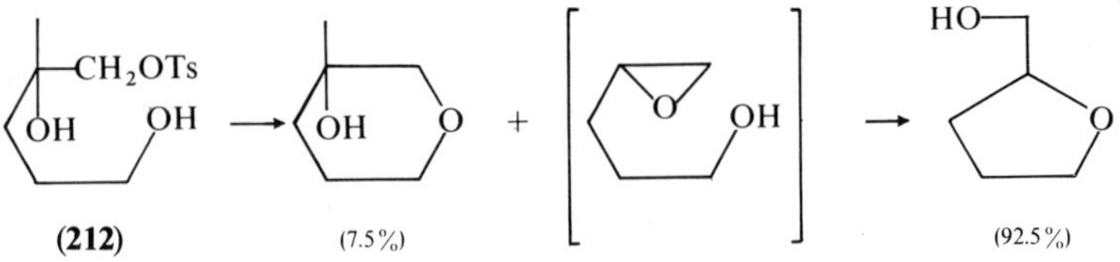

electron-withdrawing group. In (**206**), the most favored participation is probably observed since an adjacent oxygen is on every carbon, tending to increase the concentration of R—O⁻ ion at every position and thus balancing out the electronic effects. In (**211**) and (**212**) the O⁻-5 and O⁻-6 participant, respectively, has no adjacent electron-withdrawing group; therefore the results should be considered in light of these differences.

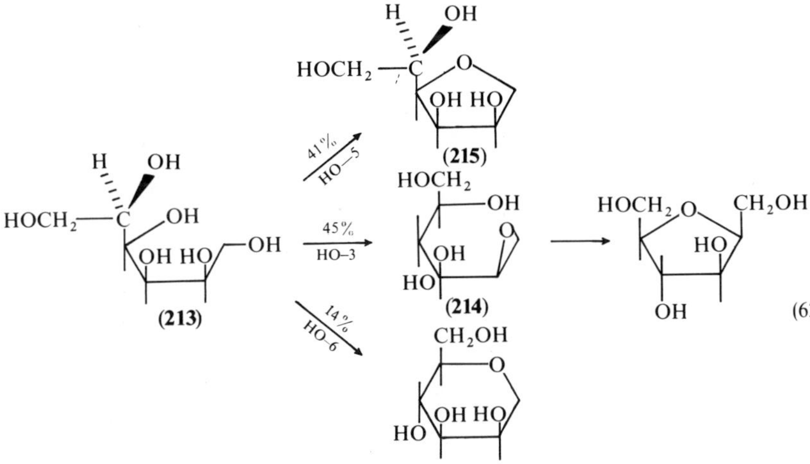

Another study pertinent to the present discussion is that of the acid-catalyzed dehydration of tetratols, pentitols, and hexitols, all of which yield tetrahydrofuran derivatives as the primary products.[109,110] These reactions are first order with respect to the alcohol and the acidity function H_0, and the direct relationship shown from plots of k_ψ versus H_0 implies that the reaction proceeds via a protonated intermediate.[110] In all cases except one, the major cyclic product is derived from HO-5 displacement of water from C-1. The exception is D-mannitol (213), where HO-3 participation leads to a slight preference for formation of 2,5-anhydro-D-glucitol (214) over 1,4-anhydro-D-mannitol (215). Inversion at C-2 occurs with HO-5 participation during ring opening of the epoxide (214).

All the hexitols and some of the pentitols gave more than one product, and the rates varied, indicating the importance of stereochemical effects. The relative rates and products from these studies are listed in Table 16. In their fairly thorough discussion of the rate and product data, Barker et al.[109,110] concluded that nonbonded interactions in the transition state are apparently very important and that a quasi-equitorial hydroxyl group at C-2 may slow the rate by interacting with the leaving water molecule. Thus the rates are rationalized on the basis that only the transition states where the hydroxyl group at C-2 is quasi-axial [i.e., (216)] are allowed.

Barker et al.[111] have also studied the deamination of 1-amino-1-deoxy-pentitols and -hexitols. Tetrahydrofuran derivatives as well as pentitols and hexitols were the products. In deaminations, the ground-state conformation was thought to be a more important factor in product formation.

The well-studied pinacol rearrangement (for a review, see Ref. 112) bears some similarities to the acid-catalyzed reactions discussed above. In this reaction, product formation generally results from migration of a hydrogen, an alkyl group, or an aryl group. Meaningful migratory aptitudes, however, are difficult to determine owing to the many factors involved. Hydroxyl group participation (migration) may occur, but it is often not observed. In part this appears to be due to two factors: (1) Migration of the hydroxyl group leads to a protonated epoxide that may reopen to the

Table 16. Anhydrization of Tetritols, Pentitols, and Hexitols at 100°C in 2 N HCl[109-110]

Polyol	Products	Relative rate
1,4-Butanediol	Tetrahydrofuran	256
1,2,4-Butanetriol	3-Hydroxytetrahydrofuran	68
Erythritol	1,4-Anydroerythritol	14
D-Threitol	1,4-Anhydro-D-threitol	33
2-O-Methyl-1,2,4-butanetriol	2-O-Methyl-3-hydroxytetrahydrofuran	68
2-Methyl-1,4-butanediol	3-Methyltetrahydrofuran	800
1,4-Pentanediol	2-Methyltetrahydrofuran	830
D,L-*erythro*-1,2,4-Pentanetriol	D,L-*erythro*-2-Methyl-4-hydroxytetrahydrofuran	250
D,L-*threo*-1,2,4-Pentanetriol	D,L-*threo*-2-Methyl-4-hydroxytetrahydrofuran	
L-1,2,5-Pentanetriol	L-2-Hydroxymethyltetrahydrofuran (9 %D)	172
Ribitol	1,4-Anhydro-D,L-ribitol	50
Xylitol	1,4-Anhydro-D,L-xylitol	29
D-Arabinitol	1,4-Anhydro-D-arabinitol (62.5%)	10
	1,4-Anhydro-L-ribitol (7.6%)	
	1,4-Anhydro-D-xylitol (11.7%)	
	1,4-Anhydro-D-Lyxitol (6.2–6.9%)	
	1,5-Anhydro-D-arabinitol (4.6%)	
	1,4-Anhydro-D-ribitol (3.3%)	
	1,4-Anhydro-L-xylitol (4.3%)	
2-Deoxy-D-ribitol	1,4-Anhydro-2-deoxy-D-ribitol	207
1-Deoxy-D-arabinitol	1,4-Anhydro-5-deoxy-D-lyxitol	58
D-Lyxitol	1,4-Anhydro-D-Lyxitol	1
Allitol	1,4-Anhydro-D,L-allitol (92%)	1700
	One unknown (8%)	
D-Talitol	1,4-Anhydro-D-talitol (68%)	930
	1,4-D-altritol (14%)	
	Two unknowns (18%)	
L-Iditol	1,4-Anhydro-L-iditol (85%)	670
	2,5-Anhydro-D-glucitol (15%)	
D-Glucitol	1,4-Anhydro-D-glucitol (85%)	510
	1,4-Anhydro-L-glucitol (13%)	
	2,5-Anhydro-L-iditol (2%)	
Galacitol	1,4-Anhydro-D,L-galacitol (97%)	173
	One unknown (3%)	
D-Mannitol	1,4-Anhydro-D-mannitol (41%)	13
	2,5-Anhydro-D-glucitol (45%)	
	1,5-Anhydro-D-mannitol (14%)	

original cation, and (2) migration of a group other than the hydroxyl group leads to a cation stabilized by the hydroxyl ion (an oxonium ion) that can lead directly to product by proton loss. The ramifications of this reaction cannot be discussed here for lack of space. Pocker and Ronald[113] have recently treated the subject of hydroxyl group participation in the pinacol reaction and have found it to be competitive as a neighboring group. In studies of the sulfuric-acid-catalyzed rearrangements of 1,2-di-*p*-anisyl-1,2-diphenylethylene glycol, tetra-*p*-anisylethylene glycol, and their respective epoxides, the order of neighboring group activity was found to be

$$\begin{array}{ccccccc} p\text{-anisyl} & > & -\mathrm{OH} & \geq & p\text{-tolyl} & > & \text{phenyl} \\ >1000 & & 37\text{–}10 & & 11 & & 1 \end{array}$$

Anchimeric assistance by the hydroxyl group has been reported in some more unusual substrates than those above, e.g., bisethylenediaminecobalt (III) complexes.[114] In aqueous hydroxide solution (**217**) reacts some 50 times faster than 4-chloro-1-butanol to form the cyclic chelate (**218**). The experimental observations support the mechanism shown. In this system HO-5 competes with N-3 participation. An analysis of equilibria and rate constants indicated that lysis of bromide by the coordinated nitrogen to produce the aziridine (**219**) is 10^2–10^3 times faster than by the coordinated oxygen.

(**217**) (**218**)

(**219**)

(**220**)

(**221**)

(**222**)

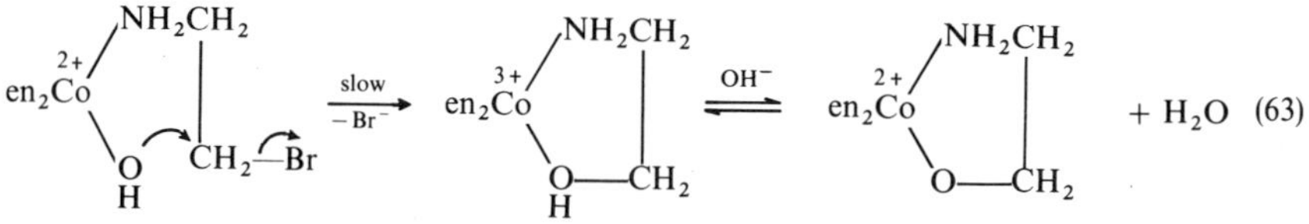

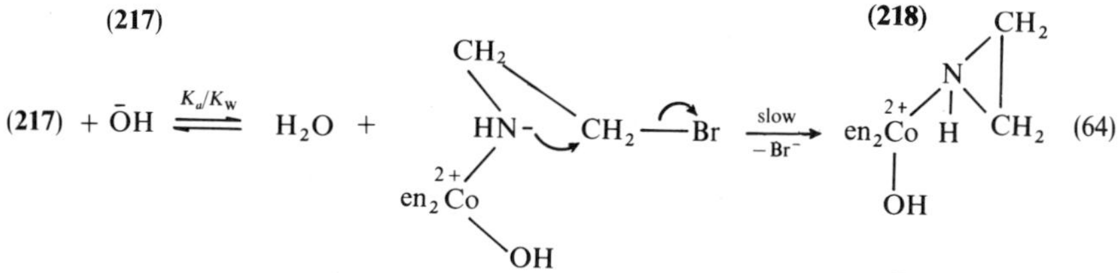

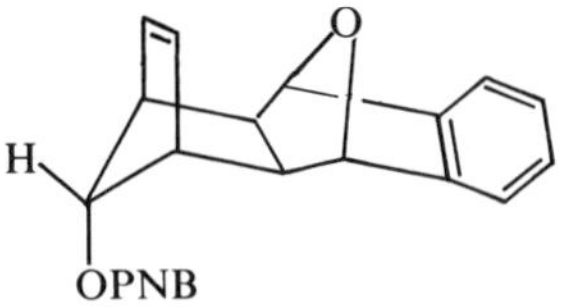

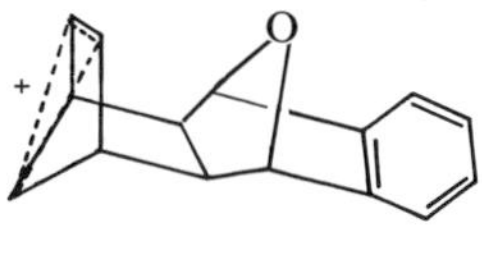

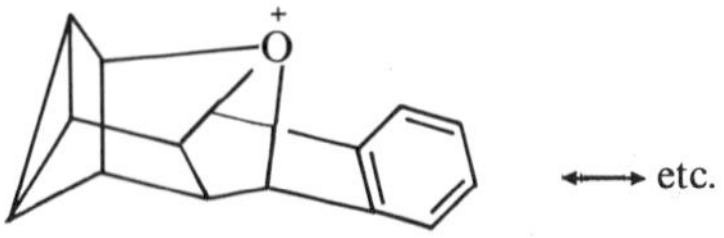

Participation by Oxygen Groups

Finally, Paquette and Dunkin[115] found no evidence indicating oxygen participation (laticyclic participation) in the solvolysis of (**220**). Thus anchimeric assistance by the π bond occurs, leading to the intermediate (**221**), and an intermediate such as (**222**), if it exists, is not indicated by kinetic or product studies.

4.2.2. In Addition Reactions*

A neighboring hydroxyl group may participate in electrophilic addition reactions. Although this type of effect is discussed in more detail in Volume 2 some pertinent examples here will serve to illustrate this effect with alcohols.

Winstein and Goodman[117] obtained the epoxide (**224**) upon treatment of the olefin (**223**) to conditions appropriate for the addition of BrOH. HO-3

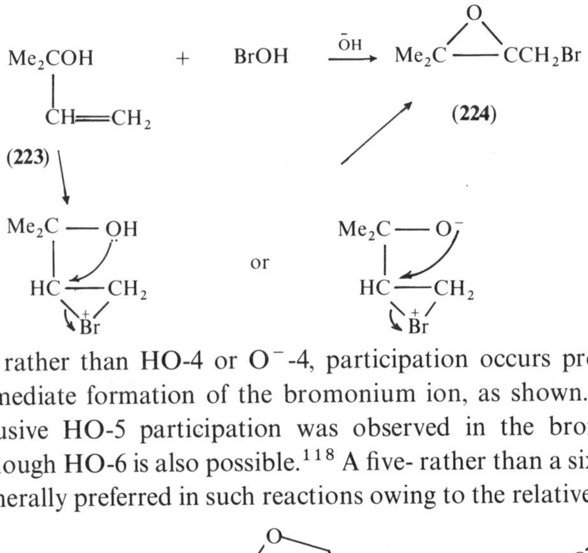

(65)

or O$^-$-3, rather than HO-4 or O$^-$-4, participation occurs probably after the intermediate formation of the bromonium ion, as shown.

Exclusive HO-5 participation was observed in the bromination of (**225**), although HO-6 is also possible.[118] A five- rather than a six-membered ring is generally preferred in such reactions owing to the relatively favorable

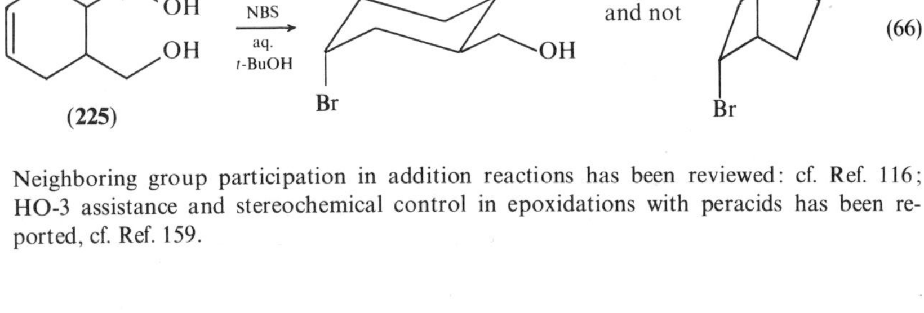

(66)

---

* Neighboring group participation in addition reactions has been reviewed: cf. Ref. 116; HO-3 assistance and stereochemical control in epoxidations with peracids has been reported, cf. Ref. 159.

entropy of activation in the formation of the five-membered ring (see Chapter 2). Nevertheless, the reaction probably proceeds by hydroxyl group attack on the reversibly formed bromonium ion.[119] Since bicyclic ring formation is involved in this example, the balance between $\Delta H^{\ddagger}$ and $\Delta S^{\ddagger}$ may be quite different from that for acyclic examples.

In studies of the bromination and iodination of $CH_2\!\!=\!\!CH(CH_2)_n OH$, an assisted process was suggested from the kinetic data (Table 17) when $n = 3$ or 4, although participation was less important in the case of bromination.[120] For $n = 3$, k_A/k_s was calculated to be 1.5 for bromination in methanol, 0.7 for bromination in water, and approximately 60 for iodination in water containing potassium iodide; for $n = 4$ the corresponding values are 0.2, 0.4, and approximately 8. Cyclic ethers were obtained (Table 17) and identified. The results of the kinetic and product studies were interpreted in terms of the formation of cyclic oxonium ions as shown in Eq. (67). The marked larger acceleration for iodination suggests that the response by the neighboring group is greatest for the more selective electrophilic reagent.

$$\underset{H-\overset{..}{O}}{\overset{E^+}{\underset{}{\searrow}}\!\!C\!\!=\!\!C\!\!\underset{}{\swarrow}(CH_2)_{3-4}} \longrightarrow \underset{H-\overset{+}{O}}{\overset{E}{\underset{}{|}}\!\!\underset{}{C-C}\!\!\underset{}{|}(CH_2)_{3-4}} \qquad (67)$$

When the series of alkenols was brominated in trifluoroacetic acid, the relative rates for $CH_2\!\!=\!\!CH(CH_2)_n OH$ were 1.00, 4.89, 11.2, and 14.8 when $n = 1-4$, respectively. The absence of anchimeric assistance was attributed to hydrogen bonding between the hydroxyl group and the solvent, which reduces the hydroxy group's nucleophilicity (Hooley and Williams[120]).

The bromocyclization of alkenes such as (226) may give a mixture of

Table 17. Reaction of $CH_2\!\!=\!\!CH(CH_2)_n OH$ with Bromine and Iodine[120]

	Bromination in MeOH		Iodination in H_2O	
n	Rel. rate	% Cyclic prod.	Rel. rate	% Cyclic prod.
1	1.0	None	1.0	None
2	1.86	None	2.0	None
3	93	50–60	200	100
4	61	5–10	35	—

isomers. In the case of (**226**) the ratio of (**227**) and (**228**) is almost 2:1.[121] Synthetic interest in this reaction has led Wong *et al.*[122] to study the factors affecting the stereochemical course of these addition–cyclization processes. An interesting consequence of their work is the finding that (**229**) gives a

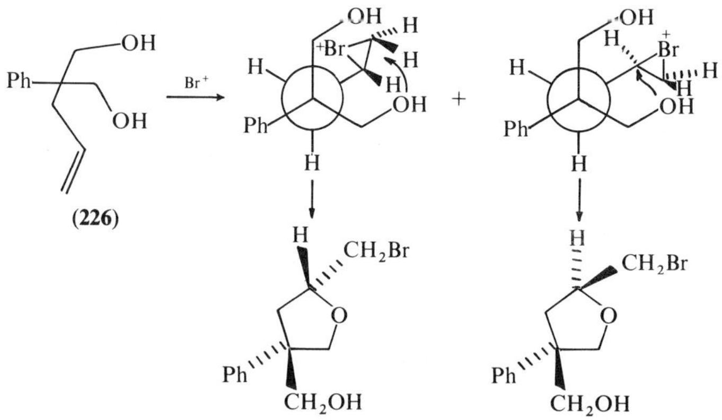

(**226**) (**227**) (**228**) (68)

1:1 mixture (71 % yield) of the isomeric HO-5 participation products. No R_3N-6 product was isolated. Additionally, bromination of (**230**) afforded an 87 % yield of a single product. Hence HO-5 participation in bromocyclization reactions appears to be superior to both R_3N-6 participation and O-5 amide participation.

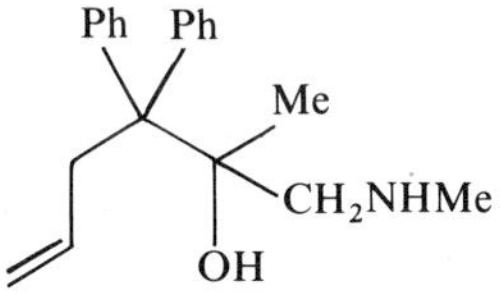

(**229**)

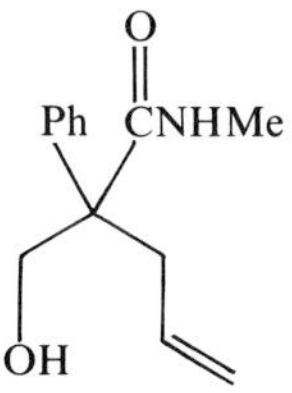

(**230**)

The hydroxyl group appears to effectively participate in the addition of metal salts to certain alkenes. Bly and Bly[123] showed that oxymercuration

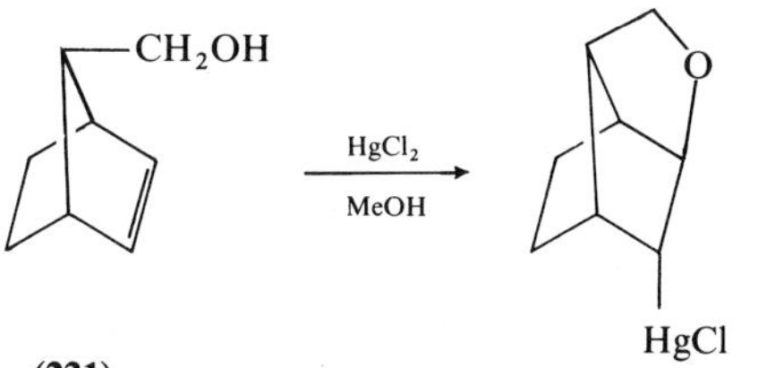

(**231**) (69)

Table 18. Rates of Hydroxymercuration[127] and Hydroxythallation[128] of Some Alkenols in 0.01 M $HClO_4$ Solution at 25°C

Alkenol	k (mole sec^{-1})	
	Hydroxymercuration	Hydroxythallation
1-Propen-3-ol	1.12×10^3	1.21
1-Buten-4-ol	8.4×10^3	15.0
1-Penten-5-ol	$>1000 \times 10^3$	317
1-Hexen-6-ol	$\sim 100 \times 10^3$	86

of syn-7-hydroxymethyl-2-norbornene (231) occurs with ring closure. A similar observation has been made in the hydroxymercuration of endo-5-hydroxymethyl-2-norbornene.[124] Since it is observed[125,126] that 100% exo-cis oxymercuration of even hindered derivatives [(232) or (233)] occurs, it seems reasonable to suggest that endo mercuric attack on (231) prevailed

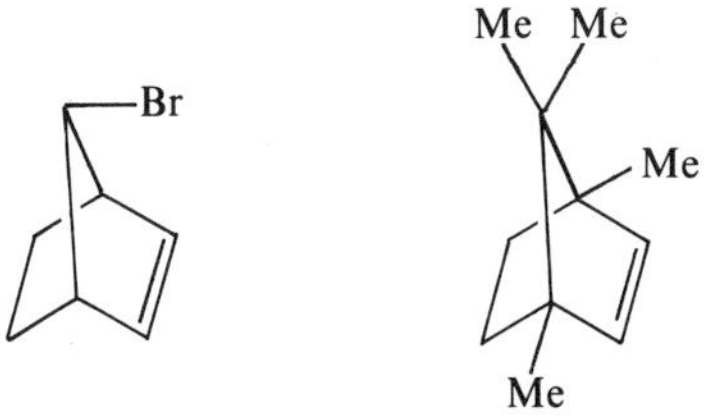

(232)   (233)

because of differences in electronic effects. Thus anchimeric assistance by the hydroxyl group is probably occurring.

Anchimeric assistance in oxymercuration[127] and oxythallation[128] reactions has been substantiated by the kinetic measurements of Halpern et al. (see Table 18). When log k values for the alkenols in each reaction were plotted versus Taft's σ^* constants (see Chapter 3) only 1-penten-5-ol and 1-hexen-6-ol fell off the lines (ρ in each case ~ 3.2) defined by $CH_2=CHR$, where R = alkyl or H. The positive deviation by 1-penten-5-ol and 1-hexen-6-ol is good evidence for anchimeric assistance. Isolation of the furan and pyran derivatives expected from HO-5 and HO-6 involvement substantiate the claim of anchimeric assistance in oxymercuration.

Interestingly, cyclic products are not found in the O-5 and O-6 assisted oxythallation reactions perhaps because of rearrangements during the dethallation step.[128] The finding of the tetrahydrofuran products (235) and (237) after dethallation of the oxidation products of 1-buten-4-ol (234)

Participation by Oxygen Groups

and 1-penten-4-ol (**235**) was attributed to hydroxy group participation, as shown in Eq. (70).[128,129]

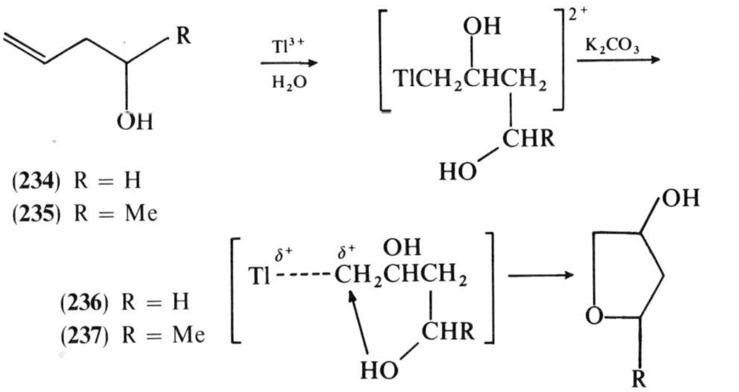

(**234**) R = H
(**235**) R = Me

(**236**) R = H
(**237**) R = Me

(70)

A novel observation of participation in a carbene reaction serves as our final example here. *endo*-5-Hydroxymethyl-2-norbornene (**238**), which Factor and Traylor[124] found to add mercuric ion with participation, affords oxohomobrendene derivatives (**239**) under dihalocarbene addition conditions.[130] The process can be envisioned to proceed as shown in Eq. (71).

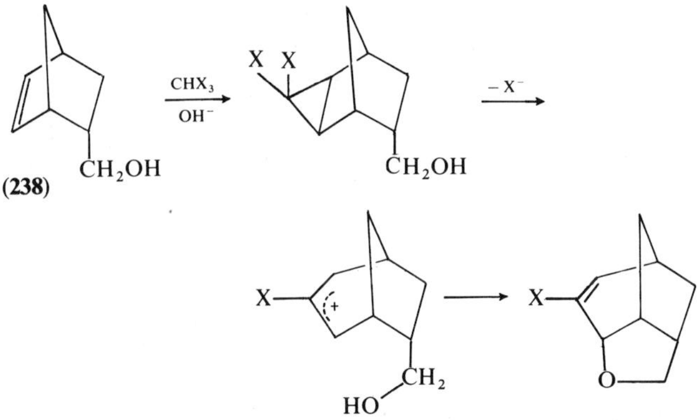

(**238**)

(71)

(**239**) X = Cl, Br

4.2.3. In Elimination Reactions

We have already discussed one way in which a hydroxyl group is involved in an elimination process (cf. Scheme 10 and related discussion). Hirsh[131] and Kovacic[132] and their co-workers have now described participation by the hydroxyl group under Hofmann elimination conditions. For example, the quaternary ammonium hydroxide (**240**) gives a mixture of products shown. Except for the normal Hofmann elimination product (**241**) nucleophilic involvement by the hydroxyl or oxide group is indicated.[131]

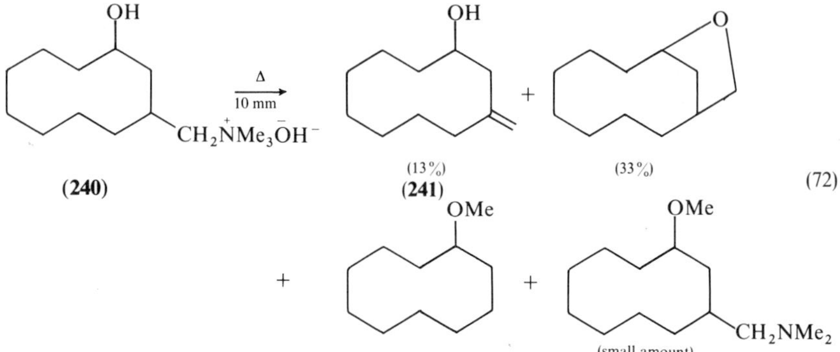

(**240**) → [Δ, 10 mm] → (**241**) (13%) + (33%) + (72)

+ (3%) + (small amount) + CH$_2$NMe$_2$

4.2.4. In Hydrolysis and Related Reactions

The literature is replete with examples of participation by ionized and un-ionized hydroxyl groups in a variety of hydrolysis reactions. Although a more complete discussion of intramolecular catalysis in hydrolytic reactions is presented in Volume 2, some examples are included here to demonstrate the extent to which this effect is observed.

Lactonization of hydroxy acids has been used as one means of gaining information on the propensity of the hydroxyl group in a certain configuration to interact with a carboxyl group. The relative rates for the acid-catalyzed lactonization of the hydroxy acids (**242**)–(**246**) are 1, 3.9, 4.2, 84, and 13,000, respectively (see the third reference in Ref. 133; some of the structures in this reference are incorrect; cf. Ref. 134). This variation, which obviously reflects differences in $\Delta S^{\ddagger}$ (see Chapter 2), may result in part

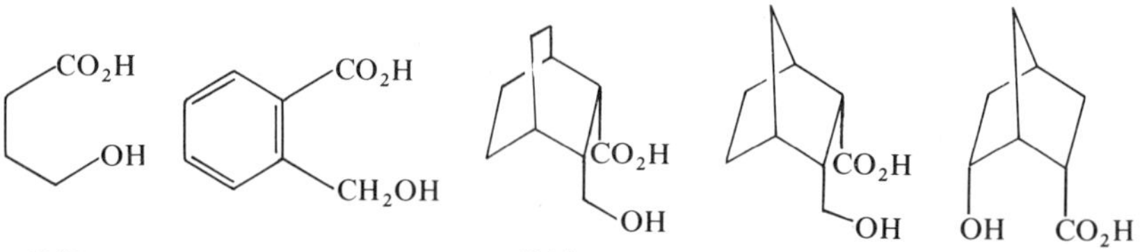

(**242**) (**243**) (**244**) (**245**) (**246**)

from differences in the conformational preferences in the ground state, differences in torsional strain in the transition state, and perhaps solvent

Participation by Oxygen Groups 183

orientation. DeTar[135] has used hydrocarbon models in calculating steric acceleration in such cyclization reactions. His results agree well with the experimental values and thus eliminate the need for special explanations such as orbital steering,[133,134,136] i.e., that enzymes and intramolecular structures can achieve accelerated rates by steering the reacting atoms into a preferred orientation.

Since lactonization of the above acids presumably (see the third reference in Ref. 133) occurs by the mechanism shown in Scheme 11, these studies should be applicable to acid-catalyzed ester hydrolysis because of the

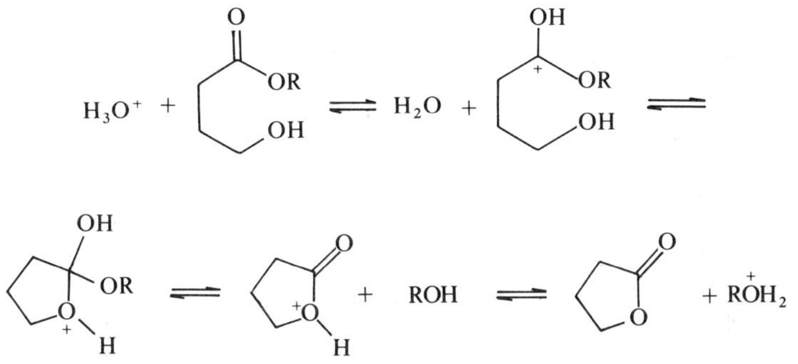

Scheme 11

similarity of the two reactions. Thus the effectiveness of a neighboring hydroxyl group in these reactions, and probably others as well, is highly dependent on the factors discussed above (See the discussion on pp. 66–69 and Ref. 137).

Nucleophilic participation by the hydroxyl group has been noted in acid-catalyzed studies of systems other than carboxyl groups. In dilute aqueous acid solution, methyl furanosides [e.g., (**248**)] were formed by ring closure during hydrolysis of the acetal bonds of the acyclic dimethyl acetals of glucose (**247**) and galactose.[138] It was estimated that the rates of these cyclizations were faster than the expected unassisted rates of hydrolysis,

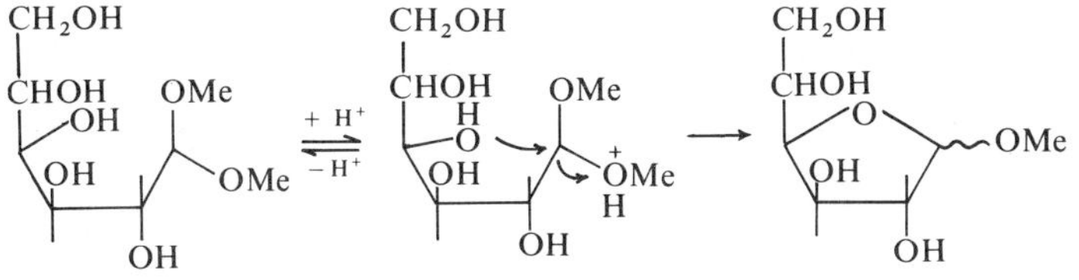

(73)

and it was therefore concluded that they were synchronous processes, the hydroxyl group providing anchimeric assistance.

Hydroxyl groups also participate as nucleophiles in a variety of other acid-catalyzed cyclizations of synthetic utility. One example is the Ritter reaction,[139] which normally produces amides upon reaction of olefins or alcohols with nitriles. Cyclization occurs with 1,3-diols, hydroxy alkenes, or epoxides.[140-142] Examples of each are given in Eqs. (74)–(76).

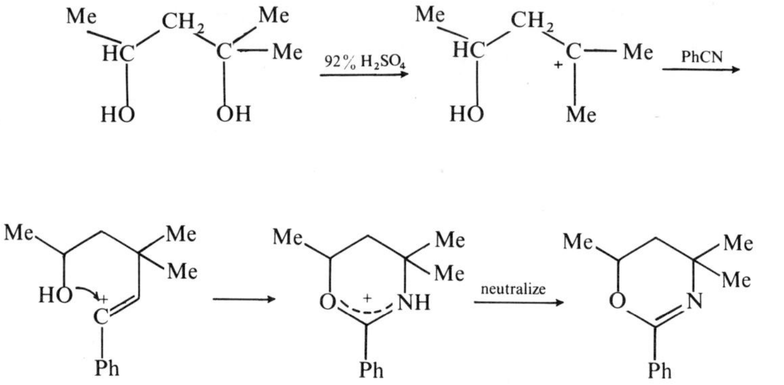

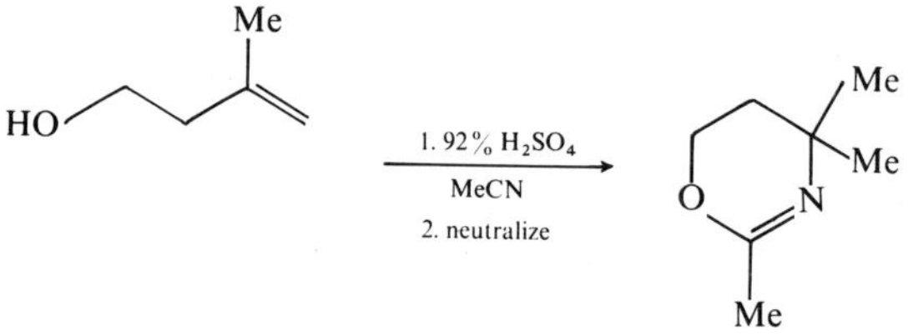

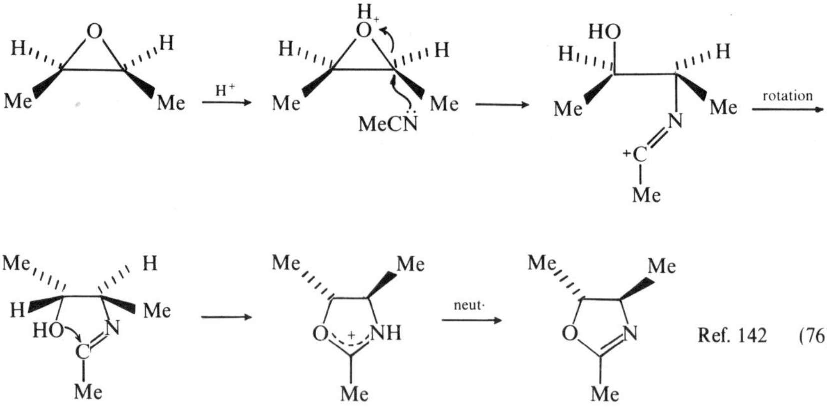

Participation by Oxygen Groups

O$^-$-7 Participation has been invoked to account for the base-catalyzed hydrolysis of 3-methyluridine (**249**).[143] The product 2-methylurea riboside (**250**) was said to result from hydroxyl participation, since methylation of the 5'-hydroxyl group inhibits this reaction.

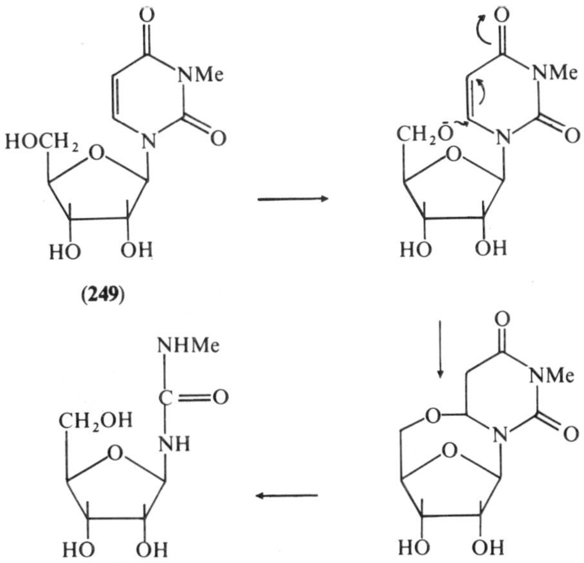

(**249**)　→　(77)

(**250**)　←　

Another interesting example of hydroxyl group participation occurs upon treatment of the hydroxyferrocene cyanide (**251**) with potassium cyanide in ethanol. The cyanoamide (**252**) is said to form by cyanide attack on the cyclic intermediate.[144] Alternatively, it seems to us that Ph$_2$C—O bond cleavage to the amide and the highly stabilized carbonium ion may precede CN$^-$ attack since the process as shown involves attack at a highly hindered carbon.

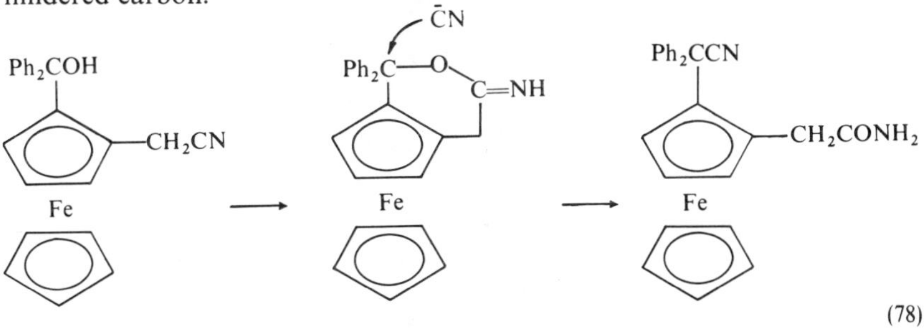

(**251**)　→　　　→　(**252**)　(78)

Kupchan and co-workers[145] have presented evidence for the argument that the hydrolysis of 1,3-diaxial hydroxyacetates is an instance of concerted general base–general acid catalysis of ester solvolysis. The rates of methanolysis of caprostane-3β,5β-diol 3-monoacetate (**253**), strophanthidol 3-acetate (**254**), and methyl strophanthidinate 3-acetate (**255**) in triethylamine buffers in methanol-chloroform are 300, 470, and 320 times faster, respectively, than that of caprostanol acetate. It was suggested that the 5β-hydroxyl group provides intramolecular assistance by hydrogen bonding to the ether oxygen of the ester group as shown. The enhanced rates for strophanthidin-3-acetate and neogermitrine, which were, respectively, 1200 and 4200 times faster than caprostanol acetate, were attributed to additional hydrogen bonding to the carbonyl oxygen atom by the hemiketal hydroxyl groups, as shown in (**256**) and (**257**), respectively.

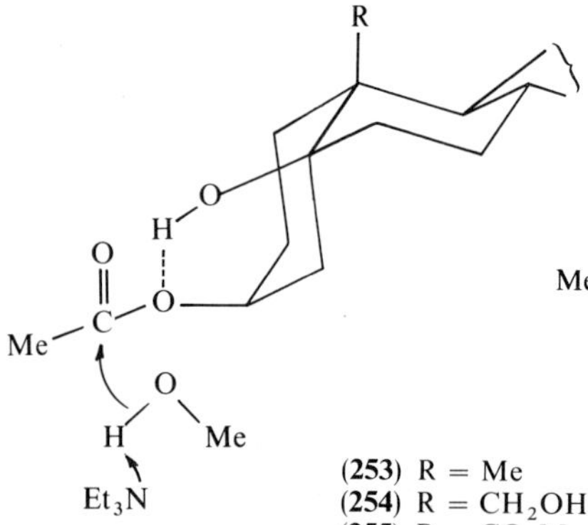

(253) R = Me
(254) R = CH$_2$OH
(255) R = CO$_2$Me

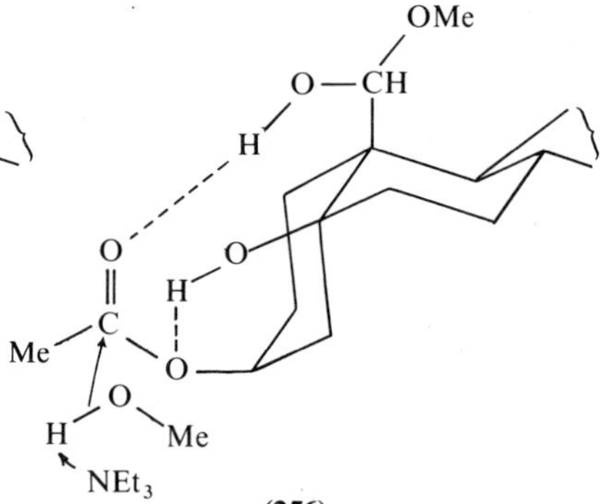

(256)

(79)

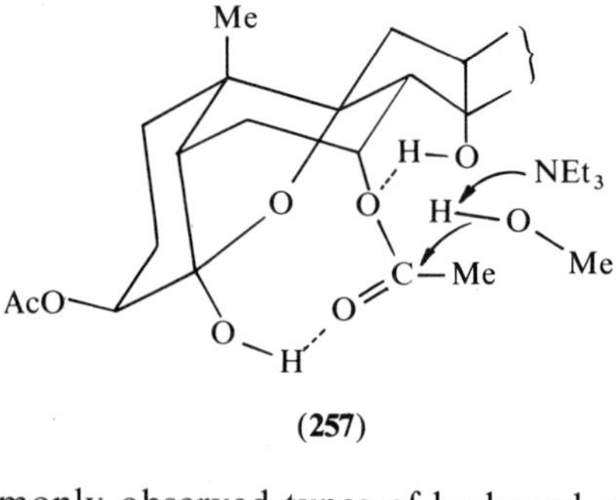

(257)

The more commonly observed types of hydroxyl group participation in the hydrolysis of carboxylic acid derivatives and some less common ones[146] are discussed in Volume 2. Intramolecular catalysis has been reviewed elsewhere.[37,147]

4.2.5. In Reduction Reactions*

There are a few examples of hydroxyl group participation in reductions by complex metal hydrides. Franzus and Snyder[148] report that reduction of norbornadiene-7-ol (258) or its acetate with lithium aluminum hydride results only in reductive formation of *anti*-7-norbornenol. Use of $LiAlD_4$ and H_2O work-up yields a single deuterium incorporation in an *exo* position. D_2O work-up leads to the incorporation of a second *exo*-deuterium. The reaction was thus formulated to occur as shown in Scheme 12. Borden[149] has published studies that indicate that the step (259)–(260) occurs with intramolecular deuteride attack.

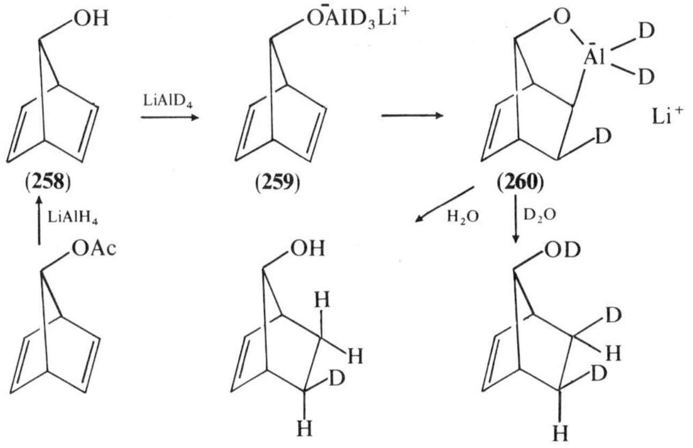

Scheme 12

(80)

4.3. Hydroperoxide Groups

The hydroperoxide group presents an interesting system for study since the hydroxyl group is attached to an atom containing nonbonding electrons. In general, such nucleophiles are more reactive than predicted from their polarizability and basicity. This is termed the α effect; for an excellent review of this phenomenon, see Ref. 150. Thus the hydroperoxide anion may be a better neighboring group than the oxide anion even

---

* Discussed in Volume 2; also see S.-C. Chen, *Synthesis*, 691 (1974).

Table 19. Comparison of the Neighboring Group Effect of the Peroxide Anion

System	k at 30°C (sec^{-1})^a	Rel. rate
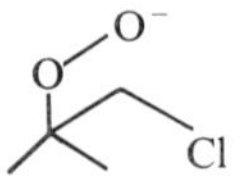 (261) [structure with O-O$^-$ and Cl]	2.35×10^{-3}	2.35×10^4
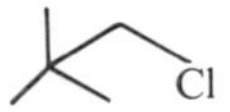 [structure with Cl] + $C_2H_5O^-$	10^{-7}	1.0
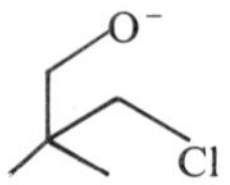 (262) [structure with O$^-$ and Cl]	6.4×10^{-5}	6.4×10^2

^{a}Hydroxide and oxide rates are based on product analysis.

though this is not predicted by a consideration of the basicity of the two. This is borne out in the studies of Richardson and co-workers.[151] Data from the rates of reaction of some halides (Table 19) indicate substantial anchimeric assistance in the formation of the 1,2-dioxetane product. Anchimeric assistance produces a 10^4 increase in rate for (261) as compared to the reaction of t-amyl chloride with ethoxide ion. By measuring the rates of ring closure of (262) under the same conditions, Richardson et al.[151] were able to calculate a value of 37 for the α effect.

4.4. Oxime Groups

The phosphate ester (263) is hydrolyzed very rapidly in the pH range 2–4 with a rate 2×10^7 times greater than that of the hydrolysis of ethyl p-nitrophenylphosphonate to give ethanol.[152] The reaction is independent of pH in the range 2–3.5, and at pH = 5 aniline is the product. This suggests a mechanism involving intramolecular general acid catalysis by the oxime group, as shown in Scheme 13.[152]

In similar hydrolytic processes in basic solution, the oxime group has been shown to afford general base catalysis.[153] Both syn- (265) and anti- (264) oximes show enhanced hydrolysis rates in hydroxide solution, but the syn isomer is faster. The anti isomer hydrolysis is 10^6 times faster than

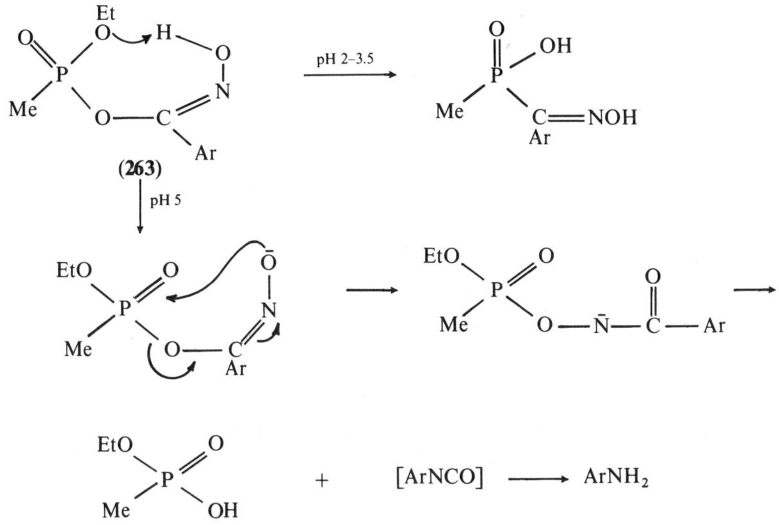

Scheme 13

the ethyl ester, and the *syn* isomer adds another factor of 120. The mechanism shown was proposed to account for these results.[153] The similarity of the *syn* and *anti* enhancements suggest a water-mediated process instead of a mechanism similar to that shown in Scheme 13.

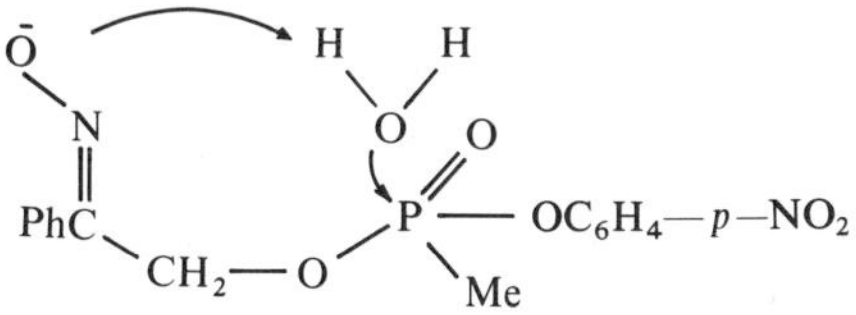

(264)

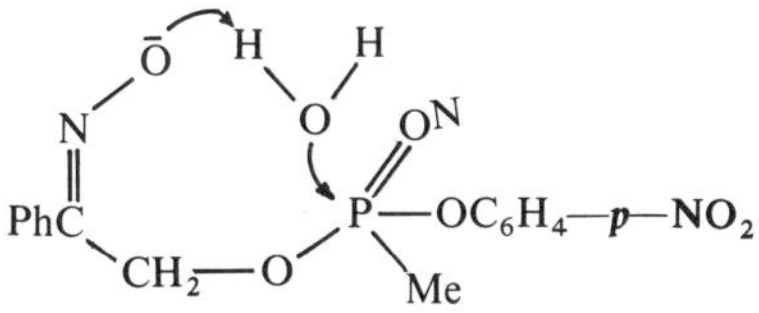

(265)

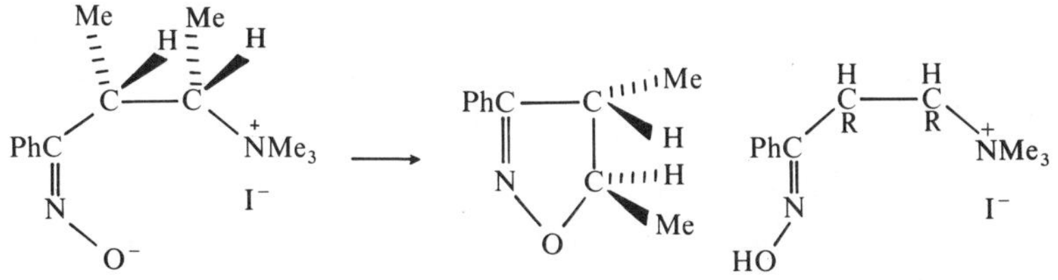

(266) **(267)** (81)

Oximes are useful in preparing isoxazolines and isoxazoles. Since some ring closures to produce these heterocycles may involve oxime anchimerism, Scott and MacConaill[154] studied the reaction with a number of Mannich base oxime methiodides. They found that the oximate anion was a powerful nucleophile and that displacement of trimethylamine, as in (**266**), occurred stereospecifically. The oxime of structure (**267**) undergoes elimination rather than cyclization.

References

1. S. Winstein, E. Allred, R. Heck, and R. Glick, *Tetrahedron*, **3**, 1 (1968).
2. E. L. Allred and S. Winstein, *J. Am. Chem. Soc.*, **89**, 3991 (1967).
3. E. L. Allred and S. Winstein, *J. Am. Chem. Soc.*, **89**, 3998 (1967).
4. E. R. Novak and D. S. Tarbell, *J. Am. Chem. Soc.*, **89**, 73, 3086 (1967); A. W. Friederang and D. S. Tarbell, *J. Org. Chem.*, **33**, 3797 (1968).
5. S. Winstein and L. L. Ingraham, *J. Am. Chem. Soc.*, **74**, 1160 (.952).
6. P. M. Henrichs and P. E. Peterson, *J. Am. Chem. Soc.*, **95**, 7449 (1973).
7. E. L. Allred and S. Winstein, *J. Am. Chem. Soc.*, **89**, 4012 (1967).
8. B. Capon, in: *Organic Reaction Mechanisms 1969* (B. Capon and C. W. Rees, eds.). pp. 81–82, John Wiley & Sons, Inc., New York (1970); see ref. 155.
9. J. R. Hazen, *Tetrahedron Lett.*, 1897 (1969),
10. D. S. Noyce, B. R. Thomas, and B. N. Bastian, *J. Am. Chem. Soc.*, **82**, 885 (1960); D. S. Noyce and B. N. Bastian, *ibid.*, **82**, 1246 (1960).
11. J. S. Brimacombe and O. A. Ching, *Chem. Commun.*, 781 (1968).
12. J. G. Buchanan, A. R. Edgar, and D. G. Large, *Chem. Commun.*, 558 (1968).
13. J. R. Hazen and D. S. Tarbell, *Tetrahedron Lett.*, 5927 (1968).
14. A. Kirrmann and N. Hamaide, *Bull. Soc. Chim. Fr.*, 789 (1957); A. Kirrmann and L. Wartski, *Compt. Rend.*, **259**, 2857 (1964); *Bull. Soc. Chim. Fr.*, 3077 (1965); *Rev. Roum. Chim.*, **10**, 1277 (1965).
15. A. Kirrmann and L. Wartski, *Bull. Soc. Chim. Fr.*, 3825 (1966).
16. H. Meerwein, in: *Methoden der organischen Chemie (Houben-Weyl)* (E. Muller, ed.), Bd/VI/3, Sauerstoffverbindugen I, p. 336, Georg Thieme Verlag, Stuttgart (1965).
17. M. J. Blandamer, H. S. Golinkin, and R. E. Robertson, *J. Am. Chem. Soc.*, **91**, 2678 (1969).
18. S. Winstein and R. B. Henderson, *J. Am. Chem. Soc.*, **65**, 2196 (1943).
19. D. D. Roberts and W. Hendrickson, *J. Org. Chem.*, **34**, 2415 (1969).
20. R. A. Evans, *Diss. Abstr. Int. B.*, **32**, 3848 (1972).
21. G. E. McCasland, M. O. Naumann, and L. J. Durham, *J. Org. Chem.*, **34**, 1382 (1969).
22. S. Winstein, C. R. Lindegren, and L. L. Ingraham, *J. Am. Chem. Soc.*, **75**, 155 (1953).
23. R. Heck, J. Corse, E. Grunwald, and S. Winstein, *J. Am. Chem. Soc.*, **79**, 3278 (1957).
24. S. Winstein, C. R. Lindegren, H. Marshall, and L. L. Ingraham, *J. Am. Chem. Soc.*, **75**, 147 (1953).
25. J. M. Harris, *Prog. Phys. Org. Chem.*, **11**, 89 (1974).
26. F. J. Lotspeich, *J. Org. Chem.*, **33**, 3316 (1968).
27. S. K. Core and F. J. Lotspeich, *J. Org. Chem.*, **36**, 399 (1971).
28. D. C. Kleinfelter and P. H. Chen, *J. Org. Chem.*, **34**, 1741 (1969).
29. V. Dauksas and I. Dembinskiene, *Zh. Vses. Khim. Obshchest*, **14**, 119 (1969); *Chem Abs.*, **70**, 114328 (1969).

30. E. L. Allred and S. Winstein, *J. Am. Chem. Soc.*, **89**, 4008 (1967).
31. A. J. Birch and E. G. Hutchinson, *J. Chem. Soc. Perkin Trans. 1*, 1546 (1972).
32. D. Guillerm-Dron, M. L. Capmau, and W. Chodkiewicz, *Tetrahedron Lett.*, 37 (1972).
33. R. M. Munavu and H. H. Szmant, *J. Org. Chem.*, **41**, 1832 (1976).
34. E. M. Kaiser and D. W. Slocum, *in: Organic Reactive Intermediates* (S. P. McManus, ed.), pp. 406–415, Academic Press, Inc., New York (1973); H. Gilman and J. W. Morton, Jr., *Org. React.*, **8**, 258 (1954).
35. P. B. Valkovich, G. W. Gokel, and I. K. Ugi, *Tetrahedron Lett.*, 2947 (1973).
36. A. I. Meyers and E. D. Mihelich, *J. Org. Chem.*, **40**, 1186 (1975); A. I. Meyers and G. Knaus, *J. Am. Chem. Soc.*, **96**, 6508 (1974); A. I. Meyers, G. Knaus, and K. Kamata, *ibid.*, **96**, 268 (1974); A. I. Meyers and M. E. Ford, *Tetrahedron Lett.*, 1341 (1974); A. I. Meyers and K. Kamata, *J. Org. Chem.*, **39**, 1603 (1974).
37. B. Capon, *Chem. Rev.*, **69**, 407 (1969); L. Goodman, *Adv. Carbohydr. Chem.*, **22**, 109 (1967).
38. G. R. Gray, F. C. Hartman, and R. Barker, *J. Org. Chem.*, **30**, 2020 (1965).
39. J. S. Brimacombe and O. A. Ching, *Carbohydr. Res.*, **5**, 239 (1967); *ibid.*, **8**, 374 (1968); *J. Chem. Soc. C*, 1642 (1968); O. A. Ching, *Bol. Soc. Quim. Peru*, **36**, 13, 45, 60 (1970); *Chem. Abstr.*, **73**, 110017 (1970); **74**, 64368, 64372 (1971).
40. N. A. Nelson, *J. Org. Chem.*, **38**, 3798 (1973).
41. G. T. Kwiatkowski, S. J. Kavarnos, and W. D. Closson, *J. Heterocycl. Chem.*, **2**, 11 (1965); see also D. Gagnaire, *Bull. Soc. Chim. Fr.*, 1813 (1960).
42. H. Morita and S. Oae, *Tetrahedron Lett.*, 1347 (1969); H. G. Richey, Jr., and D. V. Kinsman, *ibid.*, 2505 (1969).
43. H. Gilman and A. P. Hewlett, *Rec. Trav. Chem. Pays-Bas*, **51**, 93 (1932); L. A. Paquette, R. W. Begland, and P. C. Storm, *J. Am. Chem. Soc.*, **90**, 6148 (1968).
44. D. S. Tarbell and J. R. Hazen, *J. Am. Chem. Soc.*, **91**, 7657 (1969).
45. L. A. Paquette, R. W. Begland, and P. C. Storm, *J. Am. Chem. Soc.*, **92**, 1971 (1970).
46. L. A. Paquette and P. C. Storm, *J. Am. Chem. Soc.*, **92**, 4295 (1970).
47. L. A. Paquette and P. C. Storm, *J. Org. Chem.*, **35**, 3390 (1970).
48. L. A. Paquette and R. W. Begland, *J. Am. Chem. Soc.*, **87**, 3784 (1965).
49. L. A. Paquette and M. K. Scott, *J. Am. Chem. Soc.*, **94**, 6751 (1972).
50. L. A. Paquette and M. K. Scott, *J. Am. Chem. Soc.*, **94**, 6760 (1972).
51. L. A. Paquette, I. R. Dunkin, J. P. Freeman, and P. C. Storm, *J. Am. Chem Soc.*, **94**, 8124 (1972).
52. L. A. Paquette and I. R. Dunkin, *J. Am. Chem. Soc.*, **95**, 3067 (1972).
53. F. Kohen, G. Adelstein, and R. E. Counsell, *Chem. Commun.*, 770 (1970).
54. J. C. Martin and P. D. Bartlett, *J. Am. Chem. Soc.*, **79**, 2533 (1957).
55. L. A. Spurlock and R. G. Fayter, Jr., *J. Am. Chem. Soc.*, **94**, 2707 (1972).
56. J. P. Ward, *Tetrahedron Lett.*, 3905 (1965).
57. J. R. Hazen, *J. Org. Chem.*, **35**, 973 (1970).
58. P. G. Gassman and J. L. Marshall, *Tetrahedron Lett.*, 2429 (1968).
59. P. G. Gassman, J. L. Marshall, and J. G. Macmillan, *J. Am. Chem. Soc.*, **95**, 6319 (1973).
60. P. G. Gassman and J. G. Macmillan, *J. Am. Chem. Soc.*, **91**, 5527 (1969).
61. N. A. Hughes and P. R. H. Speakman, *Chem. Commun.*, 199 (1965); *J. Chem. Soc. C*, 1182 (1967).
62. C. L. Stevens, R. P. Glinski, K. G. Taylor, P. Blumbergs, and F. Sirokmann, *J. Am. Chem. Soc.*, **88**, 2073 (1966); C. L. Stevens, R. P. Glinski, G. E. Gutowski, and J. P. Dickerson, *Tetrahedron Lett.*, 649 (1967); C. L. Stevens, R. P. Glinski, K. G. Taylor, and F. Sirokman, *J. Org. Chem.*, **35**, 592 (1970); also see S. Hanessian, *Chem. Commun.*, 796 (1966).
63. N. A. Hughes, *Chem. Commun.*, 1072 (1967).
64. C. Bullock, L. Hough, and A. C. Richardson, *Chem. Commun.*, 1276 (1971).

65. R. U. Lemieux and B. Fraser-Reid, *Can. J. Chem.*, **42**, 539 (1964).
66. F. F. Blicke, W. B. Wright, Jr., and M. F. Zienty, *J. Am. Chem. Soc.*, **63**, 2488 (1941).
67. V. Prelog and S. Heimbach-Juhasz, *Ber.*, **74**, 1702 (1941).
68. D. S. Noyce and H. I. Weingarten, *J. Am. Chem. Soc.*, **79**, 3093 (1957).
69. K. B. Wiberg, *J. Am. Chem. Soc.*, **74**, 3957 (1952).
70. H. Meerwein, P. Borner, O. Fuchs, H. J. Sasse, H. Schrodt, and J. Spille, *Chem. Ber.*, **89**, 2060 (1956).
71. H. Perst, *Oxonium Ions in Organic Chemistry*, Verlag Chemie–Academic Press, Weinheim/Bergstr. (1971); C. U. Pittman, Jr., S. P. McManus, and J. W. Larsen, *Chem. Rev.*, **72**, 357 (1972).
72. W. W. Epstein and A. C. Sonntag, *J. Org. Chem.*, **32**, 3390 (1967); D. B. Denney and J. Giacin, *Tetrahedron*, **20**, 1377 (1964).
73. R. L. Cargill, T. E. Jackson, N. P. Peet, and D. M. Pond, *Acc. Chem. Res.*, **7**, 106 (1974).
74. P. E. Peterson, C. Casey, E. V. P. Tao, A. Agtarap, and G. Thompson, *J. Am. Chem. Soc.*, **87**, 5163 (1965).
75. J. L. Jernow, D. Gray, and W. D. Closson, *J. Org. Chem.*, **36**, 3511 (1971).
76. N. Heap, G. E. Green, and G. H. Whitham, *J. Chem. Soc. C*, 160 (1969).
77. H. W. Heine, A. D. Miller, W. H. Barton, and R. W. Greiner, *J. Am. Chem. Soc.*, **75**, 4778 (1953).
78. H. W. Heine and W. Siegfried, *J. Am. Chem. Soc.*, **76**, 489 (1954).
79. B. Capon and I. Farazmand, unpublished results.
80. W. H. Richardson, C. M. Golino, R. H. Wachs, and M. B. Yelvington, *J. Org. Chem.*, **36**, 943 (1971).
81. M. S. Gibson and R. W. Bradshaw, *Angew. Chem. Int. Ed. Engl.*, **7**, 919 (1968).
82. T. H. Cromartie and C. G. Swain, *J. Am. Chem. Soc.*, **97**, 232 (1975).
83. M. L. Bender, *Mechanisms of Homogeneous Catalysis from Protons to Proteins*, John Wiley & Sons, Inc. (Interscience Division), New York (1971), Chap. 4; W. P. Jencks, *Catalysis in Chemistry and Enzymology*, McGraw-Hill Book Company, New York (1969), Chap. 3.
84. C. G. Swain, D. A. Kuhn, and R. L. Schowen, *J. Am. Chem. Soc.*, **87**, 1553 (1965).
85. W. P. Jencks, I. Givot, and A. Satterthwait, unpublished results quoted by Jencks in Ref. 83, p. 170; W. P. Jencks, personal communication.
86. A. Fry, in: *Isotope Effects in Chemical Reactions* (C. J. Collins and N. S. Bowman, eds.), Chap. 6, Van Nostrand Reinhold Company, New York (1970).
87. A. H. Ewald and D. J. Ottley, *Aust. J. Chem.*, **20**, 1335 (1967).
88. G. Kohnstam and M. Penty, in: *Hydrogen-Bonded Solvent Systems* (A. K. Covington and P. Jones, eds.), p. 275, Taylor and Francis, London (1968).
89. S. Winstein and E. Grunwald, *J. Am. Chem. Soc.*, **70**, 828 (1948).
90. G. Bodennec, H. Bodot, and A. Nattaghe, *Bull. Soc. Chim. Fr.*, 876 (1967); see also H. Bodot and E. Laurent-Dieuzeide, *ibid.*, 3908 (1966); P. Crouzet, É. Laurent-Dieuzeide, and J. Wylde, *ibid.*, 4047 (1967).
91. D. D. Roberts, *J. Org. Chem.*, **33**, 118 (1968).
92. M. R. Smith, Master's thesis, The University of Alabama in Huntsville (1972); see also F. C. Haupt and M. R. Smith, *Tetrahedron Lett.*, 4141 (1974).
93. D. D. Roberts and J. G. Traynham, *J. Org. Chem.*, **32**, 3177 (1967).
94. L. N. Owen and P. N. Smith, *J. Chem. Soc.*, 4026 (1952); L. N. Owen and G. S. Saharia, *ibid.*, 2582 (1953); H. Bodot, J. Jullien, and M. Mousseron, *Bull. Soc. Chim. Fr.*, 1101, 1110 (1958).
95. H. W. Heine, *J. Am. Chem. Soc.*, **79**, 6268 (1957).
96. P. Ballinger and F. A. Long, *J. Am. Chem. Soc.*, **81**, 2347 (1959).
97. C. G. Swain, A. D. Ketley, and R. F. W. Bader, *J. Am. Chem. Soc.*, **81**, 2353 (1959).

98. G. H. Twigg, W. S. Wise, H. J. Lichtenstein, and A. R. Philpotts, *Trans. Faraday Soc.*, **48**, 699 (1952).
99. J. G. Buchanan and H. Z. Sable, *in: Selective Organic Transformations* (B. S. Thyagarajan, ed.), Vol. 2, pp. 1–96, John Wiley & Sons, Inc. (Interscience Division, New York (1972).
100. M. Černý, J. Staněk, and J. Pacák, *Collect. Czech. Chem. Commun.*, **34**, 849 (1969).
101. J. M. Coxon, M. P. Hartshorn, and A. J. Lewis, *Chem. Commun.*, 1607 (1970).
102. II. B. Henbest and B. Nicholls, *J. Chem. Soc.*, 221 (1959).
103. S. Searles, Jr., R. G. Nickerson, and W. K. Witsiepe, *J. Org. Chem.*, **24**, 1839 (1959).
104. N. G. Gaylord, J. H. Crowdle, W. A. Himmler, and H. J. Pepe, *J. Am. Chem. Soc.*, **76**, 59 (1954).
105. R. B. Clayton, H. B. Henbest, and M. Smith, *J. Chem. Soc.*, 1982 (1957).
106. H. B. Henbest and B. B. Millward, *J. Chem. Soc.*, 3575 (1960); see also C. R. Lindegren and S. Winstein, *Abstracts, 123rd American Chemical Society Meeting, 1953,* The American Chemical Society, Washington, D.C. (1953), p. 30M.
107. E. J. Corey, R. B. Mitra, and H. Uda, *J. Am. Chem. Soc.*, **86**, 485 (1964).
108. F. C. Hartman and R. Barker, *J. Org. Chem.*, **28**, 1004 (1963); **29**, 873 (1964).
109. R. Barker, *J. Org. Chem.*, **35**, 461 (1970); see also J. Jacobus, *ibid.*, **38**, 402 (1973).
110. B. G. Hudson and R. Barker, *J. Org. Chem.*, **32**, 3650 (1967).
111. D. D. Heard, B. G. Hudson, and R. Barker, *J. Org. Chem.*, **35**, 464 (1970); R. Barker, *ibid.*, **29**, 869 (1964); see also M. Chérest, F. Felkin, J. Sicher, F. Šipoš, and M. Tichý, *J. Chem. Soc.*, 2513 (1965) for a discussion on stereoelectronic control.
112. Y. Pocker, *in: Molecular Rearrangements* (P. deMayo, ed.), Part I, pp. 1–26, John Wiley & Sons, Inc. (Interscience Division), New York (1963).
113. Y. Pocker and B. P. Ronald, *J. Org. Chem.*, **35**, 3362 (1970); for an example of hydroxyl group participation in a similar process, see Ref. 101.
114. D. A. Buckingham, C. E. Davis, and A. M. Sargeson, *J. Am. Chem. Soc.*, **92**, 6159 (1970); D. A. Buckingham, D. M. Foster, and A. M. Sargeson, *ibid.*, **92**, 6151 (1970).
115. L. A. Paquette and I. R. Dunkin, *J. Am. Chem. Soc.*, **97**, 2243 (1975).
116. V. I. Staninets and E. A. Shilov, *Russ. Chem. Rev.*, **40**, 272 (1971).
117. S. Winstein and L. Goodman, *J. Am. Chem. Soc.*, **76**, 4368 4373 (1954); see Ref. 157.
118. N. L. Wendler, D. Taub, and C. H. Kuo, *J. Org. Chem.*, **34**, 1510 (1969).
119. G. Bellucci, G. Berti, G. Ingrosso, and E. Mastrorilli, *Tetrahedron Lett.*, 3911 (1973).
120. D. L. H. Williams, *Tetrahedron Lett.*, 2001 (1967); D. L. H. Williams, E. Bienvenue-Goetz, and J. E. Dubois, *J. Chem. Soc. B*, 517 (1969); S. R. Hooley and D. L. H. Williams, *J. Chem. Soc. Perkin Trans. 2*, 503 (1975).
121. I. Monkovic, Y. G. Perron, R. Martel, W. J. Simpson, and J. A. Gylys, *J. Med. Chem.*, **16**, 403 (1973).
122. H. Wong, J. Chapuis, and I. Monkovic, *J. Org. Chem.*, **39**, 1042 (1974).
123. R. K. Bly and R. S. Bly, *J. Org. Chem.*, **28**, 3165 (1963); see also R. S. Bly, R. K. Bly, A. O. Bedenbaugh, and O. R. Vail, *J. Am. Chem. Soc.*, **89**, 881 (1967), and Ref. 158.
124. A. Factor and T. G. Traylor, *J. Org. Chem.*, **33**, 2607 (1968).
125. T. T. Tidwell and T. G. Traylor, *J. Org. Chem.*, **33**, 2614 (1968).
126. H. C. Brown, J. H. Kawakami, and S. Ikegami, *J. Am. Chem. Soc.*, **89**, 1525 (1967); H. C. Brown and K.-T. Liu, *ibid.*, **89**, 3900 (1967).
127. J. Halpern and H. B. Tinker, *J. Am. Chem. Soc.*, **89**, 6427 (1967).
128. J. E. Byrd and J. Halpern, *J. Am. Chem. Soc.*, **95**, 2586 (1973).
129. P. M. Henry, *J. Am. Chem. Soc.*, **87**, 4423 (1967).
130. T. Sasaki, S. Eguchi, and T. Kiriyama, *J. Org. Chem.*, **38**, 2230 (1973).
131. J. A. Hirsch, F. J. Cross, and W. A. Meresak, *J. Org. Chem.*, **39**, 1966 (1974).
132. J.-H. Liu and P. Kovacic, *J. Org. Chem.*, **40**, 1183 (1975).
133. D. R. Storm, R. Tjian, and D. E. Koshland, *Chem. Commun.*, 854 (1971); D. R. Storm

and D. E. Koshland, *Proc. Natl. Acad. Sci. U.S.A.*, **66**, 445 (1970); *J. Am. Chem. Soc.*, **94**, 5805, 5815 (1972); D. G. Hoare, *Nature*, **236**, 437 (1972).
134. R. M. Moriarty and T. Adams, *J. Am. Chem. Soc.*, **95**, 4070, 4071 (1973).
135. D. F. DeTar, *J. Am. Chem. Soc.*, **96**, 1255 (1974).
136. B. Capon, *J. Chem. Soc. B*, 1207 (1971); M. I. Page and W. P. Jencks, *Proc. Natl. Acad. Sci. U.S.A.*, **68**, 1678 (1971); T. C. Bruice, A. Brown, and D. O. Harris, *ibid.*, **68**, 658 (1971); J. Reuben, *ibid.*, **68**, 563 (1971); G. N. Port and W. G. Richards, *Nature*, **231**, 312 (1971); G. A. Dafforn and D. E. Koshland, *Bioorg. Chem.*, **1**, 129 (1971); J. H. Block and M. A. Nunes, *J. Org. Chem.*, **35**, 3456 (1970).
137. S. Milstein and L. A. Cohen, *Proc. Natl. Acad. Sci. U.S.A.*, **67**, 1143 (1970); *J. Am. Chem. Soc.*, **94**, 9158 (1972); R. T. Borchardt and L. A. Cohen, *ibid.*, **94**, 9166 (1972); **94**, 9175 (1972); **95**, 8308, 8313, 8319 (1973).
138. B. Capon and D. Thacker, *J. Am. Chem. Soc.*, **87**, 4199 (1965).
139. L. I. Krimen and D. J. Cota, *Org. React.*, **17**, 213 (1969).
140. E. J. Tillmanns and J. J. Ritter, *J. Org. Chem.* **22**, 839 (1957).
141. S. P. McManus and J. T. Carroll, *Org. Prep. Proced.*, **2**, 71 (1970).
142. R. A. Wohl and J. Cannie, *J. Org. Chem.*, **38**, 1787 (1973).
143. Y. Kondo, J.-L. Fourner, and B. Witkop, *J. Am. Chem. Soc.*, **93**, 3527 (1971).
144. J. H. Peet and B. W. Rockett, *Chem. Commun.*, 120 (1968).
145. S. M. Kupchan, S. P. Eriksen, and M. Friedman, *J. Am. Chem. Soc.*, **88**, 343 (1966); S. M. Kupchan, S. P. Eriksen, and Y.-T. S. Liang, *ibid.*, **88**, 347 (1966); see also J. F. Baker and R. T. Blickenstaff, *J. Org. Chem.*, **40**, 1579 (1975).
146. D. A. Buckingham, F. R. Keene, and A. M. Sargeson, *J. Am. Chem. Soc.*, **96**, 4981 (1974), and references therein.
147. B. Capon, *Essays Chem.*, **3**, 127 (1972); also see Chapter 9 in Bender[83] and Chapter 10 in Jencks.[83]
148. B. Franzus and E. I. Snyder, *J. Am. Chem. Soc.*, **87**, 3423 (1965).
149. W. T. Borden, *J. Am. Chem. Soc.*, **90**, 2197 (1968).
150. R. F. Hudson, *in: Chemical Reactivity and Reaction Paths* (G. Klopman, ed.), Chap. 5, John Wiley & Sons, Inc., New York (1974).
151. W. H. Richardson, J. W. Peters, and W. P. Konopka, *Tetrahedron Lett.*, 5531 (1966); W. H. Richardson and V. F. Hodge, *ibid.*, 2271 (1970); *J. Am. Chem. Soc.*, **93**, 3996 (1971).
152. J. I. G. Cadogan and J. A. Maynard, *Chem. Commun.*, 854 (1966); J. I. G. Cadogan and D. T. Eastlick, *ibid.*, 1546 (1970); J. I. G. Cadogan, J. A. Challis, and D. T. Eastlick, *J. Chem. Soc. B*, 1988 (1971).
153. C. N. Lieske, J. W. Hovanec, and P. Blumbergs, *Chem. Commun.*, 976 (1966); C. N. Lieske, J. W. Hovanec, G. M. Steinberg, and P. Blumbergs, *ibid.*, 13 (1968).
154. F. L. Scott and R. J. MacConaill, *Tetrahedron Lett.*, 3685 (1967); R. J. MacConaill and F. L. Scott, *ibid.*, 2993 (1970); oximes may also participate through the nitrogen atom; cf E. Buehler, *J. Org. Chem.*, **32**, 261 (1967), and F. L. Scott and F. J. Lalor, *Tetrahedron Lett.*, 641 (1964), and in PdCl$_2$ catalyzed ring closure, c.f. T. Hosokawa *et al.*, *ibid.*, 383 (1976).
155. E. L. Eliel, R. O. Hutchins, R. Mebane, and R. L. Willer, *J. Org. Chem.*, **41**, 1052 (1976) have found that the preference for endocyclic or exocyclic attack by nucleophiles on five- and six-membered cyclic onium salts (O, N, or S) is relatively insensitive to a change in the charged ring atom.
156. J. K. Coward, R. Lok, and O. Takagi, *J. Am. Chem. Soc.*, **98**, 1057 (1976).
157. B. Ganem, *J. Am. Chem. Soc.*, **98**, 858 (1976), has reported on the stereochemical requirements for HO-3 chlorocyclizations.
158. R. G. Salomon and R. D. Gleim, *J. Org. Chem.*, **41**, 1529 (1976) discuss evidence for participation in the oxymercuration of *endo*-3-hydroxybicyclo[3.1.0]hexane.
159. S. Danishefsky *et al.*, *J. Am. Chem. Soc.*, **98**, 3028 (1976).

5
Participation by Sulfur Groups

5.1 Thioether Groups

5.1.1 In Solvolytic Displacement Reactions

Neighboring divalent sulfur in thioether groups is extremely active in reactions involving carbocation and free-radical intermediates. Measures of anchimeric assistance show these groups to be far superior to comparable oxygen-containing compounds. While the majority of cases reported involve S-3 participation, larger rings form as well.

The hydrolysis of 2,2′-bischloroethyl sulfide (mustard gas) was one of the first neighboring-group-accelerated reactions to get widespread attention.[1] Initially the hydrolysis of this compound is purely first order in the sulfide, and the rate is unaffected by added alkali or other nucleophiles; the increasing amount of chloride ion, however, slows the rate with time. Since formation of the primary carbocation is not expected, these data suggest an anchimerically assisted solvolysis [Eq. (1)], leading to the formation of the intermediate sulfonium ion (**1**), which may react with water, chloride ion, or any other nucleophile present. A second anchimerically assisted process that leads to displacement of the second chloro group can occur.

A series of arylthioalkyl chlorides has been studied to assess the

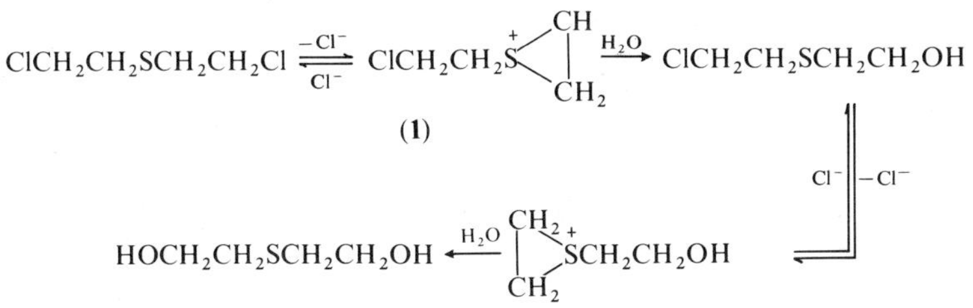

$$ClCH_2CH_2SCH_2CH_2Cl \underset{Cl^-}{\overset{-Cl^-}{\rightleftharpoons}} ClCH_2CH_2\overset{+}{S}\underset{CH_2}{\overset{CH_2}{\diagup\!\!\diagdown}} \overset{H_2O}{\longrightarrow} ClCH_2CH_2SCH_2CH_2OH$$

(**1**)

$$\Big\updownarrow Cl^- \| -Cl^-$$ (1)

$$HOCH_2CH_2SCH_2CH_2OH \overset{H_2O}{\longleftarrow} \underset{CH_2}{\overset{CH_2+}{\diagup\!\!\diagdown}} SCH_2CH_2OH \rightleftharpoons$$

Table 1. Rates of Solvolysis of Some Phenylthioalkyl Chlorides

Chloride	50% aq. acetone,[a] $k \times 10^3$ (min^{-1}, 80°C)	20 mole % aq. dioxane,[b] $k \times 10^3$ (min^{-1}, 100°C)	Methanol,[c] k_{rel}[d]
PhS(CH$_2$)$_2$Cl	—	5.5	150
PhS(CH$_2$)$_3$Cl	—	0.029	1.0
PhS(CH$_2$)$_4$Cl	65	1.1	130
PhS(CH$_2$)$_5$Cl	0.85	—	4.3
n-C$_6$H$_{13}$Cl	—	0.053	—

[a] Reference 2.
[b] Reference 3.
[c] Reference 4.
[d] No temperature was given for the comparison, and the rate data for PhS(CH$_2$)$_{3-5}$Cl were not given.

effects of ring size on sulfur participation.[2-6] The results (Table 1) show that anchimeric assistance decreases with ring size in the order 3 > 5 > 6 > 4. This is strikingly different from the order with ethers where both O-5 and O-6 anchimeric assistance were superior to O-3 participation. Bordwell and Brannen[4] also compared the effect of an α-sulfur group to those listed in Table 1. The α-sulfur group may participate in the displacement of the chloride group by direct resonance [Eq. (2)], and such participation, as expected, is quite effective (PhSCH$_2$Cl undergoes methanolysis 1900 times faster than PhSCH$_2$CH$_2$Cl at 50°C).

$$PhS-CH_2-Cl \longrightarrow Ph\overset{+}{S}=CH_2 \qquad (2)$$

In a study of S-5 versus S-6 participation in the cyclization of 4-bromobutyl p-tolyl sulfide (2) and 5-bromopentyl p-tolyl sulfide (3) in ethanol and t-butyl alcohol, it was found that equilibria between the substrates and the respective cyclic sulfonium ion resulted. In the presence of t-butoxide, however, a rapid reaction between t-butoxide and the sulfonium ions occurred, which resulted in their removal as they formed. Thus, the rate constants k_1, the reverse reaction rate constants k_{-1}, and the equilibrium constants K could be determined. Again it is seen that S-5 participation is faster than S-6 participation, but the equilibrium constant is more favorable for six-membered ring formation since the five-membered ring also opens more rapidly than the six-membered ring.

Relative to the arylthio derivatives, kinetic studies on homologous alkylthioalkyl halides have been largely ignored. It appears (see p. 51) that for ethyl ω-halogenoalkyl sulfides the rate constant for the formation of the three-membered ring sulfonium ion is much smaller than that of the

$$\underset{(2)}{\overset{Ar}{\underset{S}{\bigg|}}\diagdown\diagup\text{-Br}} \underset{k_{-1}}{\overset{k_1}{\rightleftharpoons}} \overset{Ar}{\underset{S^+}{\bigg|}}\bigtriangleup \; Br^-$$

$k_1 = 1.88 \times 10^{-4} \text{ sec}^{-1}$
$k_{-1} = 75.1 \times 10^{-4} \text{ sec}^{-1}$
$K = 2.5 \times 10^{-2}$ at 55 C

$$\underset{(3)}{\overset{Ar}{\underset{S}{\bigg|}}\diagdown\diagup\diagdown\text{-Br}} \underset{k_{-1}}{\overset{k_1}{\rightleftharpoons}} \overset{Ar}{\underset{S^+}{\bigg|}}\bigcirc \; Br^-$$

$k_1 = 4.26 \times 10^{-6} \text{ sec}^{-1}$
$k_{-1} = 1.16 \times 10^{-4} \text{ sec}^{-1}$
$K = 3.67 \times 10^{-2}$ at 55 C

five-membered ring sulfonium ion, with the cyclization to the latter being immeasurably rapid. Since this order of reactivity is the reverse of what is found in the arylthio examples, Stirling et al.[5,6] have suggested that there might be some conjugative effect of the aryl group that selectively favors formation of the three-membered ring. Such an effect, however, has not been conclusively demonstrated, although it was offered as one explanation for the observation that the $\Delta H^{\ddagger}$ for three-membered ring formation in arylthioalkyl halide is lower than the $\Delta H^{\ddagger}$ for five-membered ring formation.[5,6] It is interesting to note that an *entropy* effect and not an enthalpy accounts for the greatly accelerated ring closure of $p\text{-CH}_3\text{C}_6\text{H}_4\text{SCH}_2\cdot\text{CHPhCl}$ (Table 2).

Rearrangement by ion-pair return is common among β-halothioethers. Although this was mentioned above in connection with studies of mustard gas, a more definite observation was made in the study of the optically active derivative (4). Either optical isomer of (4) was found to racemize,

Table 2. Relative Rates of Solvolysis of Some Arylthioalkyl Halides in 80% Ethanol[6]

Halide	k_{rel} (25°C)	k_{rel} (54.2°C)	$\Delta H^{\ddagger}$ (kcal mole^{-1})	$\Delta S^{\ddagger}$ (cal mole^{-1} °K^{-1})
$p\text{-CH}_3\text{C}_6\text{H}_4\text{S(CH}_2)_2\text{Cl}$	1	1	17.8	−24
$p\text{-CH}_3\text{C}_6\text{H}_4\text{S(CH}_2)_4\text{Cl}$	0.53	0.73	19.4	−20
$p\text{-CH}_3\text{C}_6\text{H}_4\text{S(CH}_2)_2\text{Br}$	—	35	—	—
$p\text{-CH}_3\text{C}_6\text{H}_4\text{S(CH}_2)_4\text{Br}$	—	41	—	—
$p\text{-CH}_3\text{C}_6\text{H}_4\text{SCH}_2\text{CHPhCl}$	5400	—	15.4	−13
$p\text{-CH}_3\text{C}_6\text{H}_4\text{S(CH}_2)_3\text{CHPhCl}$	45	—	17.0	−19

presumably through the sulfonium ion (5) as shown.[7] In 40% aqueous acetone the rate of racemization was found to be 16.6 times the rate of hydrolysis.

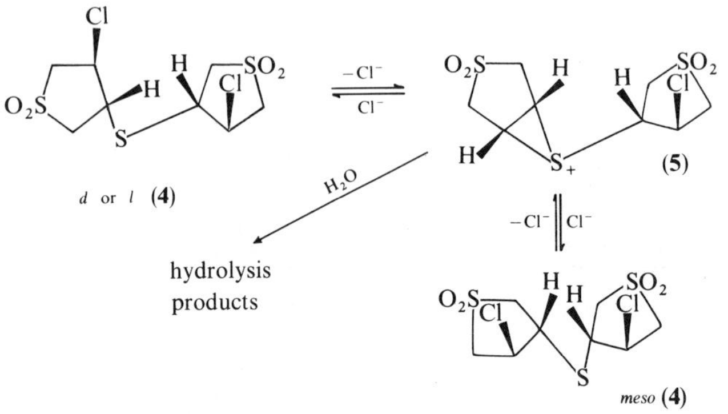

$$(3)$$

Solvolysis is also an important factor in product formation during addition of sulfenyl halides to olefins; for a review, see Ref. 8. In (6), where R is a group capable of stabilizing a carbocation, the kinetic product of addition of R'SCl is, in many cases, the product of Markovnikov addition (7), while the thermodynamic product is the anti-Markovnikov product (8). The strongly bridged intermediate (9), which forms even in styrene derivatives,[9] is said to intervene in the formation of (7) or (8). In polar media the kinetic product (7) may undergo an anchimerically assisted ionization back to (9), leading to a buildup in the concentration of (8). It is said that this phenomenon has led to some errors in assigning the exact mode of addition of alkyl and aryl sulfenyl halides to olefins.[10] Sulfuranes, such as (10), are also important intermediates in electrophilic addition reactions of sulfenyl halides[11] and may play a role in many solvolytic reactions. The

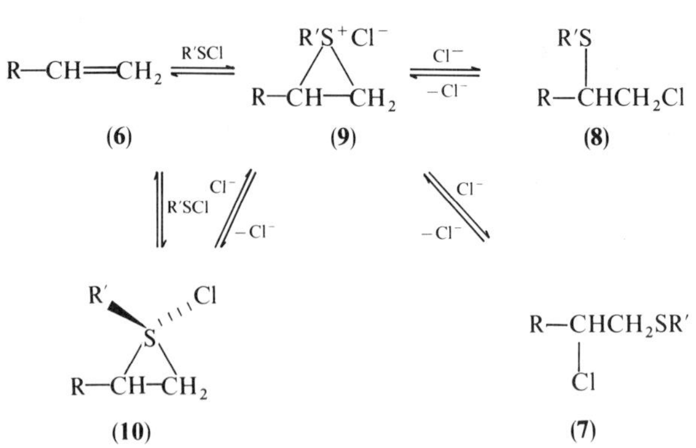

$$(4)$$

sulfurane (**10**) may form directly or indirectly in addition reactions [Eq. (4)] and by attack at sulfur rather than at carbon on sulfonium ions formed in solvolytic processes.

The ethylthio group seems to provide more anchimeric assistance than the phenylthio group, as evidenced by the observation that 2-chloroethyl ethyl sulfide solvolyses about 40 times faster than 2-chloroethyl phenyl sulfide.[2] This may be ascribed to a decreased nucleophilicity of the arylthio groups by resonance delocalization of the electrons on sulfur into the aryl ring.

trans-2-Chlorocyclohexyl and *trans*-2-chlorocyclopentyl phenyl sulfides undergo solvolysis in 80% aqueous ethanol 10^5–10^6-fold faster than their *cis* isomers.[12] Electron-releasing substituents in the *para* position of the phenyl ring enhance the rates for the *trans* compounds (ρ values are -1.43 and -1.39, respectively, at 30°C) but have no effect on the *cis* isomers. The data are in accord with anchimeric assistance by sulfur and indicates that a *trans* configuration is required for such assistance. Other studies[13] have indicated that for anchimeric assistance an antiperiplanar orbital arrangement is highly preferred if not required. Thus the rigid bicyclic derivative [(**11**), R = Ar], which can attain a *trans*-diaxial arrangement only with difficulty, solvolyzes only about four times faster than its *cis* isomer.[13] While the *cis*-methylthio isomer, with silver acetate in acetic acid, proceeded stereospecifically to yield the *syn-exo*-[3.2.1] acetate, the *trans* isomer [(**11**), R = Me] yielded a 2:1 mixture of the *syn*- and *anti*-[3.2.1] acetates.[14] The aryl derivative [(**11**), R = Ph] yielded the *syn-/anti*-acetates in a 1:1 ratio.[14] These results are explained by invoking a normal Wagner–

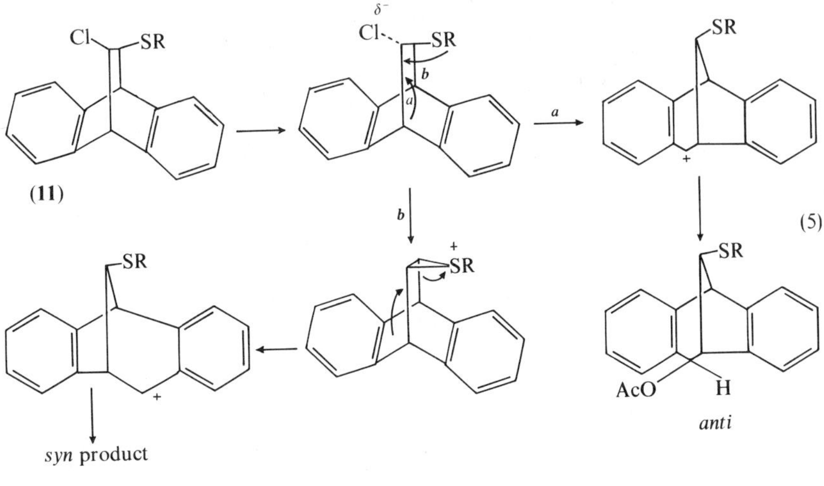

Meerwein pathway for the *cis* isomer, but neither anchimerically assisted nor unassisted processes alone account for the products from the *trans* isomers. Hence competitive processess are probably involved. It is of course possible for carbon assistance in the *cis* isomer to lead to the "low" *trans/cis* rate ratio.

The study of sulfur-bridged carbocycles[15-17] has revealed a very interesting S-3 effect that is somewhat contradictory to the above conclusions and unprecedented in studies of ether groups (see Chapter 4). Results of a study[15] of the rates of hydrolysis of the 2,5-epithio-5α-cholestan-3β-yl bromide [*endo* (12)], the 3α-yl methanesulfonate [*exo* (13)], the 3β-bromide of the corresponding sulfoxide [*endo* (14)], and of the 2α,5-epoxy-5α-cholestan-3α- and 3β-yl methanesulfonates [*exo* (15) and *endo* (15), re-

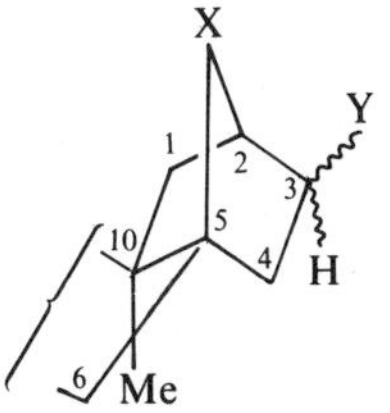

(12) X = S, Y = Br
(13) X = S, Y = OMs
(14) X = SO, Y = Br
(15) X = O, Y = OMs
(16) X = O, Y = Br

spectively], and 3α-yl bromide [*exo* (16)] in 70, 80, and 90% aqueous dioxane are shown in Table 3. With the oxygen derivatives as models for inductive retardation, the calculated anchimeric acceleration owing to sulfur in *endo* (12) is 1.1×10^{10}. The sulfur compound was also found to

Table 3. Relative Solvolytic Reactivities in Aqueous Dioxane at 25°C[15]

Substrate	k_{rel}		
	60 D[a]	70 D[a]	90 D[a]
endo (12)[b]	—	1.2×10^8	2.3×10^7
exo (13)	1.0	1.0	1.0
endo (14)[b]	0.79	—	—
endo (15)	—	0.011	—
exo (15)	—	9.0	6.4
exo-2-Norbornyl-OMs	—	1.1×10^4	6.5×10^3
endo-2-Norbornyl-OMs	—	13	36

[a] Volume % dioxane.
[b] Compared as the methanesulfonate using the leaving group conversion factor (OMs/Br = 30:1) derived from solvolyses of *exo* (15) and *exo* (16).

Table 4. Reactivity of 7-Thia- and 7-Oxa-norbornane Derivatives in Solvolysis Reactions

Substrate	Solvent[a]	k (25 °C)[b]	k_{exo}/k_{endo}
exo-2-Chloro-7-thianorbornane (17)[c]	50 D	2.60×10^{-10}	
	AcOH	3.80×10^{-14d}	$<2.11 \times 10^{-10}$
endo-2-Chloro-7-thianorbornane (18)[c]	AcOH	$>1.8 \times 10^{-4}$	
exo-2-Chloro-7-oxanorbornane (19)[e]	50 D	3.8×10^{-11}	316
endo-2-Chloro-7-oxanorbornane (20)[e]	50 D	1.2×10^{-13}	
exo-2-Chloronorbornane (21)[f]	80 E	3.9×10^{-5g}	65^g
endo-2-Chloronorbornane (22)[f]	80 E	6×10^{-7g}	

[a] 50 D = 50% aq. dioxane; 80 E = ethanol/water, 80:20 (v/v).
[b] Values at 25° are extropolated from rates at higher temperatures.
[c] Reference 16.
[d] Calculated using $m = 1.28$ in the Grunwald–Winstein equation.
[e] Reference 18.
[f] Reference 19.
[g] Rates at 85 °C.

yield a product with 100% retention of stereochemistry. It is interesting to note that the oxygen compound also gave a high degree (95%) of retained stereochemistry, indicating participation by oxygen.

Studies of the parent 7-thianorbornyl derivatives (17) and (18) proved to be even more interesting[16] (Table 4). While the acetolysis of (17) appears to have proceeded in much the same manner as the solvolysis of the exo-7-oxa derivative (19) (see p. 153)[18], the endo-7-thia derivative (18) behaves remark-

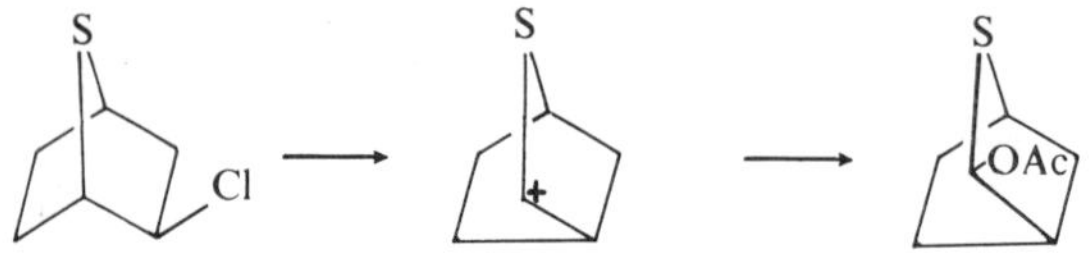

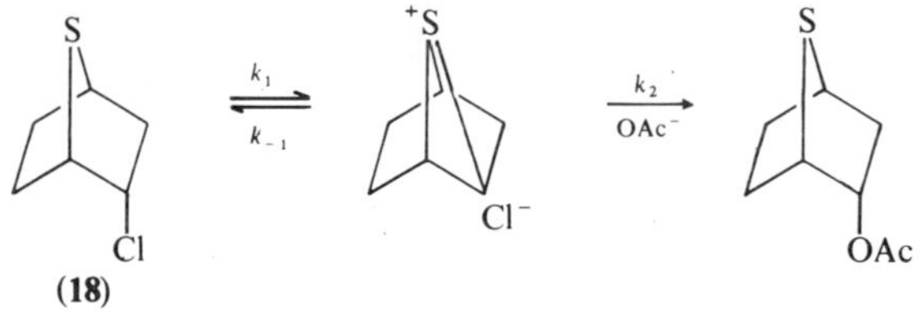

ably differently from its oxygen analog (20). The rates of solvolysis of the epimeric heterobicyclic chlorides are compared with the epimeric 2-

norbornyl chlorides (**21**) and (**22**) in Table 4. Instead of a $k_{exo}/k_{endo} \gg 1$, as is typical with norbornyl derivatives and the 7-oxa derivatives, the acetolysis of (**17**) and (**18**), like the steroidal derivatives above, gives an *exo/endo* rate ratio indicative of a strongly S-3-assisted chloride displacement in the *endo* isomer (**18**). The sulfur particpation is so efficient in (**18**) that the kinetics upon acetolysis were found to be second order, $v = k_2[\text{RCl}]$ [NaOAc]. Hence the *endo* derivative (**18**) solvolyzes at least 4.7×10^{-9} times faster than the *exo* derivative (**17**).[16] The course of the acetolysis of (**18**) is shown in Eq. (7); the rate-determining step is destruction of the sulfonium ion (**23**) by acetate ion with ion-pair return (k_{-1}) competing. At molar concentrations of sodium acetate up to 1.0 M it was found[16] that $k_{-1}[\text{Cl}^-] > k_1 > k_2[\text{OAc}^-]$ [Eq. (7)].

The driving force for participation also more than offsets the transition-state strain in the case of 2-chloromethylthiirane (**24**). The acetolysis of (**24**) proceeds 1000 times faster than cyclopropylmethyl chloride and gives, as one product, 3-acetoxythietane[20] *via* the sulfonium ion (**25**) [Eq. (8)]. The oxygen analog of (**24**), epichlorohydrin, reacts 3 times as fast as cyclopropylmethyl chloride,[21] which is already accelerated relative to primary derivatives (see Chapter 2).

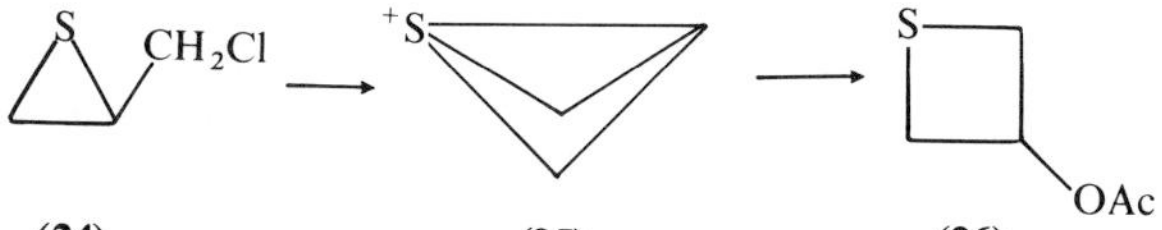

The interesting observation of isotope scrambling in the transformation of (**27**) → (**28**) is a final example of solvolytic S-3 participation in thioethers.[22] Such scrambling probably results from rearrangement *via*

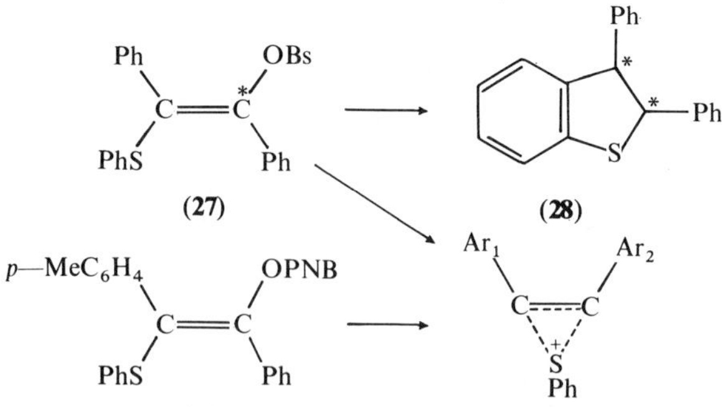

transition state (**29**), $Ar_1 = Ar_2 = Ph$, since (**29**), $Ar_1 = p$-tolyl and $Ar_2 = Ph$, has been implicated from solvolysis of (**30**).[23] An intermediate such as (**29**) is not consistent with the data.

While S-3 participation in carbocation reactions is common, cyclopropylcarbinyl cations such as (**31**) preferentially lead to ring enlargement, and a synthetic procedure for obtaining cyclobutanones has been devised based on this observation.[24] Thus the tertiary alcohol (**32**) is ether, upon treatment with aqueous fluoboric acid, gives the spiro ketone (**33**) quantitatively.

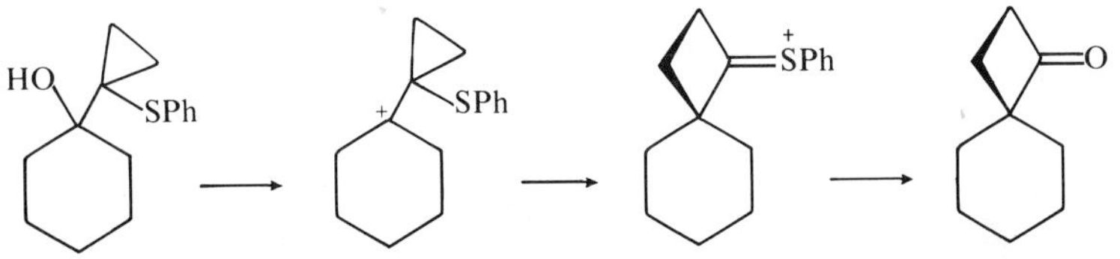

(10)

(**32**) (**31**) (**33**)

The kinetic data in Table 1 demonstrates that S-4 anchimeric assistance is not important in the hydrolysis of acyclic alkyl chlorides [e.g., n-hexyl chloride solvolyzes 1.8 times faster than $PhS(CH_2)_3Cl$ in 20 mole % aqueous dioxane at $100°C^3$]. Some data on cyclic systems will be presented here to demonstrate that proximity effects can drastically affect S-4 participation. The formation of (**35**) and (**36**) from treatment of 3β-methanesulfonyloxy-5-α-methylthiocholestane (**34**) with tetrapropylammonium acetate was rationalized in terms of S-4 participation as shown.[25] Ireland and Smith[26] found that the rates of solvolysis of (**37**) and (**38**) were slower [$k_{rel}(77°C) = 0.46$ and 0.21, respectively] than the rate for *trans*-4-t-butyl-

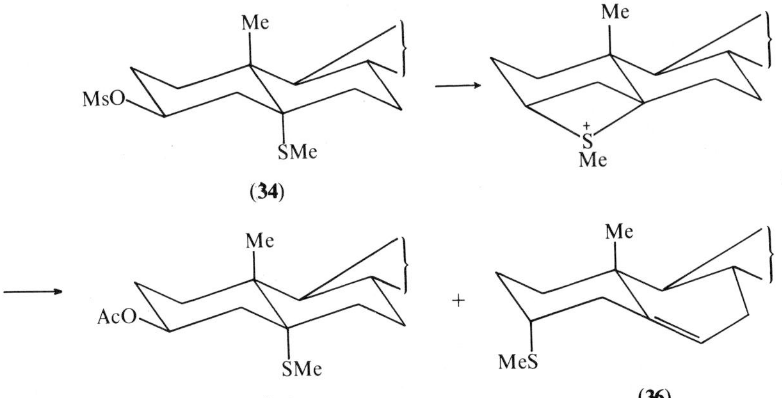

(11)

(**34**)

(**35**) (**36**)

Table 5. Relative Rates of Solvolysis in
Buffered Acetic Acid at 25°C[27]

Compound	(40)	(41)	(42)	(43)	(44)
Relative rate	5564	1.8	32.2	8.4	1

cyclohexyl toluene-*p*-sulfonate $[k_{rel}(77°C) = 1.00]$. Although anchimeric assistance by sulfur is not indicated by the kinetic results, the product studies (only *endo*-alcohol from either toluene-*p*-sulfonate) were interpreted as implying that sulfonium ion (**39**) probably intervened after the transition state for ionization.

While systems (**34**) and (**37**) may contain sufficient conformational mobility to prevent significant anchimeric assistance, Paquette and his co-workers[27] have investigated the rigid systems (**40**)–(**43**) with interesting results. No significant differences were seen in the relative rates (see Table 5) of acetolysis of (**41**)–(**43**) as compared to the bishomocubyl derivative

(**44**). These were judged to react with carbon participation. The *exo*-ester in (**40**), however, shows a considerably enhanced rate relative to the others and gives the acetate with retained configuration [i.e., (**40**). Y = H, X = OAc]. This was attributed to anchimeric assistance by sulfur, leading directly to the sulfonium ion (**45**). The addition of a methylene group to (**42**), i.e., changing (**42**) to (**40**), is said to move the sulfur atom closer to the carbon

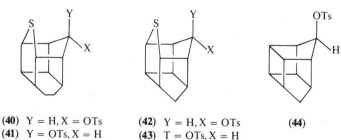

(**40**) Y = H, X = OTs
(**41**) Y = OTs, X = H

(**42**) Y = H, X = OTs
(**43**) T = OTs, X = H

(**44**)

Participation by Sulfur Groups

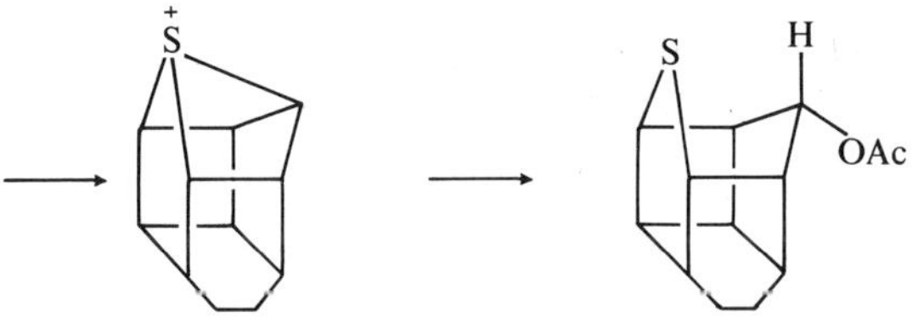

$$(40) \longrightarrow \underset{(45)}{[\text{structure}]} \longrightarrow [\text{structure with OAc}] \qquad (12)$$

bearing the leaving group.[27] Thus the differences between the anchimeric assistance afforded by sulfur in (40) and (42) were attributed to the internuclear distance between the neighboring group and the developing electron-deficient center. The product from (42) could be accounted for by invoking only carbon participation. One may conclude either that carbon participation in (42), relative to (40), has increased in importance or that sulfur participation has decreased in importance. Since the driving force for sulfur participation is generally much greater than that for carbon participation, it seems that the latter explanation is more than likely correct.

S-4 participation has also been implicated in the racemization of (46a) upon heating in methyl ethyl ketone.[28] Since the deuterated derivative (46b) produced a mixture of (46b) and (47b) under the same conditions, the reaction may involve either ion-pair return from (48) or the four-centered transition state (49).

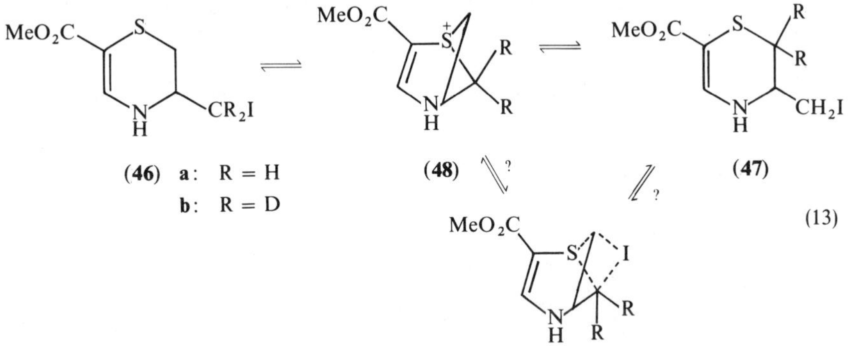

$$[\text{structures of (46), (48), (47), (49)}] \qquad (13)$$

(46) a: R = H
     b: R = D

R_2S-5 anchimeric assistance has now been documented in a variety of systems. Although (50a) has a flexible sulfur-containing ring, Gratz and Wilder[29] observed that it solvolyzes 955 times faster than its isomer (52a) and 750 times faster than the carbon analog (50b). The product is (50c), indicating that the sulfonium ion (51a) is formed. The lack of products of rearrangement was also taken as evidence against competing carbon participation. By comparison, the oxygen derivative (**50d**) solvolyzes 50

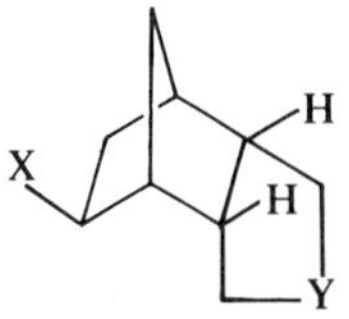

(**50**) a: X = OPNB, Y = S
     b: X = OPNB, Y = CH_2
     c: X = OH, Y = S
     d: X = OMs, Y = O

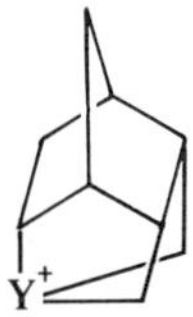

(**51**) a: Y = S
     b: Y = O

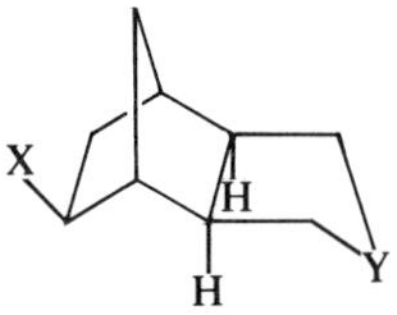

(**52**) a: X = OPNB, Y = S
     b: X = OMs, Y = O

times faster than the *exo* isomer (**52b**) and gives predominant retention with *ca.* 10% rearrangement.[30] Hence (**51b**) is likely formed by oxygen participation, but carbon participation may be competing. An intermediate similar to (**51a**) also forms in certain electrophilic addition reactions (see Section 5.1.2).

Transannular sulfur participation occurs in certain solvolytic rearrangement reactions. Hence under pinacol rearrangement conditions either glycol (**53**) or (**54**) is converted to the bicyclic derivative (**55**).[31] A possible mechanism for the formation of (**55**) from (**53**) is shown in Scheme 1. There appears to be no simple pathway to (**55**) from (**54**). Three possible pathways leading to the necessary ring expansion are shown in Scheme 1.

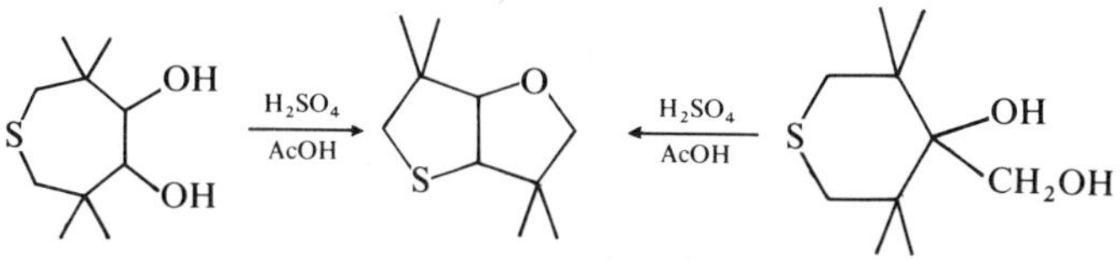

$$(53) \xrightarrow[\text{AcOH}]{H_2SO_4} (55) \xleftarrow[\text{AcOH}]{H_2SO_4} (54) \quad (14)$$

Breslow and his co-workers[32] have investigated multiple neighboring groups about benzylic cations. Unlike the corresponding oxygen compound, which was noncyclic, the triarylmethanol (**56**) yields the cyclic structure (**57**) in liquid SO_2. The derivative with two neighboring sulfur groups also cyclized. At 10°C the NMR data revealed one coordinated and one uncoordinated sulfur group as in (**58**). At higher temperatures, however, the methyl and methylene peaks coalesce, indicating a fluctional structure. An S_N1 mechanism for the process was proposed. A donating group on one

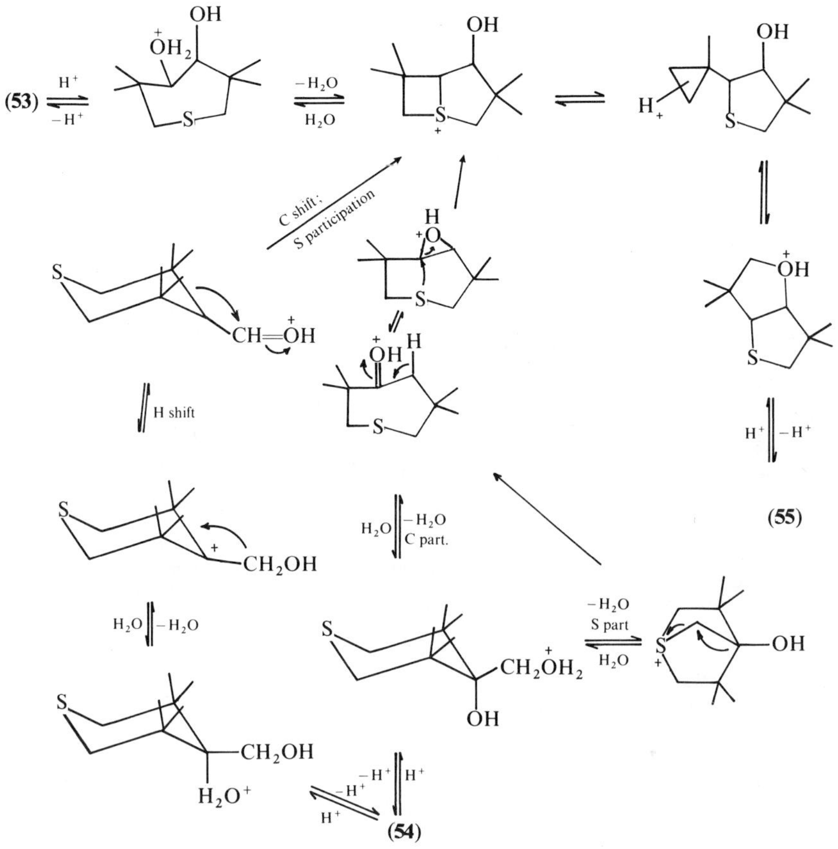

Scheme 1

of the rings, such as *p*-methoxy, stabilized the triarylcarbonium ion relative to the sulfonium ion.

Martin and Basalay[33] have studied similar degenerate rearrangements with somewhat different conclusions. 2-Thioniaaceanthrenes (**60**) formed upon generation of a benzylic cation (**59**). These sulfonium ions were

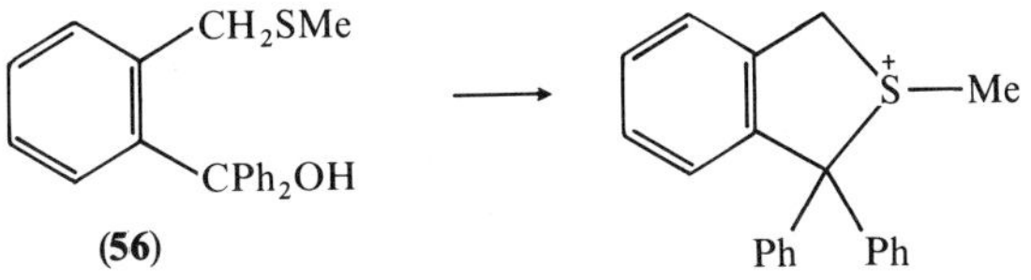

(15)

found to undergo a degenerate rearrangement, called the *bell-clapper rearrangement*. Support for an S_N2 mechanism with the transition state

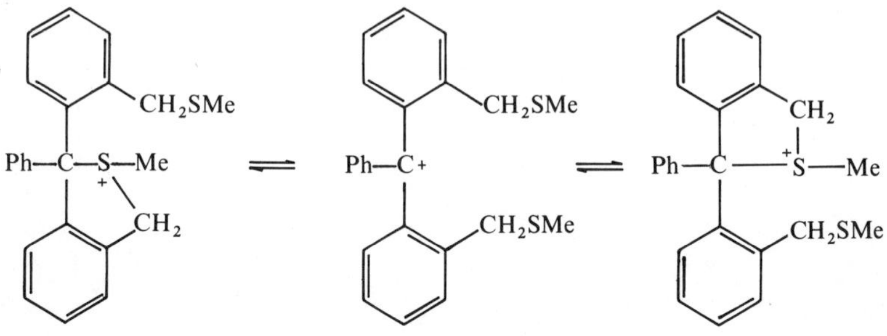

(58)   (16)

(**61**) was supported by the observations that the rearrangement rate was insensitive to substitutuion for R_1 or R_2 in (**59**), to gegenion nucleophilicity, and to substitution for R_3 in (**59**).

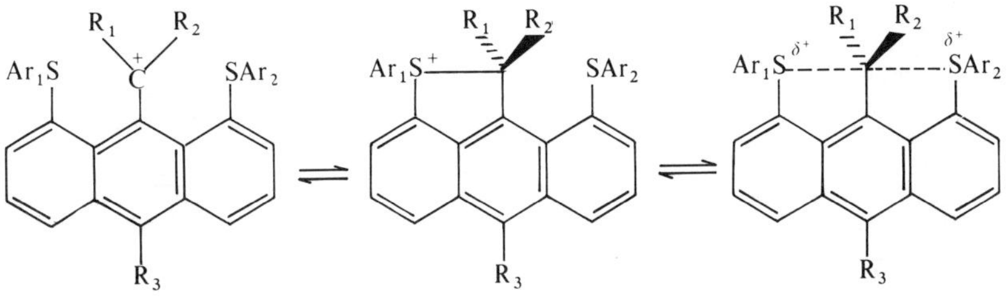

(**59**)   (**60**)   (**61**)

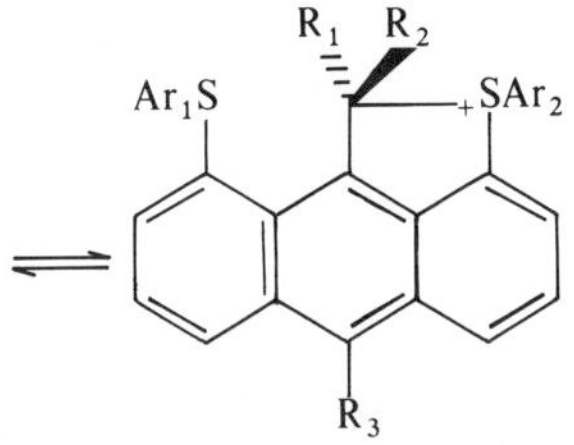

(17)

A number of examples of sulfur participation by dithioacetals and similar derivatives have been reported. The available examples suggest that such groups are quite efficient. The benzylic chloride (**62a**) undergoes acetolysis 220 times faster than its *para* isomer and gives the acetate (**62c**) as the product. S-5 participation is suggested, and structures (**63**) or (**64**) were proposed as intermediates.[34] When the toluene-*p*-sulfonate (**62b**) was acetolyzed, a stable salt of unknown structure was formed. Although the NMR spectrum of the salt was said to be independent of the anion, no NMR data were given. Presumably the salt is the sulfonium toluene-*p*-sulfonate. Participation by an *o*-MeS (S-4) or *o*-MeO (O-4) was not observed on the acetolysis of the respective substituted benzyl chlorides.[34]

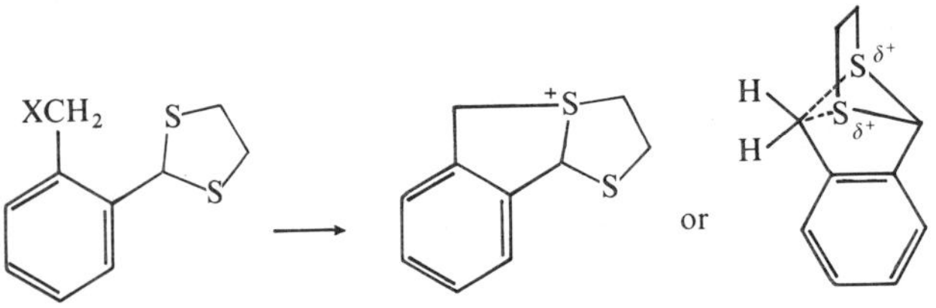

(62) → (63) or (64)   (18)

a: X = Cl
b: X = OTs
c: X = OAc

Hughes and his co-workers[35] have examined S-6 participation in carbohydrate derivatives. This type of effect readily occurs in the solvolysis of 5-O-toluene-*p*-sulfonyl arabinose diethyl thioacetal (65) in aqueous acetone with added barium carbonate as an acid scavenger. Since note has been made earlier of hydroxyl group participation under similar conditions, the S-6 assistance observed in these examples should be quite significant owing to the stronger driving force for sulfur participation.

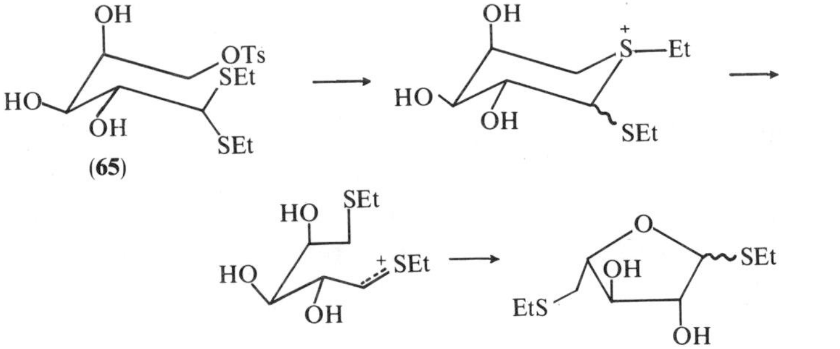

(65) →   (19)

Participation by sulfur has also been noted in attempts to hydrolyze the imino chloride functionality of penicillin intermediates, e.g., as in (66), to amide groups. Treatment of (66) with silver nitrate or aqueous acid resulted in the formation of three products, (67)–(69). Their formation can be rationalized as shown in Eq. (20).[36]

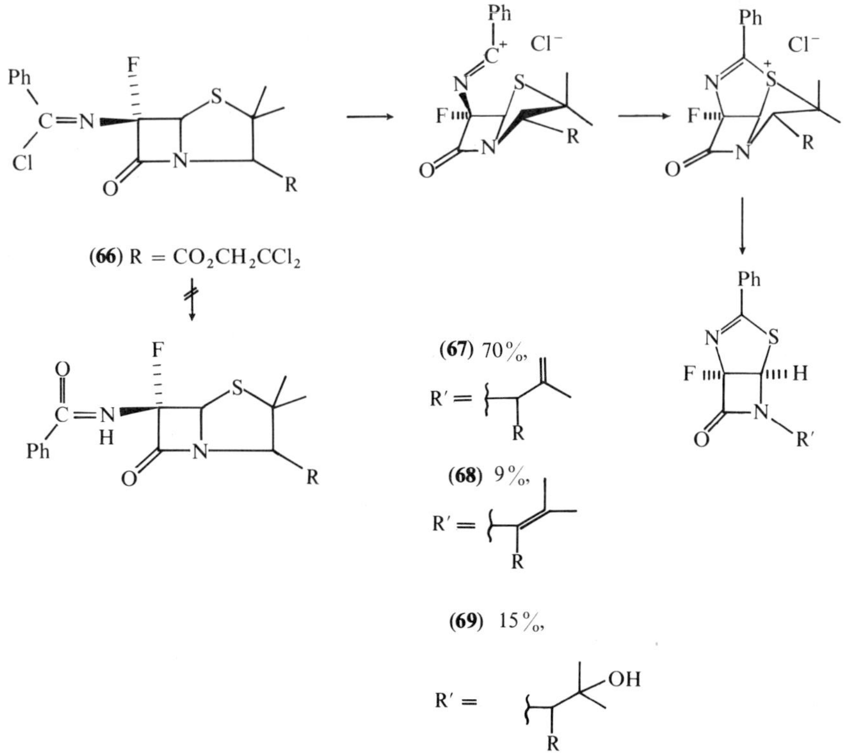

(66) R = CO$_2$CH$_2$CCl$_2$

(67) 70%, R' = [structure]

(68) 9%, R' = [structure]

(69) 15%, R' = [structure with OH]

5.1.2 In Electrophilic Addition Reactions

Although there are no unequivocal examples of anchimeric assistance by neighboring sulfur groups in radical addition reactions,[37,38] sulfur participation in electrophilic addition reactions is well known. With few exceptions, the cyclization reactions most often observed involve S-5 participation.

Kwart and Drayer[38] have studied the addition of several electrophilic reagents to *o*-allylthioanisole and conclude that the extent of MeS-5 anchimeric assistance varies with the reactant/solvent in the following qualitative order: iodination in acetic acid ≈ hydrobromination in aprotic solvent ≫ formic acid addition. No evidence was found for MeS-5 anchimeric assistance in the addition of arylsulfenyl chlorides to *o*-allylthioanisole.[38] This is contrary to some studies discussed below but consistent with the view that the ArSCl additions occur with little or no charge distribution to carbon.[8,39,40]

Iodination of unactivated double bonds is normally not observed and is not observed with o-allylanisole in acetic acid. In sharp contrast, within 3 min after admixture of an acetic acid solution of o-allylthioanisole (**70**) and iodine both reactants were nearly quantitatively consumed[38]; for a review of iodination–cyclization reactions, see Ref. 41. Hence anchimeric assistance by the neighboring sulfur group must be strong. The product of iodination is the sulfonium salt (**71**) which slowly reverts to (**70**) and iodine upon standing. Staninets et al.[42,43] who observed sulfur participation upon iodination of 4-methyl-thio-1-pentene (**72a**) [Eq. (22)], reported that the rate of iodination of (**72a**) at 20°C is some 8 powers of 10 slower than that of the

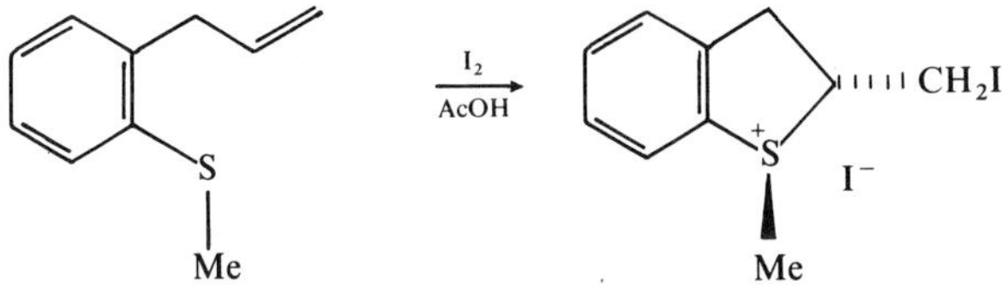

$$\text{(70)} \xrightarrow[\text{AcOH}]{I_2} \text{(71)} \tag{21}$$

analogous amine (**72b**). While no direct rate comparison with the analogous ether is available 4-penten-1-ol reacts nearly 500 times faster with iodine than

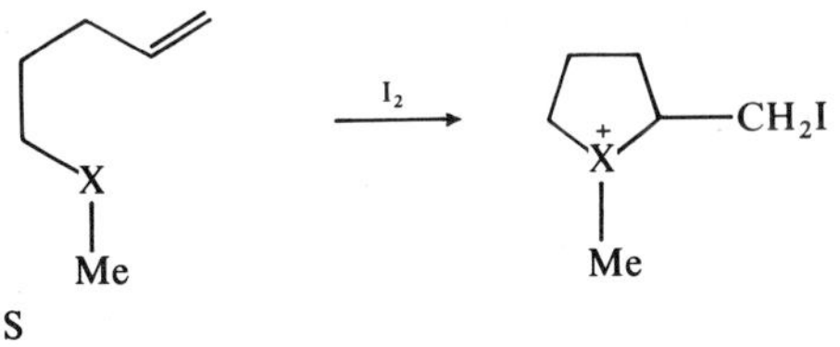

$$\xrightarrow{I_2} \tag{22}$$

(**72a**) X = S
(**72b**) X = NH

does (**72a**). Hence a wide variation in reactivity exists in these reactions, and it is apparent that factors other than nucleophilicity are important here.[41]

There is also evidence for sulfur participation in the addition of hydrogen halides and other acids. The products of reaction of the enolthioether (**73**) with aqueous perchloric acid suggest the intervention of the sulfonium ion (**75**) and its reaction (hydride abstraction) with excess (**73**).[44] Sulfur participation most likely occurs after the formation of the resonance-stabilized benzylic cation (**74**).

The addition of formic acid to *o*-allylthioanisole (**70**) proceeds with mild anchimeric assistance on the basis of rate and product studies. The major product is that of cyclization by MeS-5 participation (9:1, *trans/cis*).

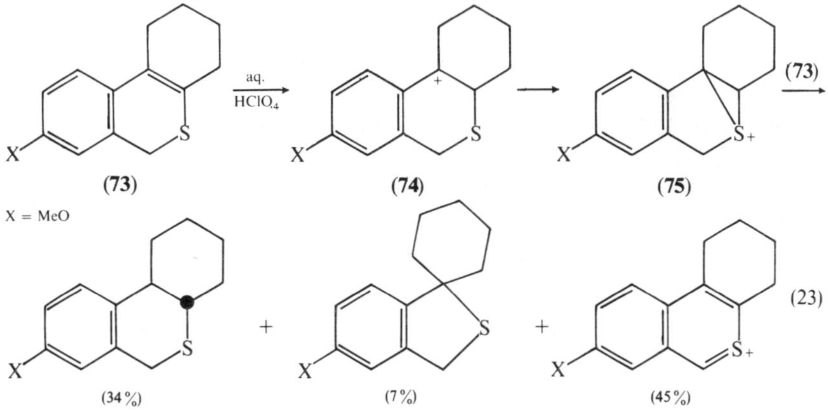

(**73**) (**74**) (**75**) (**73**)

X = MeO

(34%) (7%) (45%) (**23**)

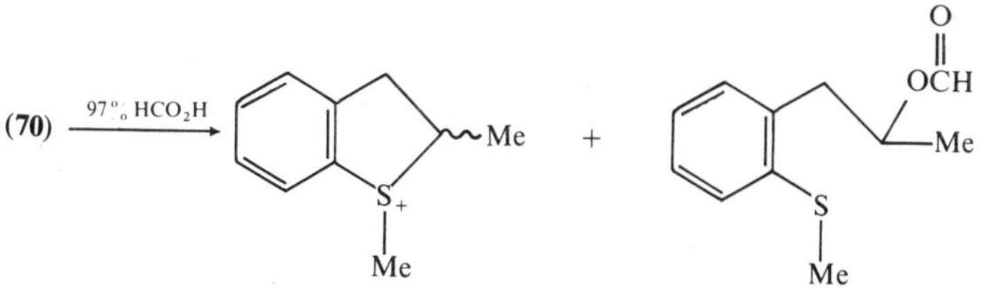

(**70**) 97% HCO₂H → ...—Me + ...—Me (**24**)

A small amount of the formate ester also resulted [Eq. (24)].[38] Interestingly, *o*-allylanisole reacts faster than (**70**), but it gives only the noncyclic addition product. This observation is said to result because of the superiority of Ar_1-3 participation as compared to MeO-5 participation.[38,45] In the case of (**70**), however, MeS-5 participation dominates. The reversal in reactivity with (**70**) can be accounted for by the greater nucleophilicity of sulfur relative to oxygen, thus favoring MeS-5 over MeO-5, and by the greater relative π-bonding ability of oxygen, which increases Ar_1-3 relative to MeO-5. The poor π-bonding ability of sulfur decreases the effectiveness of Ar_1-3 in (**70**).

In the polar hydrobromination of (**70**), only a mixture of the cyclic sulfonium bromides was formed [Eq. (25)].[38] It was not determined if anchimeric assistance was involved, but the finding of apparent simple addition[46] in the case of *o*-allylanisole suggests a difference. Alternatively, the cyclic sulfonium salts would be expected to survive longer than the analogous oxonium ions, and hence the difference may be fortuitous; that

Participation by Sulfur Groups

is, both reactions may involve either anchimerically assisted or unassisted addition–cyclization, but only the oxonium ion [e.g., (**76**)] suffers destruction by the bromide ion.

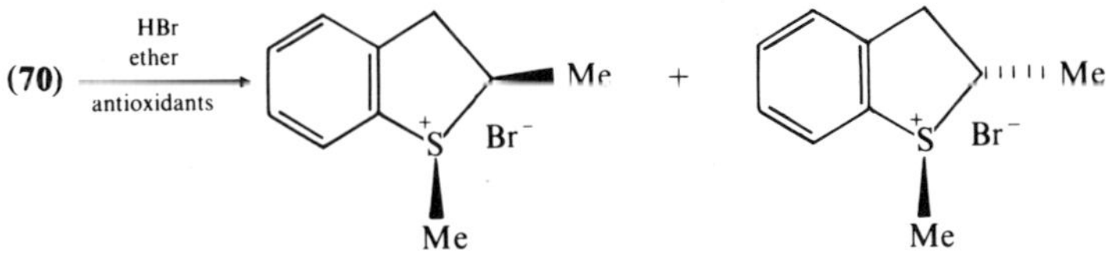

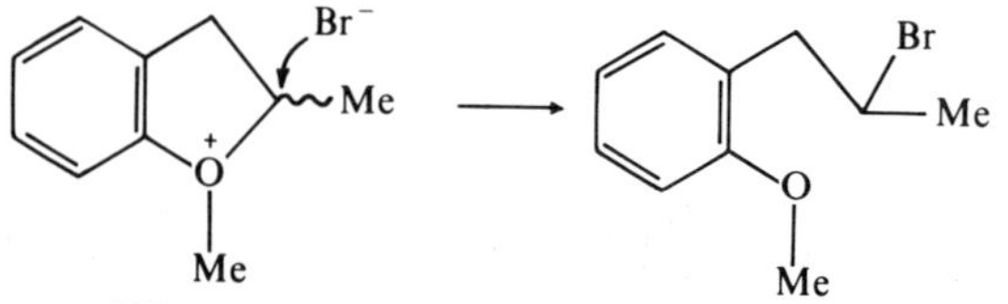

$$(70) \xrightarrow[\text{antioxidants}]{\text{HBr, ether}} \text{[sulfonium intermediate]} \longrightarrow \text{Me} + \text{Me} \quad (25)$$

$$(76) \longrightarrow \text{products} \quad (26)$$

The hydrobromination of (**77**)[47] under somewhat different conditions may involve the free carbocation (**78**) since S-6 participation occurs. Although a one-step process can be envisioned, the transition state for a concerted S-6 process would appear to be of high energy (see pp. 49–58). Alternatively, a loosely coordinated carbocation (weak anchimeric assistance) could possibly be involved.

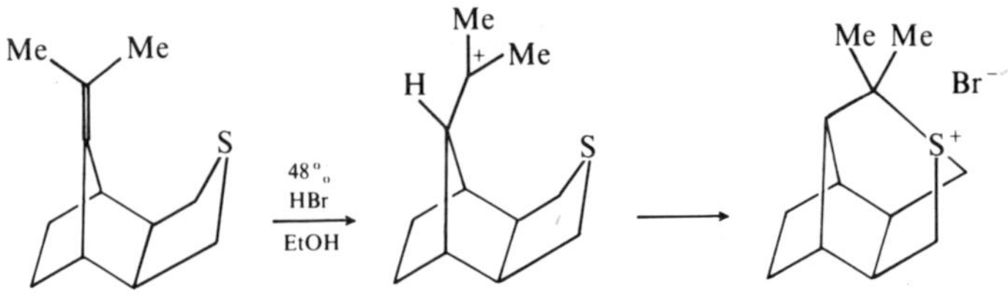

$$(77) \xrightarrow[\text{EtOH}]{48\% \text{ HBr}} (78) \longrightarrow (62\%) \quad (27)$$

Wilder and his co-workers have extensively used addition–cyclization reactions in the synthesis of heteroatom-substituted polycyclic systems. The reaction in Eq. (27) is an example. Bromination, as with 2-thia-1,2-dihydro-*endo*-dicyclopentadiene (**79**), has also been utilized. The polar bromination of (**79**) yields the isolable bromosulfonium bromide (**80**) by S-5, -6 participation.[48] When (**80**) is treated with lithium carbonate in aqueous methanol, the *exo-cis*-diol (**81**) is formed, probably by the sequence of reactions shown in Eq. (28).[49]

Raasch[40] has completed an interesting comparative study of 2-oxa-, 2-aza-, and 2-thianorbornenes in addition reactions. As a criterion for

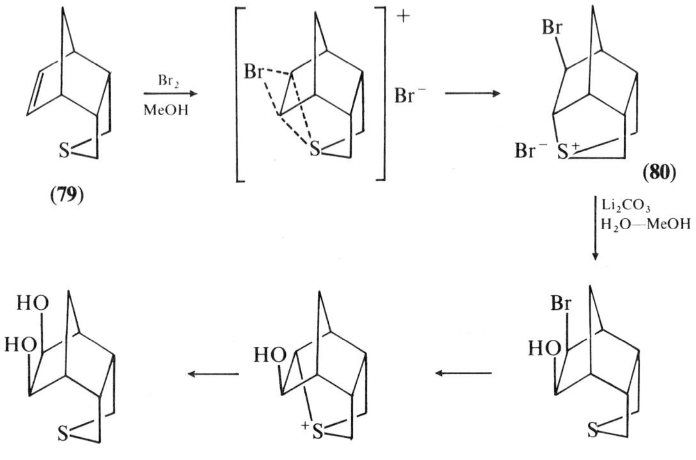

(28)

participation Raasch analyzed for rearranged versus unrearranged addition products. He reasoned that rearranged products would result, e.g., Eq. (29), if sulfur bridging occurs. Thus, instead of the *trans*-dibromide from back-side attack of bromide on (**82**), products (**82**) and (**84**), which are the result of bromide attack at C-6 and C-1, respectively, would be expected.

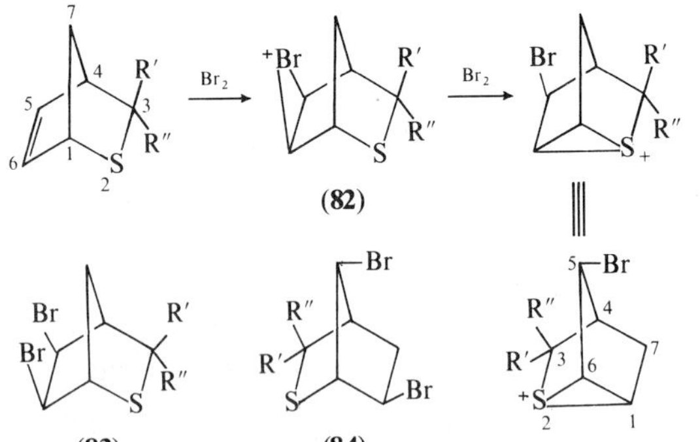

(29)

Actually, attack at C-1 is favored from both electronic and steric considerations, and hence the 6,7-dibromide should be the major product. Bromination of the 2-thia- and 2-azanorbornenes, (**85**) and (**86**), respectively, proceeded with rearrangement to the 6-7-dibromides, while the 2-oxa- (**87**) and 2-thia-2,2-dioxide (**88**) reacted, presumably, by a free-radical pathway.[40]

Participation by Sulfur Groups 215

Although studies have shown that oxygen, sulfur, and nitrogen may greatly affect the rates of solvolysis reactions, these results indicate that the electron-withdrawing groups at C-3 apparently level the participatory power of

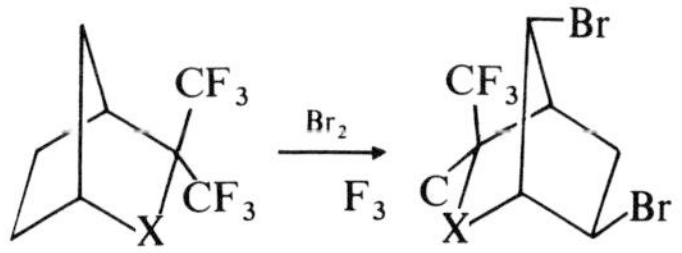

(30)

(85) X = S
(86) X = NH

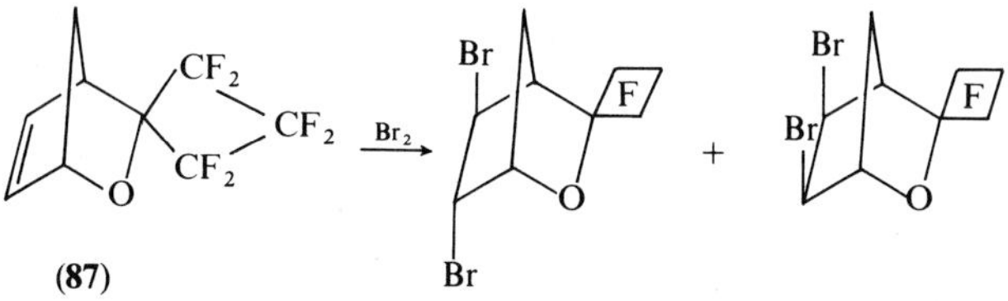

(31)

(87)

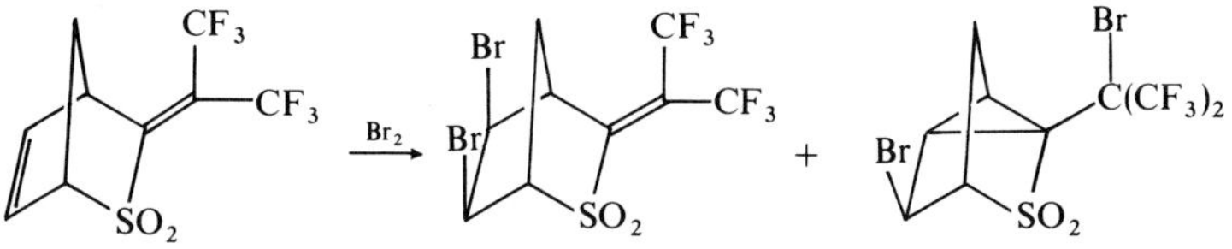

(32)

(88)

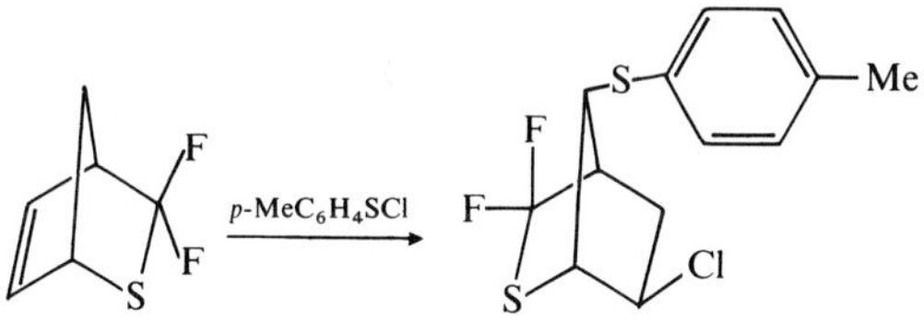

(33)

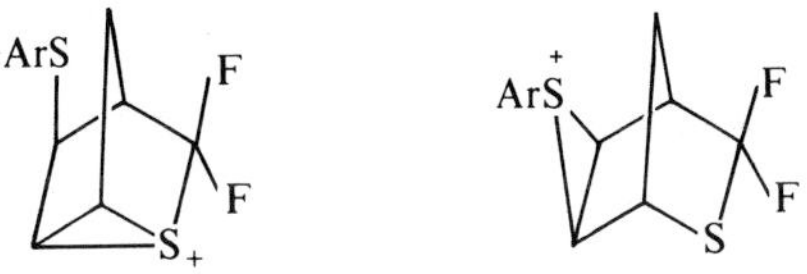

(89)   **(90)**

the oxygen atom in **(87)** but not that of the sulfur in **(85)** or of the nitrogen in **(86)**.

The addition of sulfenyl halides to 2-thianorbornenes such as **(85)** is analogous to bromination in that rearranged products are obtained, e.g.,

Eq. (33). Hence the bridged ion (**89**) and not (**90**) is product directing.

Transannular sulfur participation has also been demonstrated in a number of examples of sulfenyl halide additions to dienes. Normally,

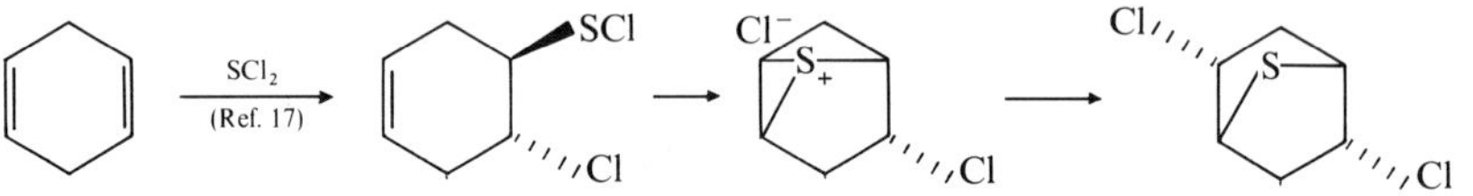

$$(34)$$

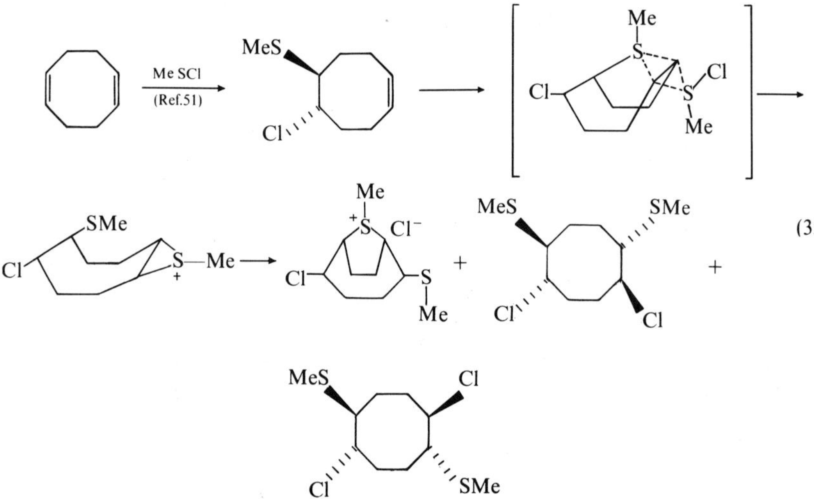

$$(35)$$

addition to the second double bond of a diene, especially conjugated dienes,[50] is retarded. Some dienes, however, appear to be activated, and transannular sulfur attack, e.g., Eq. (34), or transannular sulfur participation, e.g., Eq. (35), may occur.[8]

5.1.3 In Elimination Reactions

While there is a paucity of information on neighboring sulfur involvement in elimination reactions, the results of McCabe and Livingston[52] suggest participation in the solvolytic elimination of hydrogen chloride from the chloroalkene (**91**) and in the pyrolytic elimination of adipic acid from the adipate ester (**92**).

On the basis of an *ortho/para* rate ratio of 1.1×10^4, anchimeric assistance to elimination is also proposed in the case of the oxime ester

Participation by Sulfur Groups

(93).[53] It seems most likely that the neighboring sulfur group provides nucleophilic assistance in breaking the N—O bond. This may lead to the ion pair (94), which undergoes proton loss in a subsequent fast step [path (a), Eq. (37)]. Alternatively, extension of the N—O bond may sufficiently strengthen the basicity of the carbonyl group that the cyclic process is facilitated [path (b), Eq. (37)].

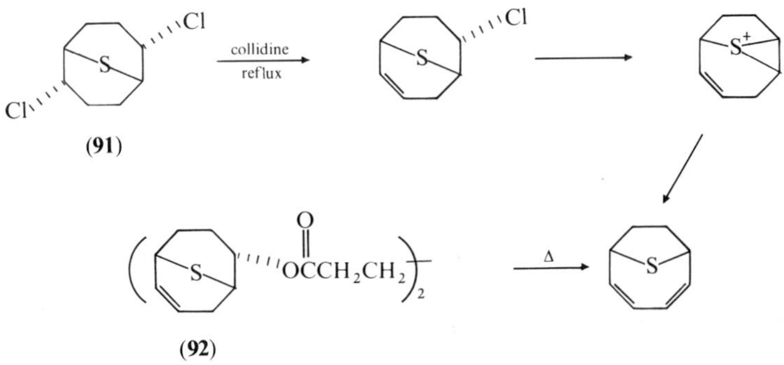

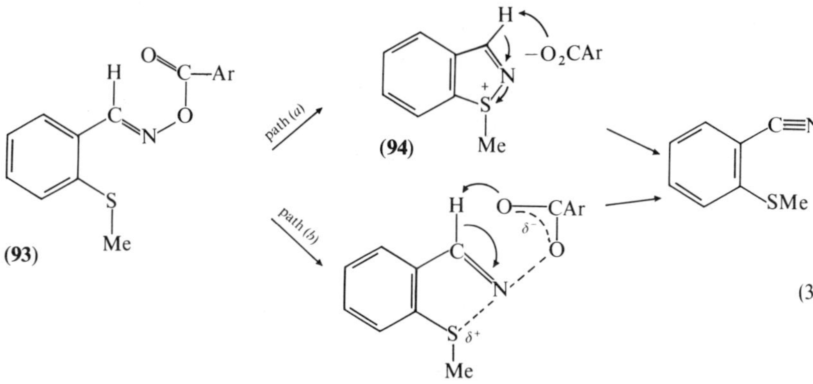

The reversibility of sulfenyl chloride addition allows exchange with another alkene, leading to elimination from the initial adduct.[54] The rates of the exchange reactions vary considerably, and when an excess of an olefin that undergoes slow elimination is present with an adduct that eliminates rapidly near-quantitative exchange may occur. An appropriate exchange pair was found to be 2-chlorocyclooctyl 4-chlorophenyl sulfide (elimination half-life = 1.3 hr) and cyclopentene (elimination half-life = 400 hr) [Eq. (38)].

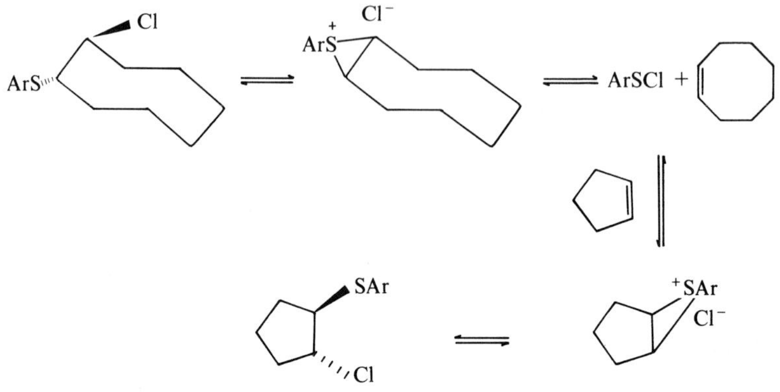

$$\text{(38)}$$

5.1.4. In Free-Radical, Carbene, and Photochemical Reactions

Martin and his co-workers have provided a firm experimental foundation for anchimeric assistance in the homolytic decomposition of perbenzoate derivatives[55-57] (Table 6). A single o-phenylthio group was found to increase the rate of decomposition of t-butyl perbenzoate by a factor of ca. 45 thousand.[55] The effect is not steric since an o-t-butyl group hardly affects the rate of perester decomposition [compare (**95**) versus (**99**), Table 6]. A noticeable but weak effect is observed with the homologous sulfide (**96**). In addition to sulfur, anchimeric assistance by iodine and vinyl groups in (**97**) and (**98**), respectively, was observed. The bridged canonical species

Table 6. Relative Rates and Activation Parameters for the Decomposition of Mono- and Disubstituted t-Butyl Perbenzoates in Chlorobenzene[55-57]

Substituents	k_{rel} (40°C)	$\Delta H^{\ddagger}$ (kcal mole^{-1})	$\Delta S^{\ddagger}$ (e.u.)
None	1	34.1	10.0
o-Me$_3$C (**95**)	2.4	34.4	12.5
o-PhSCH$_2$ (**96**)	5.6	32.2	7.2
o-I (**97**)	80.5	28.0	−0.8
o-Ph$_2$C=CH (**98**)	1.53 × 10^2	27.3	−5.0
o-PhS (**99**)	6.53 × 10^4	23.0	−3.4
2,6-Di-PhS (**100**)	4.31 × 10^4	—	—
2-PhS, 3-Me$_3$COOCO (**101**)	1.88 × 10^6	17.6	−13.9
2-I, 3-Me$_3$COOCO (**102**)	2.52 × 10^3	22.4	−11.9

Participation by Sulfur Groups 219

(**103**) and (**104**) were thought to be involved in the decomposition of (**99**).[55,56]

Although it was found that a second *o*-phenylthio group [e.g., (**100**)] added no net stablization to the decomposition transition state,[56] the

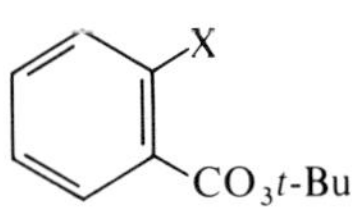

(**95**) X = *t*-Bu
(**96**) X = PhSCH$_2$
(**97**) X = I
(**98**) X = Ph$_2$C=CH
(**99**) X = PhS

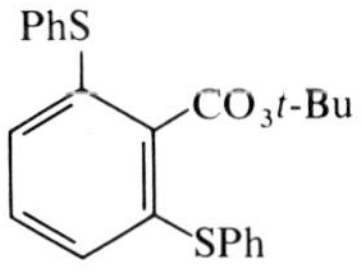

(**100**)

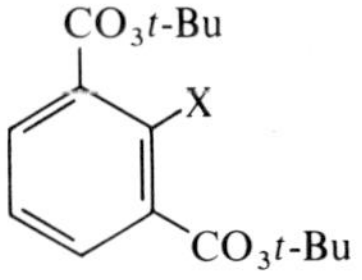

(**101**) X = PhS
(**102**) X = I

interesting observation has been made[57] that a single phenylthio or iodo group *ortho* to two perbenzoate groups simultaneously accelerates the decomposition of both perester groups. Not only is there an enhanced degree of anchimeric assistance [compare (**97**), (**99**), (**101**), and (**102**), Table 6], but Martin and Chau[57] have isolated the sulfurane (**105**) and the iodosodilactone (**106**). The mechanism shown in Eq. (40) was proposed.

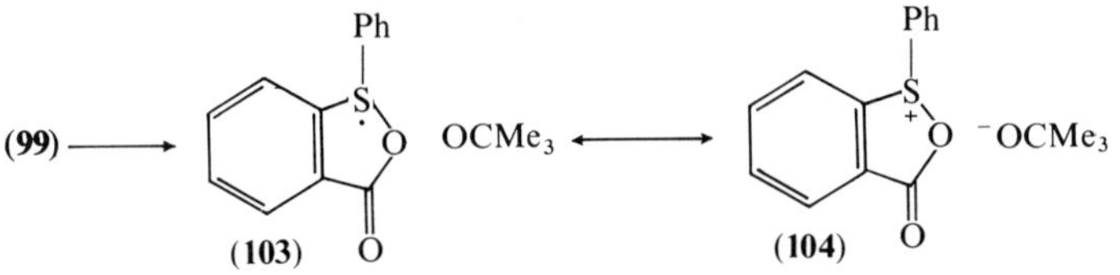

(**99**) ⟶ (**103**) ⇌ (**104**)   (39)

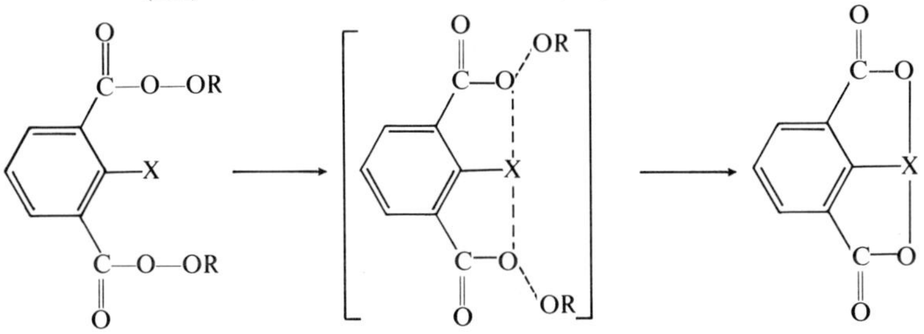

(**101**) X = PhS, R = *t*-Bu
(**102**) X = I, R = *t*-Bu

(**105**) X = PhS
(**106**) X = I

(40)

Sulfonium ions have been suggested as intermediates in the photolysis of cyclic γ-keto sulfides on the basis of product studies.[58] For example, the irradiation of thiacyclohexan-4-one (**107**) in *t*-butyl alcohol gave the products shown in Eq. (41). The formation of thiacyclobutan-2-one and ethylene can be accounted for by invoking S-4 participation as shown. Participation has also been observed in the photolytic reactions of α-dione sulfides,[59] but a search for RS-4 interactions with the carbene formed upon pyrolysis of an alkyl sulfide tosylhydrazone salt failed to provide any positive evidence.[60]

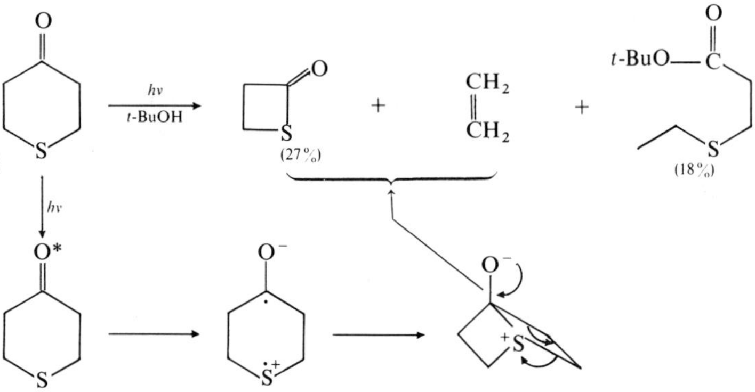

(41)

5.2 Thiol Groups

Ionized and un-ionized thiol groups generally give the same reactions as their oxygen analogs, although sulfur appears to be a more powerful neighboring group, as expected. The alkaline hydrolysis of (108) is an interesting example. The reaction apparently proceeds with S-3 displacement to form (109), which follows to (111) presumably via (110).[61]

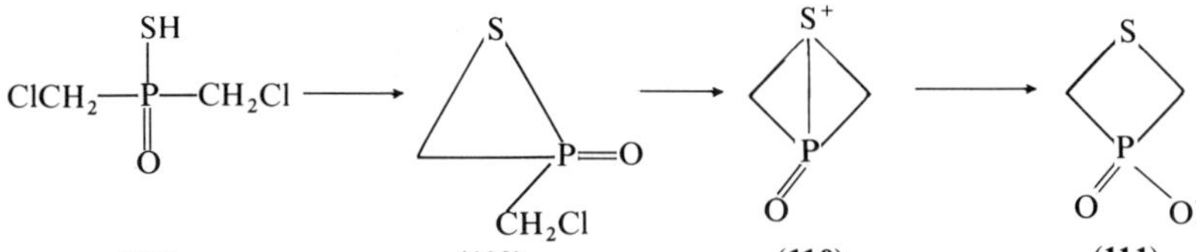

(42)

HS-3 participation was found to be more effective than HO-3 participation by a comparison of the results of solvolysis of cis- and trans-2-chlorocyclohexanols and 2-chlorohexanethiols.[62] As with the oxygen compounds, both cis- and trans-2-chlorocyclohexanethiol were found to react with enhanced rates, presumably by hydrogen participation [Eq. (43)], and sulfur participation [Eq. (44)]. The *cis* isomer reacts less rapidly than *cis*-2-chlorocyclohexanol, suggesting that the driving force for hydrogen participation (or elimination) is reduced. The *trans* isomer, however, is considerably faster than the corresponding oxygen analog (see p. 167).

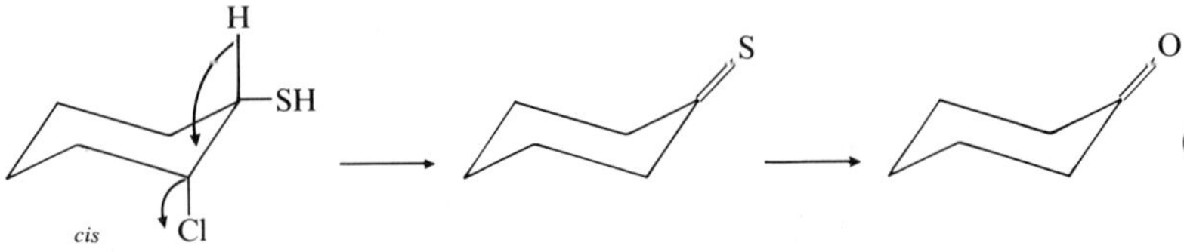

(43)

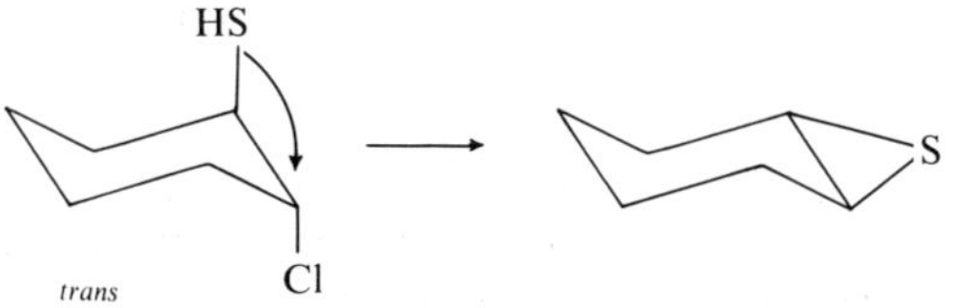

(44)

2-Mercaptoethanol was incorporated into the Boc removal solution for tryptophan protection in a Merrifield solid-phase synthesis of hen egg white lysozyme, but the reagent proved unsatisfactory.[63] The problem with the reagent was traced to its polymerization, presumably by HS-3 participation [Eq. (45) and Scheme 2].

$$n\text{-HOCH}_2\text{CH}_2\text{SH} \xrightarrow{H^+} -(CH_2CH_2S)_n + (n-1)\text{-H}_2O \quad (45)$$

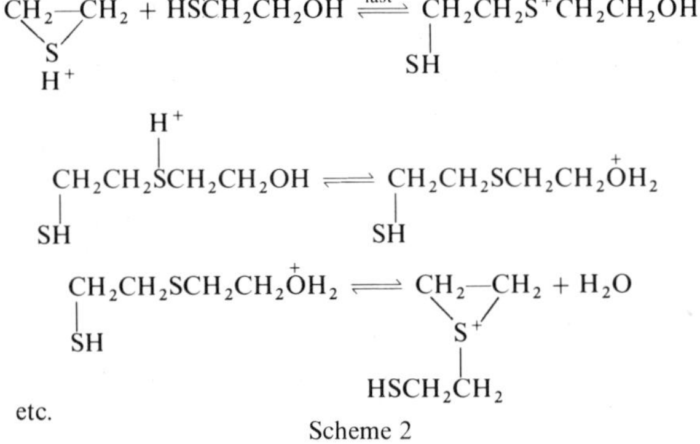

$$\text{HSCH}_2\text{CH}_2\text{OH} \rightleftharpoons \text{HSCH}_2\text{CH}_2\overset{+}{\text{OH}}_2$$

$$\underset{\text{HS}}{\overset{\text{CH}_2-\text{CH}_2}{\diagup\diagdown}}\overset{+}{\text{OH}_2} \overset{\text{slow}}{\rightleftharpoons} \underset{\underset{H^+}{S}}{\overset{\text{CH}_2-\text{CH}_2}{\diagup\diagdown}} + \text{H}_2\text{O}$$

$$\underset{\underset{H^+}{S}}{\overset{\text{CH}_2-\text{CH}_2}{\diagup\diagdown}} + \text{HSCH}_2\text{CH}_2\text{OH} \overset{\text{fast}}{\rightleftharpoons} \underset{\underset{\text{SH}}{|}}{\overset{\overset{H}{|}}{\text{CH}_2\text{CH}_2\overset{+}{\text{S}}\text{CH}_2\text{CH}_2\text{OH}}}$$

$$\underset{\underset{\text{SH}}{|}}{\overset{\overset{H^+}{|}}{\text{CH}_2\text{CH}_2\text{SCH}_2\text{CH}_2\text{OH}}} \rightleftharpoons \underset{\underset{\text{SH}}{|}}{\text{CH}_2\text{CH}_2\text{SCH}_2\text{CH}_2\overset{+}{\text{OH}}_2}$$

$$\underset{\underset{\text{SH}}{|}}{\text{CH}_2\text{CH}_2\text{SCH}_2\text{CH}_2\overset{+}{\text{OH}}_2} \rightleftharpoons \underset{\underset{\text{HSCH}_2\text{CH}_2}{|}}{\overset{\text{CH}_2-\text{CH}_2}{\diagup\diagdown}\overset{+}{S}} + \text{H}_2\text{O}$$

etc.

Scheme 2

Thietanes can be prepared by reactions involving neighboring sulfur groups. For example, the D-mannitol derivative (112) is converted into the interesting tricyclic D-iditol derivative (113) with methanolic sodium methoxide.[64]

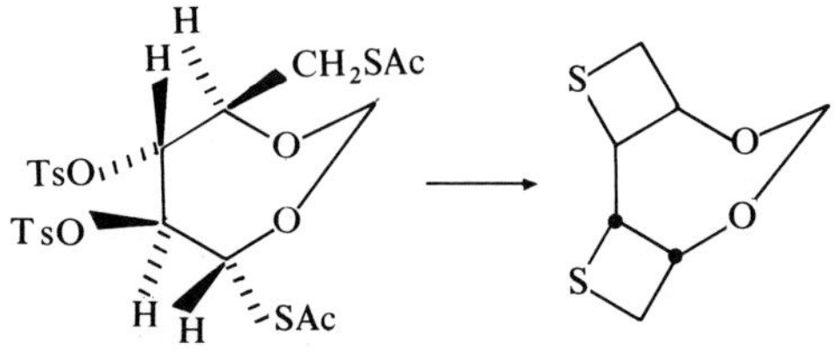

(46)

S-5 participation also occurs readily. Thus a sulfur bridge results on treatment of 6-O-toluene-p-sulfonyl-2,3,4-tri-O-acetyl-β-D-glucopyranosyl ethyl xanthate or thioacetate (114) with sodium methoxide in methanol, although O^--5 participation is also possible.[65]

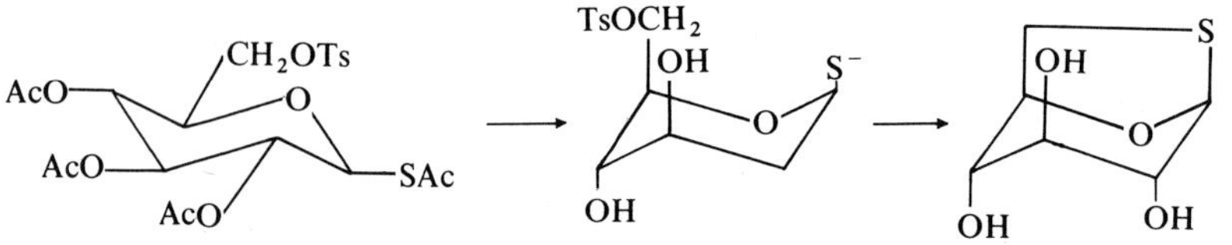

(47)

Johnson and Kingsbury[66] have used S-5 participation in their recently devised syntheses for some rigid tricyclic sulfur compounds. The p-bromobenzenesulfonates (115) were converted to (116) by the sequence shown. Esters of (116) should make intriguing models for solvolytic studies.

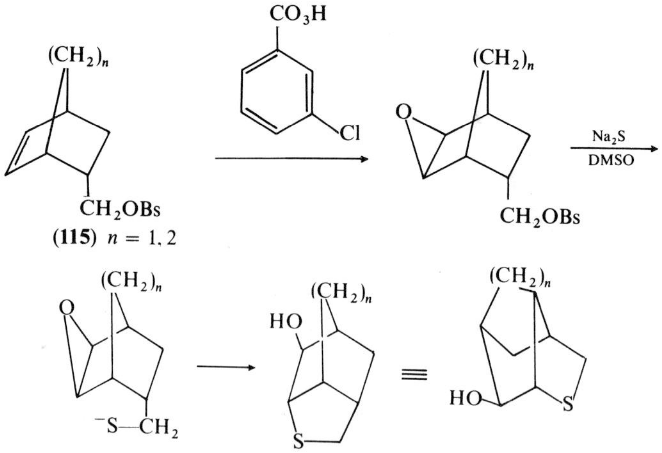

(48)

One interesting example of S^--6 participation occurs on treatment of 6-chloropenicillanic acid **(117)** with sodium methoxide.[67] The pathway through esters **(118)** and **(119)** was proposed as a route to the formation of **(120)**. Spectrophotometric evidence was gained to support the intermediacy of **(119)**.

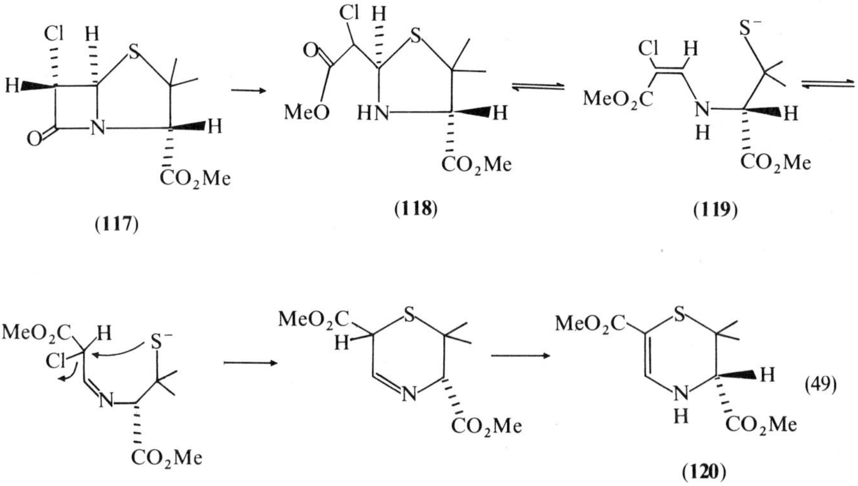

References

1. R. A. Peters and E. Walker, *Biochem. J.*, **17**, 260 (1923); A. G. Ogstom, E. R. Holiday, J. St. L. Philpot, and L. A. Stocken, *Trans. Faraday Soc.*, **44**, 45 (1948); P. D. Bartlett and C. G. Swain, *J. Am. Chem. Soc.*, **71**, 1406 (1949).
2. H. Böhme and K. Sell, *Chem. Ber.*, **81**, 123 (1948).
3. G. M. Bennett, F. Heathcoat, and A. N. Mosses, *J. Chem. Soc.*, 2567 (1929).
4. F. G. Bordwell and W. T. Brannen, Jr., *J. Am. Chem. Soc.*, **86**, 4645 (1964); but see E. Block, *J. Org. Chem.*, **39**, 734 (1974).
5. A. C. Knipe and C. J. M. Stirling, *J. Chem. Soc. B*, 1218 (1968).
6. R. Bird and C. J. M. Stirling, *J. Chem. Soc. Perkin Trans. 2*, 1221 (1973).
7. S. N. Lewis and W. D. Emmons, *J. Org. Chem.*, **31**, 3572 (1966).
8. W. H. Mueller, *Angew. Chem. Int. Ed. Engl.*, **8**, 482 (1969).
9. C. L. Wilkins and T. W. Regulski, *J. Am. Chem. Soc.*, **94**, 6016 (1972).
10. W. A. Thaler, W. H. Mueller, and P. E. Butler, *J. Am. Chem. Soc.*, **90**, 2069 (1968); see also G. M. Beverly, D. R. Hogg, and J. H. Smith, *Chem. Ind. (London)*, 1403 (1968).
11. W. A. Smit, M. Z. Krimer, and E. A. Vorob'eva, *Tetrahedron Lett.*, 2451 (1975); D. C. Owsley, G. K. Helmkamp, and S. N. Spurlock, *J. Am. Chem. Soc.*, **91**, 3606 (1969); D. C. Owsley, G. K. Helmkamp, and M. F. Rettig, *ibid.*, **91**, 5239 (1969); J. F. King, K. Abikar, D. M. Deaken, and R. G. Pews, *Can. J. Chem.*, **46**, 1 (1968); J. F. King and K. Abikar, *ibid.*, **46**, 9 (1968).

12. H. L. Goering and K. L. Howe, *J. Am. Chem. Soc.,* **79**, 6542 (1957).
13. S. J. Cristol and R. P. Arganbright, *J. Am. Chem. Soc.,* **79**, 3441 (1957).
14. S. J. Cristol, R. Caple, R. M. Sequeira, and L. O. Smith, *J. Am. Chem. Soc.,* **87**, 5679 (1965).
15. T. Tsuji, T. Komeno, H. Itani, and H. Tanida, *J. Org. Chem.,* **36**, 1648 (1971).
16. I. Tabushi, Y. Tamaru, Z. Yoshida, and T. Sugimoto, *J. Am. Chem. Soc.,* **97**, 2886 (1975).
17. E. J. Corey and E. Block, *J. Org. Chem.,* **31**, 1663 (1966); see also M. B. Dines and W. Mueller, *ibid.,* **35**, 1720 (1970).
18. J. C. Martin and P. D. Bartlett, *J. Am. Chem. Soc.,* **79**, 2533 (1957).
19. J. D. Roberts, W. Bennett, and R. Armstrong, *J. Am. Chem. Soc.,* **72**, 3329 (1950).
20. H. Morita and S. Oae, *Tetrahedron Lett.,* 1347 (1969).
21. H. G. Richey and D. V. Kinsman, *Tetrahedron Lett.,* 2505 (1969).
22. G. Capozzi, G. Melloni, G. Modena, and U. Tonellato, *Chem. Commun.,* 1520 (1969).
23. G. Modena and U. Tonellato, *J. Chem. Soc. B,* 374, 381, 1569 (1971).
24. B. M. Trost, D. Keeley, and M. J. Bogdanowicz, *J. Am. Chem. Soc.,* **95**, 3068 (1973).
25. D. N. Jones, M. J. Green, M. A. Saeed, and R. D. Whitehouse, *J. Chem. Soc. C,* 1362 (1968).
26. R. E. Ireland and H. A. Smith, *Chem. Ind. (London),* 1252 (1959).
27. L. A. Paquette, G. V. Meehan, and L. D. Wise, *J. Am. Chem. Soc.,* **91**, 3231 (1969).
28. A. R. Dunn and R. J. Stoodley, *Chem. Commun.,* 1169 (1969).
29. R. F. Gratz and P. Wilder, Jr., *Chem. Commun.,* 1449 (1970).
30. P. Wilder, Jr., and C. V. A. Drinnan, *J. Org. Chem.,* **39**, 414 (1974).
31. A. DeGroot, J. A. Boerma, and H. Wynberg, *Tetrahedron Lett.,* 2365 (1968); *Recl. Trav. Chim. Pays-Bas,* **88**, 994 (1969); M. B. Dines and W. H. Mueller, *J. Org. Chem.,* **35**, 1720 (1970).
32. R. Breslow, S. Garratt, L. Kaplan, and D. LaFollette, *J. Am. Chem. Soc.,* **90**, 4051, 4056 (1968).
33. J. C. Martin and R. J. Basalay, *J. Am. Chem. Soc.,* **95**, 2572 (1973).
34. M. Hojo, T. Ichi, Y. Tamaru, and Z. Yoshida, *J. Am. Chem. Soc.,* **91**, 5170 (1969).
35. N. A. Hughes and R. Robson, *J. Chem. Soc. C,* 2366 (1966); J. A. Harness and N. A. Hughes, *Chem. Commun.,* 811 (1971).
36. W. A. Spitzer, T. Goodson, Jr., M. O. Chaney, and N. D. Jones, *Tetrahedron Lett.,* 4311 (1971); see also D. C. Lankin, R. C. Petterson, and R. A. Velazquez, *J. Org. Chem.,* **39**, 2801 (1974).
37. L. Kaplan, *Bridged Free Radicals,* Marcel Dekker, Inc., New York (1972), p. 305.
38. H. Kwart and D. Drayer, *J. Org. Chem.,* **39**, 2157 (1974).
39. H. Kwart and R. K. Miller, *J. Am. Chem. Soc.,* **78**, 5678 (1956); S. J. Cristol, R. P. Arganbright, G. D. Brindell, and R. M. Heitz, *ibid.,* **79**, 6035 (1957); H. C. Brown, J. H. Kawikami, and K.-T. Liu, *ibid.,* **95**, 2209 (1973).
40. M. S. Raasch, *J. Org. Chem.,* **40**, 161 (1975).
41. V. I. Staninets and E. A. Shilov, *Russ. Chem. Rev.,* **40**, 272 (1971).
42. V. I. Staninets and E. A. Shilov, *Dokl. Akad. Nauk. Ukrain. SSR,* 1474 (1962).
43. E. N. Rengevich, V. I. Staninets, and E. A. Shilov, *Dokl. Akad. Nauk. SSSR,* **146**, 111 (1962).
44. E. R. deWaard, W. J. Vloon, and H. O. Huisman, *Chem. Commun.,* 841 (1970).
45. R. Heck, J. Corse, E. Grunwald, and S. Winstein, *J. Am. Chem. Soc.,* **79**, 3278 (1957).
46. Z. Horri, J. Tsuji, and T. Inoi, *Yakugaku Zasshi,* **77**, 248 (1957); *Chem. Astr.,* **51**, 8671a (1957).
47. P. Wilder, Jr., L. A. Feliu-Otero, and G. A. Diegnan, *J. Org. Chem.,* **39**, 2153 (1974).
48. P. Wilder, Jr., and L. A. Feliu-Otero, *J. Org. Chem.,* **30**, 2560 (1965).
49. P. Wilder, Jr., and L. A. Feliu-Otero, *J. Org. Chem.,* **31**, 4264 (1966).
50. W. H. Mueller and P. E. Butler, *J. Org. Chem.,* **33**, 2642 (1968).
51. W. H. Mueller, *J. Am. Chem. Soc.,* **91**, 1223 (1969).

52. P. H. McCabe and C. M. Livingston, *Tetrahedron Lett.*, 3029 (1973).
53. R. J. Crawford and C. Woo, *Can. J. Chem.*, **43**, 3178 (1965).
54. G. H. Schmid and P. H. Fitzgerald, *J. Am. Chem. Soc.*, **93**, 2547 (1971); see also P. H. Fitzgerald, *Diss. Abstr., Int. B*, **34**, 4855 (1974).
55. W. G. Bentrude and J. C. Martin, *J. Am. Chem. Soc.*, **84**, 1561 (1962).
56. T. H. Fisher and J. C. Martin, *J. Am. Chem. Soc.*, **88**, 3382 (1966).
57. J. C. Martin and M. M. Chau, *J. Am. Chem. Soc.*, **96**, 3319 (1974).
58. P. Y. Johnson and G. A. Berchtold, *J. Org. Chem.*, **35**, 584 (1970).
59. P. Y. Johnson, *Tetrahedron Lett.*, 1991 (1972).
60. P. Y. Johnson, E. Koza, and R. E. Kohrman, *J. Org. Chem.*, **38**, 2967 (1973).
61. V. E. Bel'skii, N. V. Isvasyuk, S. V. Povarenkina, and I. M. Shermergorn, *Izv. Akad. Nauk SSSR Ser. Khim.*, 1407 (1970).
62. P. Crouzet, E. Laurent-Dieuzeide, and J. Wylde, *Bull. Soc. Chim. Fr.*, 1463 (1968).
63. J. J. Sharp, A. B. Robinson, and M. D. Kamen, *J. Am. Chem. Soc.*, **95**, 6097 (1973).
64. M. Akagi, S. Tejima, and M. Haga, *Chem. Pharm. Bull. (Tokyo)*, **11**, 58 (1963).
65. A. M. Creighton and L. N. Owen, *J. Chem. Soc.*, 1024 (1960).
66. C. R. Johnson and W. D. Kingsbury, *J. Org. Chem.*, **38**, 1803 (1973).
67. I. McMillan and R. J. Stoodley, *Tetrahedron Lett.*, 1205 (1966).

6
Participation by Nitrogen Groups

6.1. Amino Groups

6.1.1. Anchimeric Assistance in Ring Closure Reactions

In view of the well-known nucleophilicity of the amino group it is not surprising that a considerable number of examples of participation by amines have appeared. Yet the assignment of a mechanism is not always simple because the interpretation of kinetic studies is often complicated by the appearance of competitive reactions. Competing with intramolecular displacement by the amino group (ring closure) are dimerization, polymerization, elimination, fragmentation, and solvent displacement, although all these processes would generally not occur in a single system.

Freundlich and Kroepelin[1] measured the rates of reaction of a series of ω-bromoalkylamines in water and found that their order of reactivity with ring size was $5 > 6 > 3 > 7 > 4$ (Table 1). This order parallels that of analogous ω-hydroxy and ω-methoxy compounds. In 1936 Salomon[2] discussed some of the above-mentioned complications that affect the interpretation of the rate data. He derived the relative rates for ring closure (Table 1) from a combination of kinetic and preparative experiments. Undoubltedly, modern analytical methods would have been beneficial to Salomon in dissecting the various rate processes.

A similar ring closure order is observed when an arylamino group is the nucleophile (Table 2).[3] The precursor for formation of the five-membered ring, $PhNH(CH_2)_4Br$, could not be obtained for rate studies. This may be a consequence of its fast rate of ring closure under the conditions of its synthesis.

Table 1. Ring Closure Rates of ω-Bromoalkylamines in Water at 25°C

$H_2N-(CH_2)_{n-1}-Br$, n = ring size	Observed rate, $10^3 k$ (min^{-1})	k_{rel} (obs.), reaction[a]	k_{rel}, ring closure[b]
3	36	72	70
4	0.5	1.0	1.0
5	3×10^4	6×10^4	6×10^4
6	500	1×10^3	1×10^3
7	1.0	2.0	17

[a]Reference 1.
[b]Reference 2.

It is interesting to note that it is apparently an enthalpy of activation effect that results in the relative slow rate of formation of the azetidine from N-γ-bromopropylaniline (see Table 2) since the measured $\Delta S^\ddagger$ values for three- and four-membered ring formation are identical. In each case a high yield of cyclic product is obtained, and hence the activation parameters are for the ring closure reaction.

The results from a study of the cyclization of a series of ω-chloroalkylamines[3] in 50% dioxane–water are shown in Table 3. As with the bromides, the yield of cyclic product is high in all cases, but none of the ring closures is quantitative. The experimental $\Delta S^\ddagger$ values for N-phenyl substrates undergoing either three- or five-membered ring formation are slightly more negative than their respective unsubstituted counterparts. To the contrary the $\Delta S^\ddagger$ values for the reaction of the substrates with a phenyl group attached to the carbon bearing the amino group were slightly less negative than their unsubstituted counterparts. From the data in Table 3, it seems apparent that the differences in the rates of closure to three- and five-membered rings are primarily attributable to differences in $\Delta H^\ddagger$.

Table 2. Ring Closure of ω-Bromoalkylamines in 60% Ethanol–Water[a]

$PhNH-(CH_2)_{n-1}-Br$, n = ring size	k_{rel} (25°C)	$\Delta H^\ddagger$ (kcal mole^{-1})	$\Delta S^\ddagger$ (cal mole^{-1} °K^{-1})	%cyclic product[b]
3	52.6	19.4	−11	91
4	1.0	21.7	−11	98
6	503	16.2	−17	100

[a]Reference 3; solution contained K_2CO_3.
[b]GLC analysis.

Participation by Nitrogen Groups

Table 3. Ring Closure of ω-Chloroalkylamines in 50% Dioxane–Water [a]

Substrate	k_{rel} (25°C)	$\Delta H^{\ddagger}$ (kcal mole^{-1})	$\Delta S^{\ddagger}$ (cal mole^{-1} °K^{-1})	%cyclic product[b]
$NH_2(CH_2)_2Cl$	5.3	19.3	−15	86
$NH_2(CH_2)_4Cl$	1777	16.5	−13	92
$PhNH(CH_2)_2Cl$	1.0	19.8	−17	96
$PhNH(CH_2)_4Cl$	833	16.4	−15	92
$2,6\text{-}Me_2C_6H_4NH(CH_2)_2Cl$	1.7	20.7	−13	86
$NH_2CHPhCH_2Cl$	5.5	21.9	−7	91
$NH_2CHPh(CH_2)_3Cl$	1680	17.5	−10	95

[a] Reference 3.
[b] GLC analysis.

The phenyl substituents to nitrogen appear to slightly increase the activation enthalpy for three-membered but not for five-membered ring formation. An increase of about 1 kcal is observed for both ring sizes when the phenyl group is on the carbon bearing the amino group.

The ring closure reactions are thought to proceed as shown in Eq. (1), although in basic solution the pathway shown in Eq. (2) is possible. Measurements of rates at different pH values allow the calculation of the rate constants for ring closure of the free base (Tables 2 and 3.)[3]

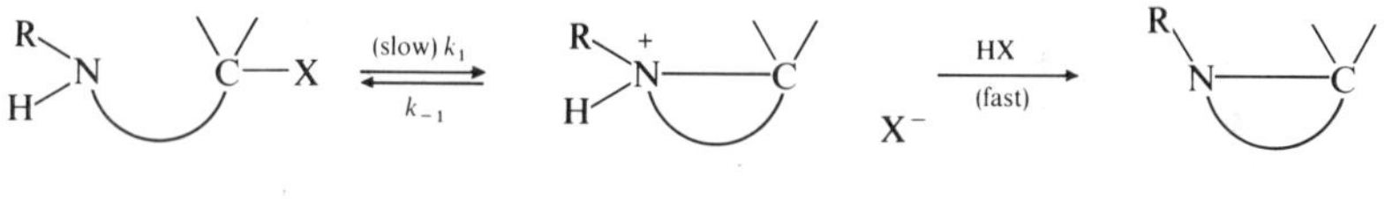

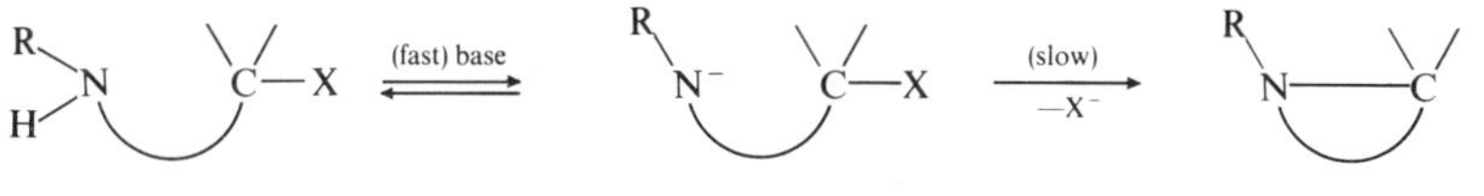

Kinetic studies of the base-catalyzed ring closure of ω-aminoalkyl hydrogen sulfates have also been carried out (Table 4). The relative rates reveal an order of 5 > 3 > 4, which is the same as that of the ω-haloalkylamines. These cyclizations, like those of the haloalkylamines, were first order in substrate and zero order in base. Hence the reaction sequence probably occurs as shown in Eq. (3).

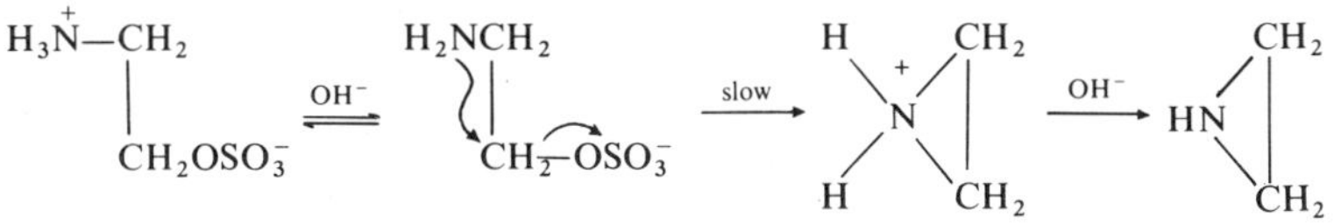

Table 4. Ring Closure of ω-Aminoalkyl Hydrogen Sulfates in Aqueous Potassium Hydroxide at 75°C[4]

$NH_3^+-(CH_2)_{n-1}-OSO_3^-$, n = ring size	$10^3 k$ (min^{-1})	k_{rel}
3	0.48	12.3
4	0.039	1.0
5	213	5.5×10^3

Table 5. Substituent Effects on the Rate of Ring Closure of β-Aminoalkyl Hydrogen Sulfates in Aqueous Potassium Hydroxide at 75°C[4]

$H_3N^+-\overset{\beta}{C}-\overset{\alpha}{C}-OSO_3^-$

substituent	$10^4 k$ (min^{-1})	k_{rel}
None	4.80	1.0
α-Methyl	9.45	1.97
α-Ethyl	9.62	2.01
α-Isopropyl	4.97	1.03
α-t-Butyl	0.34	0.072
α-Phenyl	286	59.2
α-Benzyl	2.38	0.50
β-Methyl	31.0	6.45
β-Ethyl	44.1	9.21
β-Isopropyl	55.3	11.5
β-Phenyl	5.87	1.22
threo-α,β-Dimethyl	64.3	13.4
erythro-α,β-Dimethyl	81.8	17.0
β,β-Dimethyl	195	40.5

Table 6. Ring Closure of trans-2-Aminocycloalkyl Hydrogen Sulfates in Aqueous Potassium Hydroxide at 75°C[4]

Substrate	n	$10^4 k$ (min^{-1})	k_{rel}
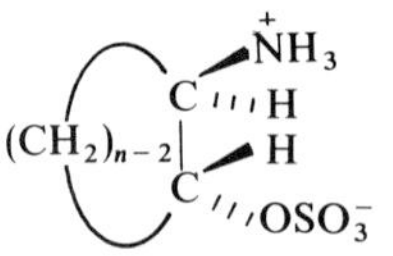 $(CH_2)_{n-2}$ ring with NH$_3^+$ and OSO$_3^-$	5	105	1.28
	6	11.4	0.139
	7	41.0	0.501
	8	13.1	0.160
erythro-MeCHCHMe, H_3N^+ OSO_3^-		81.8	1.0

Because of the utility of β-aminoalkyl sulfates in the synthesis of aziridines, Dewey and Bafford[4] have evaluated the effects of substituents on the β-aminoalkyl sulfate cyclization reaction (Table 5 and 6). From the data in Table 5 it is seen that β substituents are more rate enhancing than α substituents. In fact, as the α substituents increase in bulk, the rate enhancement vanishes, and with the α-t-butyl a rate retardation exists. This is steric retardation since an α-phenyl group greatly accelerates the reaction, presumably by stabilizing the partial positive charge that forms at carbon upon C—O bond breaking. The rate data in Table 6 closely parallel in order those of other S_N2-type reactions of cycloalkyl substrates.[5]

6.1.2. R_2N-3 Participation

Because of their relationship to mustard gas (*bis*-2-chloroethyl sulfide), β-aminoalkyl halides were studied intensely during the war years of the 1940s[6]; for a review of the early literature, see Ref. 6. Their hydrolysis reactions occur *via* isolable yet reactive aziridinium ions [Eq. (4)] and may be quite complex owing to the availability of competing reations with

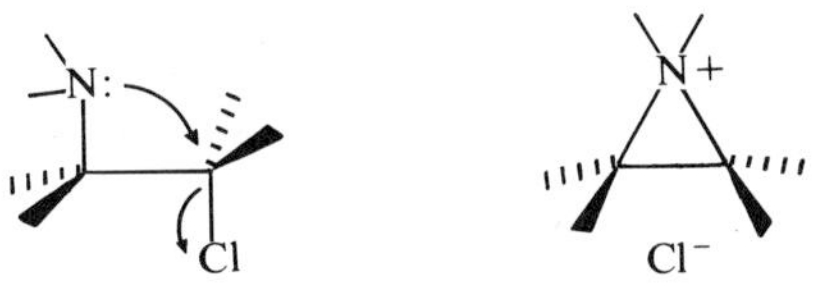

$$\tag{4}$$

reasonable probabilities. Thus aqueous solvolysis does not always provide, at least as the major product, the β-aminoethanol derivative one might expect. We shall now briefly explore the chemistry of some substrates containing the β-amino group.

6.1.2.1. Dimerization

Bartlett et al.[7] showed that, in aqueous acetone solution, nitrogen mustard (**1**) dimerizes probably by the sequence of reactions shown in Scheme 1. The regeneration of the original amino chloride in the first reversible step no doubt contributed to the difficulties in obtaining reliable rate measurements.[2]

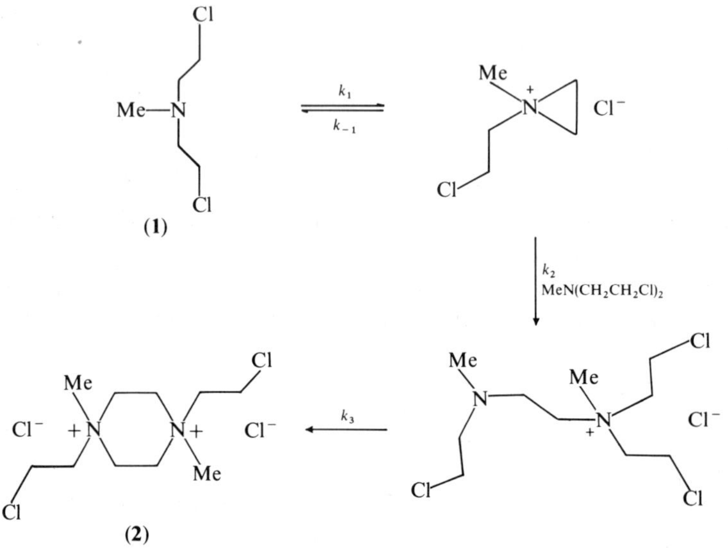

Scheme 1

6.1.2.2. Rearrangement and Cyclization Reactions

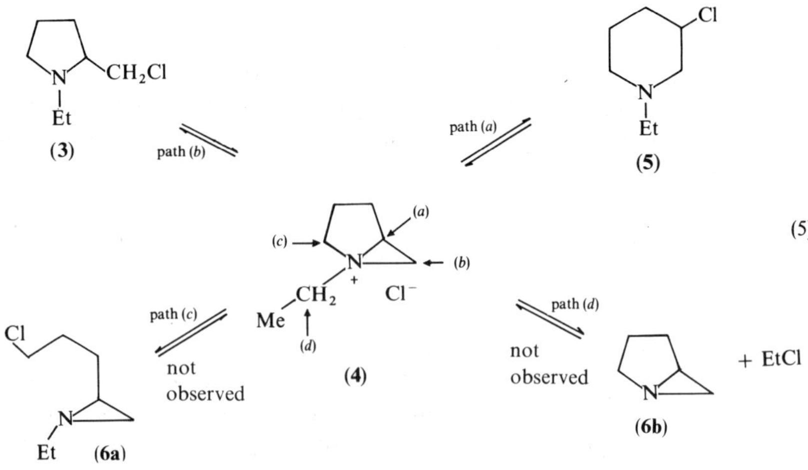

When the intermediate aziridinium ion is unsymmetrical, rearrangement may result.[8-10] Such is the case with the pyrrolidine derivative (**3**), studied by Fuson and Zirkle.[9] Four products are possible from nucleophilic attack on the intermediate (**4**), yet the secondary chloride (**5**) is the major product of rearrangement. It is well known[11] that primary positions are more reactive than secondary positions by the S_N2 mechanism and that the reverse is true for the S_N1 mechanism (i.e., secondary > primary), although both primary and secondary substrates generally prefer an S_N2-like mechanism if possible.[12] Thus it is likely that (3) undergoes nitrogen-assisted displacement to give (4). Since the rate of reaction of (3) is several hundred times that of model primary chlorides, the anchimerically assisted route to (4) seems assured. The formation of (5) as the product does not mean that the rate of nucleophilic attack by path (a) is faster than that of paths (b)–(d). To the contrary, the rate of attack to return (3) should be important since (1) that attack relieves a large amount of strain [strain decreases (6b) > (6a) > (3) > (5)] and (2) paths (b)–(d) involve attack at primary positions while path (a) involves attack at a secondary position. Hence the formation of (5) must infer thermodynamic control; that is, the piperidine derivative is assumed to be more stable than (3), (6a), or (6b) under equilibrium conditions.

An alternative mechanism[13] accounting for formation of (5) is shown in Eq. (6). This mechanism has been rejected[14] since it violates the principle of microscopic reversibility. Additional evidence against this mechanism

(6)

(5)

has been provided by Hammer and Heller.[10] The rate of reaction of 3-chloropiperidine (5) with 14 nucleophiles is independent of the nucleophile and occurs more than 10^4 times faster than the reaction of cyclohexyl chloride with hydroxide ion. Furthermore, (4) is implicated as the product-forming intermediate by the observation that optically active 3-substituted piperidines formed from optically active 3-chloropiperidine [Eq. (7)]. Thus, if 3-chloropiperidine undergoes anchimerically assisted displacement to give an intermediate that gives optically active products, formation of (5) from (3) cannot occur, as is shown in Eq. (6).

Before proceeding to other reactions of aziridinium ions, it is interesting to point out here the differences between oxygen and nitrogen as neighboring

(7)

groups. We included in Chapter 4 a discussion of the studies of compounds **(7)** (see pp. 145–146) and **(8)** (see p. 146). No unusual reactivity was

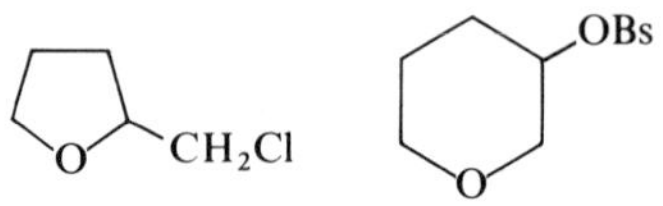

(7)   (8)

observed upon solvolysis of **(7)**, and no products of rearrangement, e.g., Eq. (8), were observed.[15] In the case of **(8)**, Tarbell and Hazen[16] found

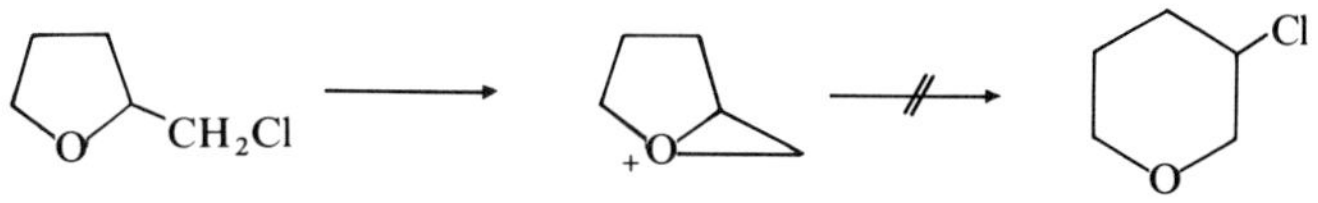

(7) (8)

no rearranged products but suggested, on the basis of rate studies, that transannular participation by oxygen was occurring to give an encumbered ion [Eq. (9)]. Stereochemical studies such as those carried out with 3-

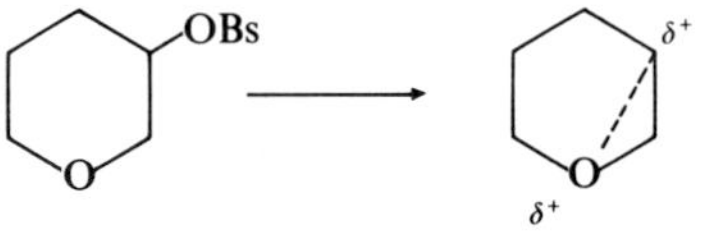

(9)

chloropiperidine ought to provide a better understanding of the pyran system and will allow a better comparison of oxygen versus nitrogen as neighboring groups; transannular participation in O-, S-, and N-substituted ring systems has been estimated by the use of comparative carbonyl stretching values; cf. Refs. 16 and 17.

Participation by nitrogen in cyclic systems is apparently absent or severely retarded when the intermediate aziridinium ion is highly strained. This is presumed to be the case with 8-chloromethylpyrrolizidine **(9)** since no rearrangement occurs.[18] It is possible, however, that the intermediate **(A)** forms but that **(9)**, and not its isomer **(B)**, is favored both kinetically and thermodynamically; hence, rearrangement is not observed.

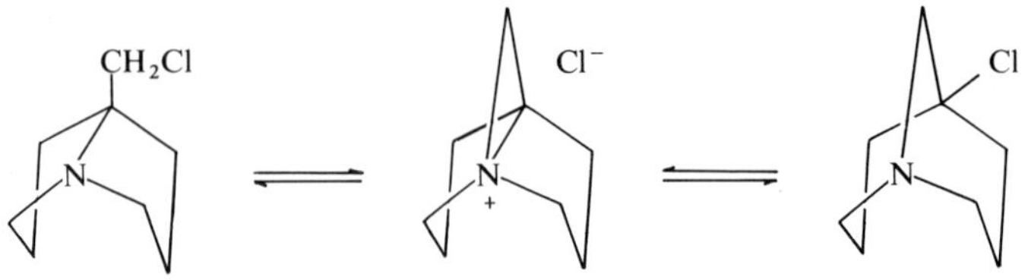

(9)   (A)   (B)   (10)

Although the 1-*t*-butyl-1-azabicyclo[1.1.0]butonium ion (**10**) should also be highly strained, it and other such derivatives have been implicated as intermediates upon solvolysis of azetidine and aziridine precursors. Gaertner[19] claimed that (**11**) and (**12**) are among the least reactive β-aminoalkyl chlorides known, yet isomerization of (**11**) ⇌ (**12**) occurred in either carbon tetrachloride or acetonitrile. With nucleophiles present, both gave products of simple displacement, hydrolysis, and rearrangement. The rate constants for solvolysis (in 50% aqueous ethanol) of (**11**) and (**12**) were

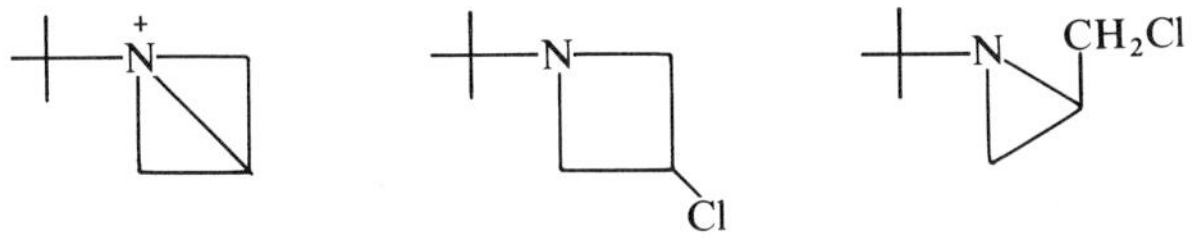

(**10**)   (**11**)   (**12**)

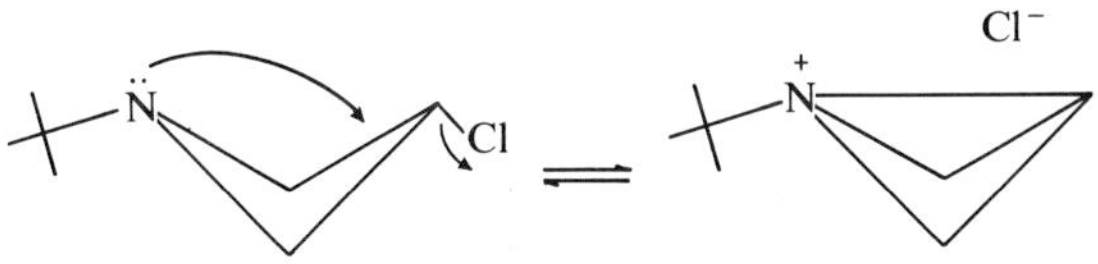

(**13**)   (**11**)

0.50 hr^{-1} (50°C) and 0.015 hr^{-1} (70°C), respectively. Both these compounds solvolyzed faster than the corresponding cyclobutyl and cyclopropylcarbinyl chlorides; however, their relative rate order is reversed. Anchimeric assistance by nitrogen seems to account for the results, with (**10**) as an intermediate, although strain may cause the N—C_3 bond of (**10**) to be weak. Apparently (**10**) forms less readily from (**12**) than from (**11**). This may be primarily an entropy effect since (**11**) may be in a favorable conformation, e.g., (**13**), for anchimeric assistance, while (**12**) must rotate to attain a proper arrangement. The solvolysis of (**14**) in aqueous ethanol also probably proceeds via (**10**) as an intermediate since the azetidinol (**15**) and its ether (**16**) are among the reaction products.[20]

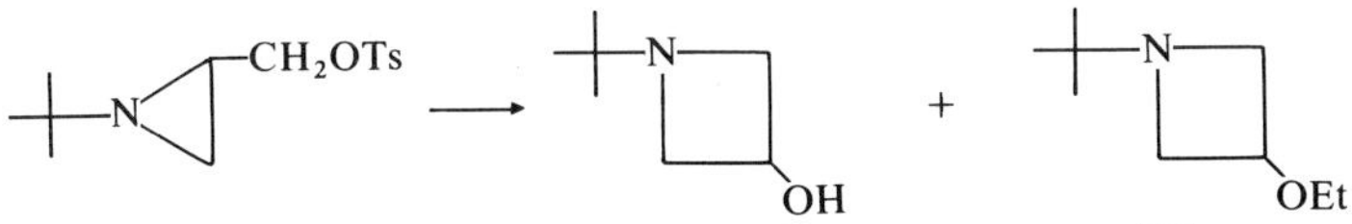

(**14**)   (**15**)   (**16**)   (**12**)

Cromwell *et al.*[21,22] and Hortmann and Robertson[23] have reported studies that partially overlap yet basically complement the earlier studies. Cromwell *et al.*[21] reported that the rate of reaction of the toluene-*p*-sulfonate

(17) with methanolic potassium cyanide is independent of the cyanide concentration and is equal to the rate of methanolysis of (17), suggesting that both reactions proceed through either the bridged or unbridged cations, e.g., (10) or (18), respectively.

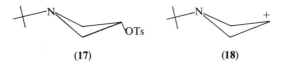

(17)   (18)

Upon substituting a methyl group at C-2 of the starting azetidine, Higgins and Cromwell[22] were able to obtain definitive stereochemical and rate data supporting azabicyclobutonium ion formation as an intermediate. Hydrolysis of both the *cis* and *trans* derivatives in 60% aqueous acetone provided stereospecific retention of configuration. Thus (19) provided exclusively (20), while (21) yielded only (22). Hortmann and Robertson[23]

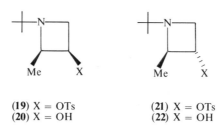

(19) X = OTs   (21) X = OTs
(20) X = OH   (22) X = OH

showed that substituted 1-azabicyclobutanes react with acids to yield 3-substituted azetidines [Eq. (13)].

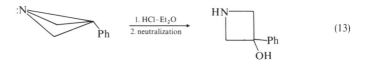   (13)

The relative rate data (Table 7) for solvolysis of (17), (19), and (21) are consistent with anchimeric assistance in (19), but the mechanims for (21) is in doubt. The Arrhenius plot for the solvolysis of (21) in 60% aqueous acetone is nonlinear, and its solvolysis rate is considerably less than that for either (17) or (19). On the basis of the rate data, one may agree that the *cis* derivative (19) may be accelerated owing only to anchimeric assistance by nitrogen. The relative rate data may also reflect steric effects. Schleyer et al.[24] found that *cis*- and *trans*-3-*t*-butylcyclobutyl toluene-*p*-sulfonate undergo hydrolysis in 60% aqueous acetone considerably slower than cyclobutyl

Table 7. Rate Data and Activation Parameters for the Hydrolysis of Azetidinyl Toluene-p-sulfonates in 60% Aqueous Acetone

Compound	k_1 (sec^{-1})(30°C)	$\Delta H^{\ddagger}$ (kcal mole^{-1})	k_{rel}
(17)[a]	2.69×10^{-4}	22.1	3.20
(19)[b]	7.67×10^{-3}	23.7	91.3
(21)[b]	2.23×10^{-5}	[c]	0.265
Cyclobutyl-OTs[d]	8.40×10^{-5}	21.9	1.0

[a] Reference 21.
[b] Reference 22.
[c] The Arrhenius plot is nonlinear.
[d] Reference 24.

toluene-p-sulfonate* (respectively, they react at rates 0.0076 and 0.17 times that of the unsubstituted ester at 30°C). The additional cis-methyl group in (19) could accelerate hydrolysis, but it seems likely that a mechanism involving steric assistance would lead to overall trans (inverted) and not 100% cis product, e.g., Eq. (14), owing to the relief of the unfavorable 1,2 interactions that exist in the cis derivative (19) [or (20)]. Hence a mechanism involving anchimeric assistance appears more consistent with the data.

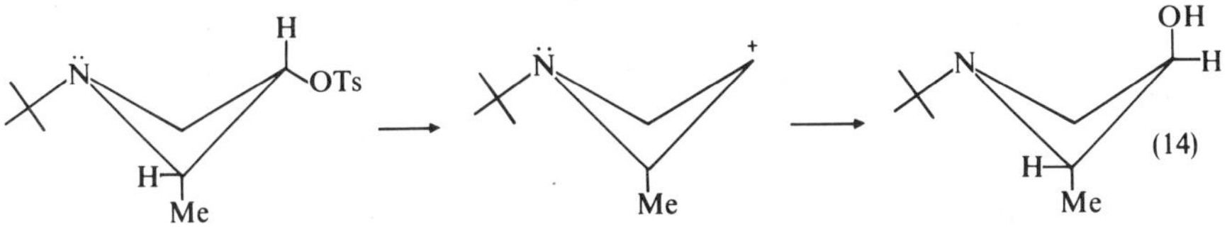

(19)   (14)   Less hindered than (20)

As compared to (19), the trans derivative (21) should suffer fewer nonbonded interactions in the ground state, but as the anchimerically assisted transition state is approached the bulky t-butyl group must move into a conformation where nonbonded interactions increase, e.g., Eq (15). The opposite is true with (19); that is, the nonbonded interactions between

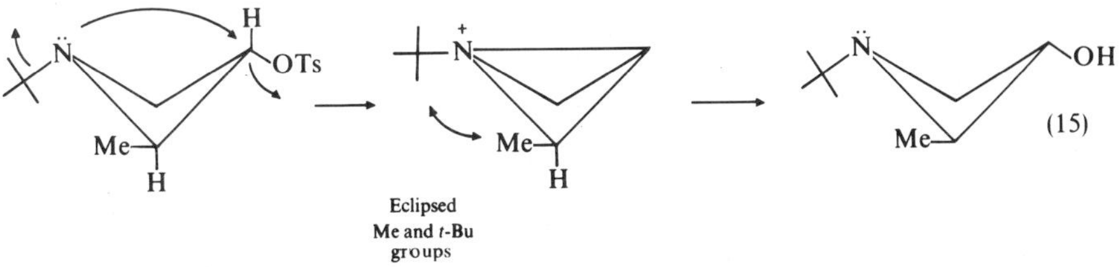

Eclipsed Me and t-Bu groups   (15)

* The cyclobutyl derivatives give cyclopropylcarbinyl products.

the *t*-butyl and methyl group should decrease on going to the anchimerically assisted transition state.

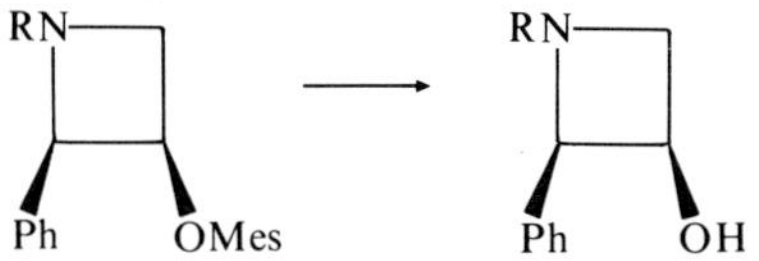

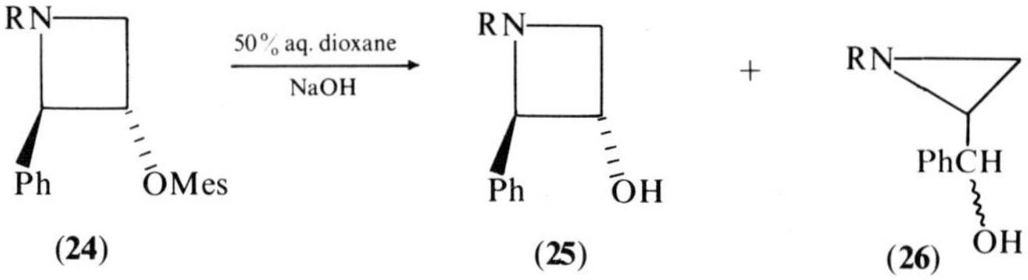

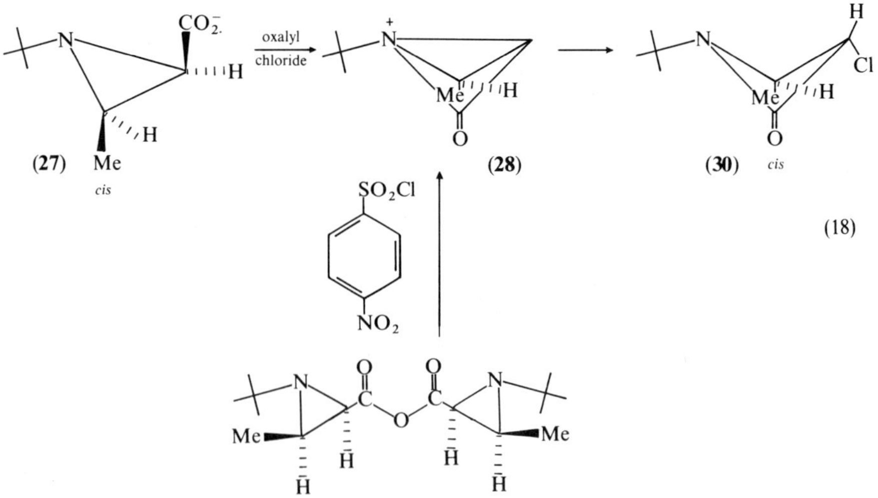

The above argument does not, however, account for the nonlinearity of the Arrhenius plot. Higgins and Cromwell[22] logically assumed that solvolysis may be occurring by two mechanisms. What is unusual is that a single product is found. In a similar study Okutani *et al.*[25] found that while the *cis*-methanesulfonate (**23a**) undergoes hydrolysis to give only *cis*-alcohol (**23b**), the *trans* derivative (**24**) gives a mixture of the retained alcohol (**25**) and the rearranged *erythro*- and *threo*-alcohols (**26**). It is inviting to suggest that the *trans* derivatives are reacting simultaneously

(**23a**) (**23b**)

R = cyclohexyl

(16)

(**24**) (**25**) (**26**)

(17)

(**27**) (**28**) (**30**) (**29**)

(18)

by nitrogen assistance *and* by a mechanism similar to that followed by cyclobutyl derivatives. Of course, other plausible explanations can be proposed.[22,25]

A final example of a reaction that may involve an azabicyclobutonium ion comes from studies of Deyrup and Clough[26] on the interconversions of aziridine carboxylates and β-lactams. The stereoselectivity of the reaction of certain substrates, e.g., (**27**), suggests the intermediacy of cations such as (**28**). The NMR spectrum of the anhydride (**29**) in liquid SO_2 with added *p*-nitrobenzenesulfonyl chloride gave evidence of an intermediate. Based on the chemical shift data, the presence of (**28**) was claimed. The simplicity of the reported spectral data and the fact that (**30**) apparently did not ionize to (**28**) upon dissolution into a saturated solution of antimony pentafluoride in liquid sulfur dioxide suggests caution in accepting the conclusion that (**28**) was observed in the NMR experiments. It was pointed out[26] that concerted reaction pathways are available[27] that could account for the conversion of β-lactams to aziridine carboxylates. Since a concerted mechanism can also be written to explain the stereospecific conversion of (**27**) to (**30**) [Eq. (19)], the intermediacy of (**28**) remains questionable.

The racemization of L-(+)-2-α-tropinol (**31**) upon refluxing overnight with acetic anhydride was taken as evidence of the possible involvement

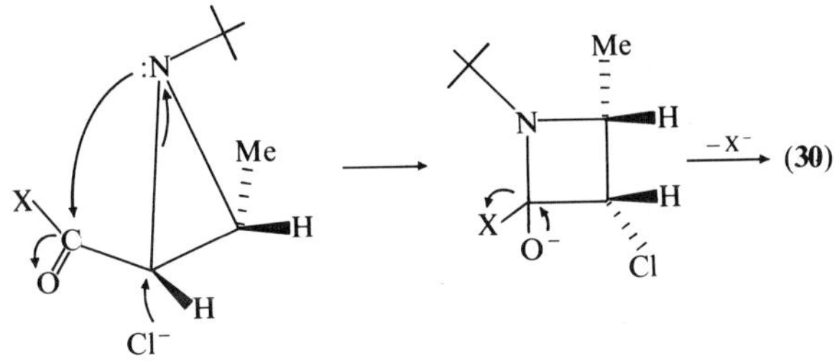

(19)

of the neighboring nitrogen group in displacement of the acetate group in (**32**).[28] Evidence for this mechanism included isolation of the optically

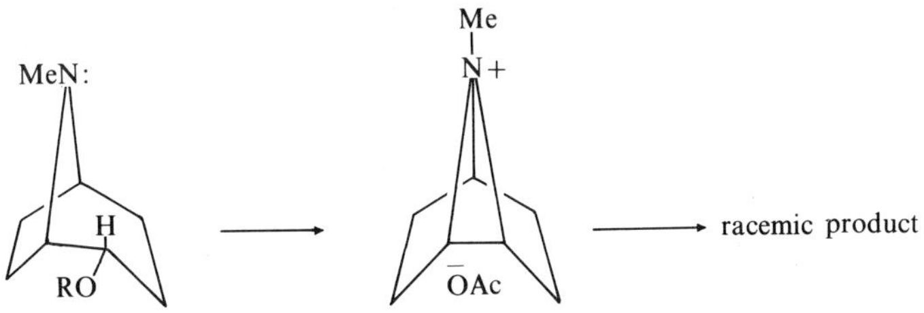

→ racemic product (20)

(**31**) R = H
(**32**) R = Ac

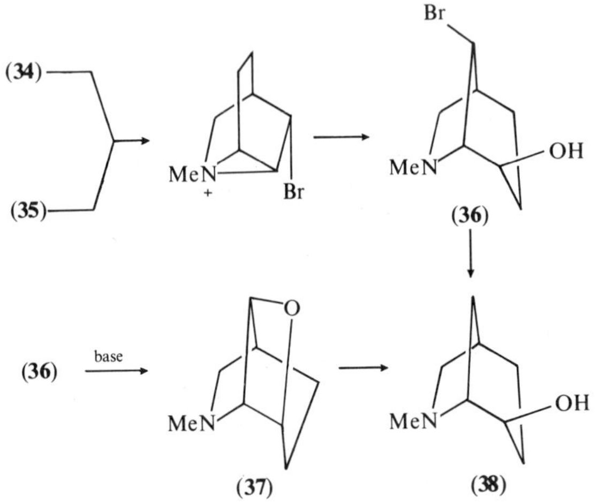

$$\begin{array}{c}(34)\\(35)\end{array}\Bigg\} \longrightarrow \underset{(35)}{\text{MeN}^+}\text{—Br} \longrightarrow \underset{(36)}{\text{MeN}}\text{—OH} \quad (21)$$

$$(36) \xrightarrow{\text{base}} \underset{(37)}{\text{MeN}}\text{—O} \longrightarrow \underset{(38)}{\text{MeN}}\text{—OH}$$

active acetate (as the HCl salt) after short reaction periods and observation that L-(−)-2β-tropinol [the *exo* isomer of (31)], fails to racemize under conditions where the racemization of (31) was rapid.[28]

The finding of substantial driving force for anchimeric assistance by nitrogen (the strong tendency for participation is even shared by the nitrogen group of amides[29]) in bicyclic compounds is not surprising in view of the propensity for assistance by oxygen (see pp. 150–153), which is considerably less active as a neighboring group. Hutchins and Rua[30] have verified the extraordinary reactivity of azabicyclic derivatives.* Their relative rate data are shown in Table 8. The amino halides are so reactive as the free amines that they were only isolated at hydrohalide salts. The relative rate data illustrate that the β-amino group can lend tremendous assistance to ionization of a halide in either the azabicyclo[2.2.2]octane, e.g., (33) or

---

* There have been a considerable number of recent studies (for leading references, see Ref. 31) of the ionization of azabicyclic compounds where the nitrogen (or the leaving group) is at a bridgehead position and the leaving group (or the nitrogen) is attached to the α-carbon, e.g.,

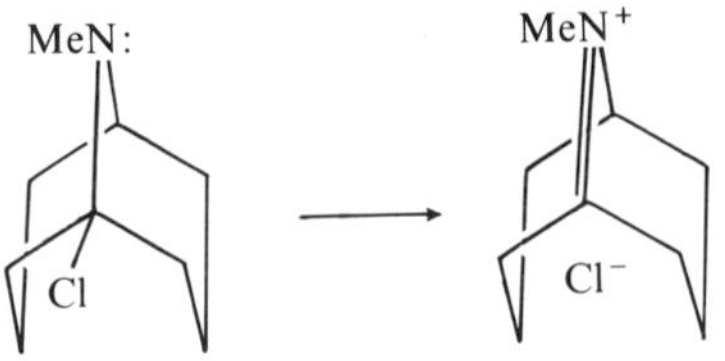

$$\text{MeN:} \underset{\text{Cl}}{\diagup\!\!\!\diagdown} \longrightarrow \text{MeN}^+ \underset{\text{Cl}^-}{\diagup\!\!\!\diagdown}$$

Any participation by nitrogen can be considered a resonance effect; hence we shall not treat such examples here; for a review of azacarbocations, see Ref. 32.

Table 8. *Relative Rates of Solvolysis of Azabicyclic and Related Compounds at $0°C$ in 80% Aqueous Ethanol*

Compound	Relative rate
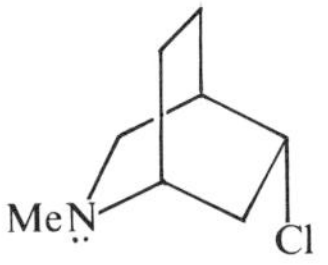 (MeN, Cl) azabicyclic	1.0
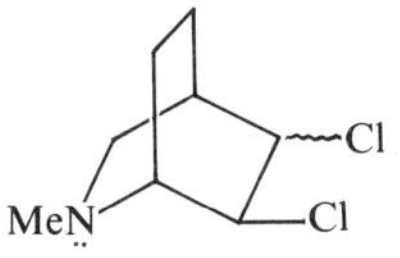 (33) MeN, Cl, Cl	4×10^8
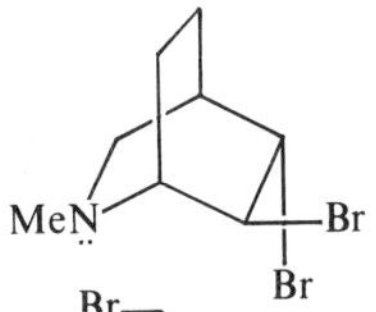 (34) MeN, Br, Br	$\sim 1.7 \times 10^9$
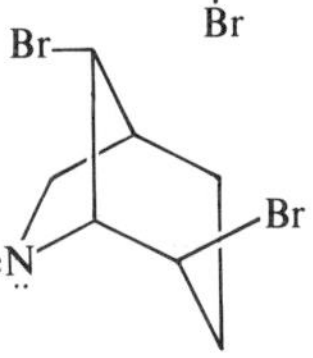 (35) MeN, Br, Br	$\sim 1.7 \times 10^9$
$Me_2NCH_2CH_2Cl$	2.4×10^3
exo-2-Norbornyl Chloride	2.8×10^2

(34), or the azabicyclo[3.2.1]octane system, e.g., (35). The acceleration of about 5 powers of 10 over the already assisted β-N,N-dimethylaminoethyl chloride is remarkable. Both (34) and (35) gave the hydroxy derivative (36). Further reaction of (36) in the presence of base afforded the amino ether (37), by O^--4 cyclization, which could be converted to the known amino alcohol (38) by reduction.

While the majority of studies of β-amino group participation have centered on alkyl amino groups in the types of systems mentioned above, some other system types have received attention. Mihalic et al.[33] have reported on the regio- and stereoselectivity of bromide displacement in derivatives of the aryl amine (39). Some examples of amino group participation have also appeared in the carbohydrate literature.[34] For example, the azido group has been used as a precursor for the amino group in the preparation of the aziridine (**40**).[35] Finally, MeO-5 and R_2N-3 participation were

(**39**)

(22)

(**40**)

(23)

(**41a**) $R_1 = R_2 = H$

(**41b**)

(**41c**)

R_1 = Me or H; R_2 = OMe or OH

(24)

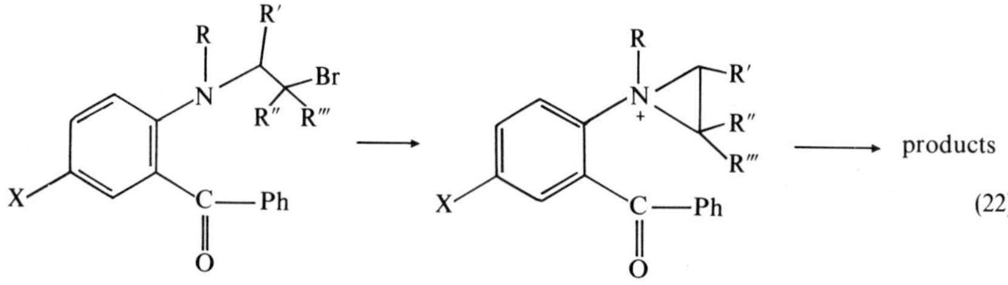

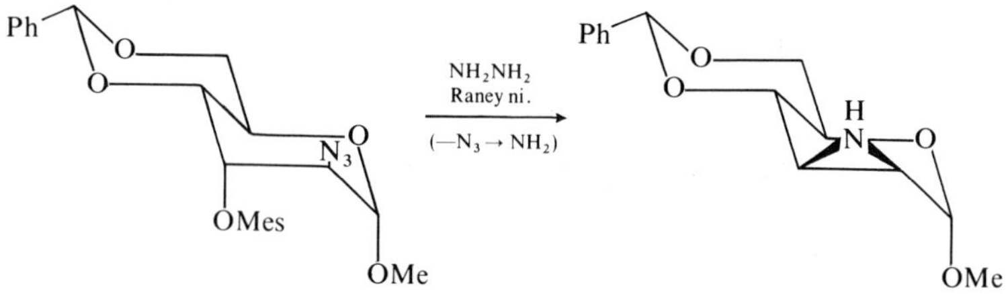

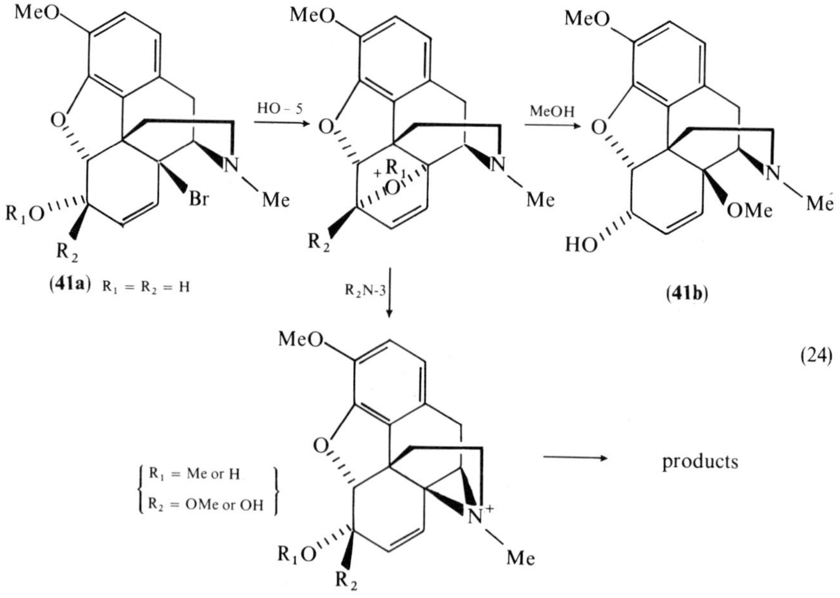

Participation by Nitrogen Groups

put forth to account for the formation of the isolable aziridinium ion (**41c**) upon solvolysis of acetals of halogeno-codeinone and -neopinone. The perchlorate salt of (**41c**) was isolated upon treatment of the 8β-iodoneopinone derivative with $AgClO_4$ in benzene.[36] In the presence of the free 6α-hydroxy group, e.g., the codeine derivative (**41a**), HO-5 participation is confirmed by the isolation of the methanol substitution product of retained configuration, i.e., (**41b**).[37]

6.1.2.3. Fragmentation

As we shall see below, framentation is a major reaction pathway for γ-aminoalkyl halides and esters; for a review of these and related fragmentations, see Ref. 40. It is uncommon in other aminoalkyl homologs owing to the specific requirements for fragmentation. An interesting example of fragmentation has been shown to occur in a β-amino mixed anhydride [Eq. (25)].[38]

$$Ph_2C-\underset{\underset{Ph}{\underset{|}{\underset{NH}{|}}}}{\overset{\overset{O}{\|}}{C}}-OTs \longrightarrow Ph_2C=\overset{+}{N}Ph + CO \quad (25)$$
$$H$$
$$OTs^-$$

6.1.3. R_2N-4 Participation

As shown from the kinetic data earlier in this chapter, ring closure to four-membered azacarbocyclic rings (N-4 participation) is quite slow relative to ring closure to higher or lower homologs. Nevertheless, R_2N-4 participation does occur. In fact, this mode of participation is shown to be more efficient than HO-3 participation in 1-alkylamino-3-chloro-2-propanols as 1-alkyl-3-azetidinols are the exclusive products [Eq. (26)].[39]

$$RNHCH_2\overset{\overset{O}{\diagup\ \diagdown}}{CH}-CH_2 \xleftarrow{\ /\!\!/\ } RNHCH_2\overset{\overset{OH}{|}}{C}HCH_2Cl \longrightarrow RN\!\!\begin{array}{c}\rule{1.2cm}{0.4pt}\\ \ \\ \rule{1.2cm}{0.4pt}\end{array}\!\!{-}OH \quad (26)$$

Grob[40] and his co-workers have extensively studied reactions of the neighboring amino group in γ-substituted aminoalkyl halides. Four processes are found to occur in simple acyclic examples (see Table 9 and

Table 9. Influence of Structure on the Rate and Course of the Reaction of γ-Chloroamines in 80% Ethanol at 56°C[40]

Compound		$k \times 10^5$	$k/k_h{}^a$	Reaction path (%)[b]
$Me_2NCH_2CH_2CH_2Cl$	(42)	10.8	4.3×10^3	$100R^c$
$Me_2NCH_2\underset{Me}{\overset{Me}{C}}CH_2Cl$	(43)	98	3.4×10^6	$100R^c$
$Me_2NCH_2CH_2CHCl\text{\textbackslash}Me$	(44)	1.87	520	$100R$
$Me_2NCH_2\underset{Me}{\overset{Me\ Me}{C}}CHCl$	(45)	28.2	4100	$72R, 19F$
$Me_2NCH_2\underset{Me\ Me}{\overset{Me\ Me}{C}}CCl$	(46)	14,500	125	$70F, 30E$
$Me_2N\underset{Me\ Me}{\overset{Me\ Me}{C}}CH_2CCl$	(47)	41,200	24	$80F, 20E$
$Me_2NCH_2CH_2\underset{Me\ Me}{C}Cl$	(48)	35.5	0.52	$38F, 37E, 23S, 2R$

[a] The rate constant k_h is the rate for the compound where i-Pr is substituted for NMe_2 in each case.
[b] R = ring closure; F = fragmentation; E = elimination; S = substitution.
[c] Elimination N-6 is impossible.

Scheme 2). In addition to solvent participation (k_s) and normal nucleophilic nitrogen participation (k_Δ), the observed reactions are elimination (k_e) (Scheme 3) and a fragmentation (k_f) process sometimes called the *Grob fragmentation* (Scheme 4).

It is interesting that k_s products were not found in the primary or secondary substrates, i.e., (42)–(45), but only in the tertiary substrate (48). Thus the substitution product most likely results from attack on the tertiary cation (49) or on the cyclic cation (50). The slow rate of reaction of (48) compared to its model, $Me_2CH(CH_2)_2CMe_2Cl$, suggests that it is under-

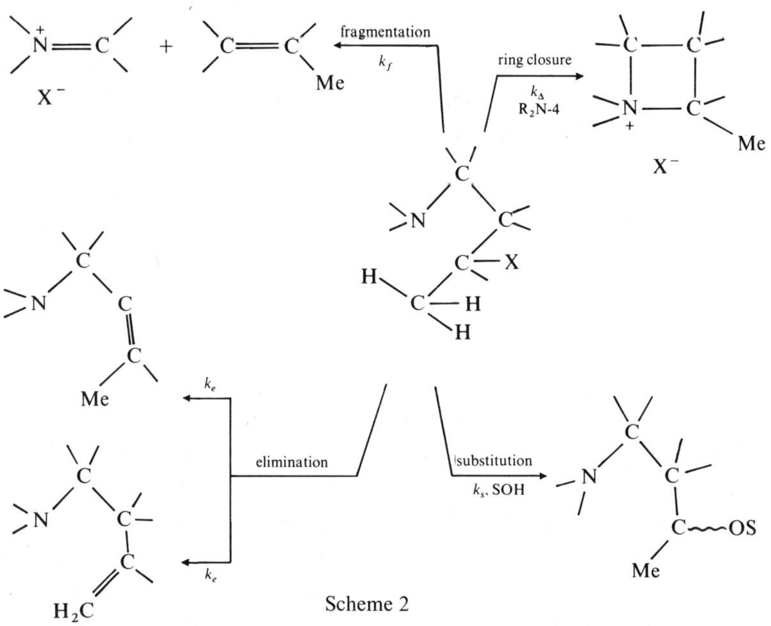

Scheme 2

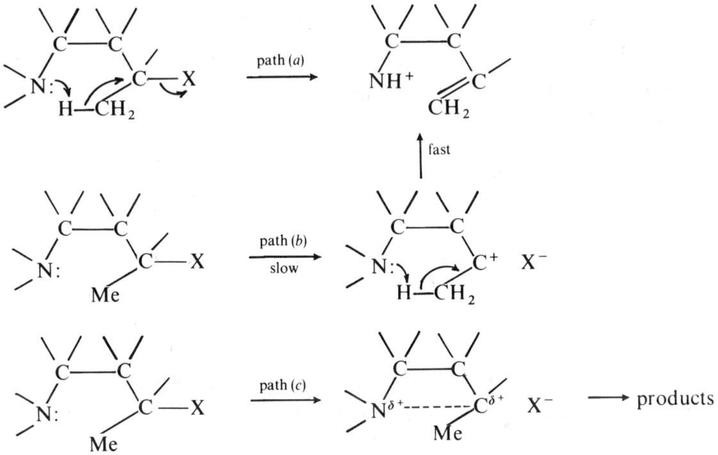

Scheme 3

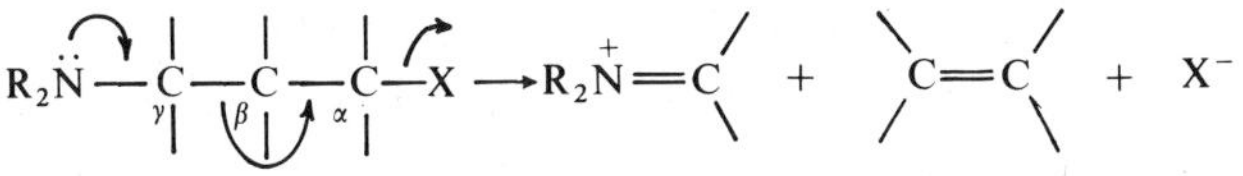

Scheme 4

going rate-limiting ionization. Thus, the nonbonding interactions of the methyl groups apparently are destabilizing (50) and the transition state leading to it. Hence its formation will be slow and its destruction rapid. As a consequence, (49) may be more stable than (50). The other substrates

$$\begin{array}{cc} \text{Me} & \text{Me} \\ & \diagup \\ \diagdown & \text{NCH}_2\text{CH}_2\overset{+}{\text{C}} \\ \diagup & \diagdown \\ \text{Me} & \text{Me} \end{array} \qquad \text{Me}-\overset{+}{\underset{\text{Me}}{\text{N}}}\kern-0.5em\square\kern-0.5em\begin{array}{c}\text{Me}\\\text{Me}\end{array}$$

(49)   (50)

in Table 9 are indicated to be anchimerically assisted by nitrogen both from rate comparisons and from the use of other probes.[40]

Elimination from (46) and (47) is readily visualized according to two mechanisms, pathways (a) and (b) in Scheme 3. Unfortunately the experimental facts are inconsistent with either mechanism. First, anchimeric assistance is indicated from relative rate comparisons as stated above; this could eliminate pathway (b). Second, there is no primary isotope effect upon substituting deuterium in the reactive positions of (46), i.e., $Me_2NCH_2 \cdot CMe_2C(CD_3)_2Cl$. Grob and Jenny[41] have suggested that nitrogen assistance may occur to give an internally solvated carbocation (51) [Eq. (27)]. It is possible then to conceive of elimination (and fragmentation) products arising from an intermediate such as (51), e.g., pathway (c) in Scheme 3.

(27)

(51)

Unlike the exceptions noted above, the large majority of Grob fragmentations are synchronous and accelerated without direct nitrogen participation at the carbon bearing the leaving group. Grob[40,42] has said that a reaction proceeds with *frangomeric assistance* when the rate constant for fragmentation, k_f, is greater than that for normal ionization, k_i.[40] The rate constant k_f is equivalent to k_Δ if the nitrogen is said to be anchimerically assisting fragmentation. The term k_i is perhaps misleading since the large majority of chemists use k_c or k_s to describe the normal reaction (see pp. 13–14 and 114).

Participation by Nitrogen Groups

The Grob fragmentation exhibits remarkable stereochemistry. The structural and stereoelectronic requirements for concertedness are demonstrated with the substrate in Scheme 4. The C_α—X bond and the orbital of the lone pair of electrons on nitrogen must be parallel antiperiplanar[42] to the C_β—C_α bond. The stereoisomeric 3-chlorotropones (**52**) and (**53**) serve to illustrate the reaction stereochemistry and frangomeric assistance.[44] The 3β-chloride (**52**) meets the stereochemical requirements and thus should fragment by the concerted pathway at an enhanced rate. On the other hand, the 3α-chloride (**53**) does not have the C—Cl antiparallel to the C_β—C_α bond and thus should react by a mechanism different from (**52**). Grob et al.[44] found that the 3β-chloride (**52**) reacts 13,500 times as fast as the model (**55a**) and gives only the expected product of fragmentation

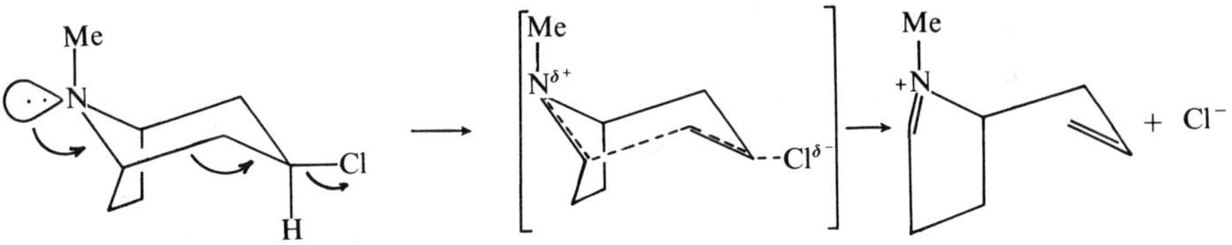

(28)

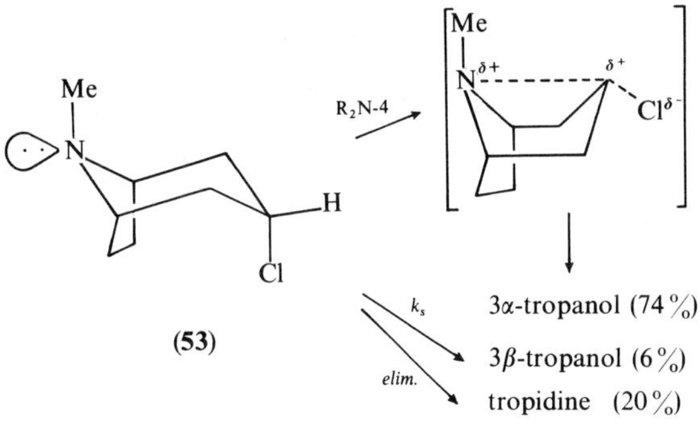

3α-tropanol (74%)
3β-tropanol (6%)
tropidine (20%)

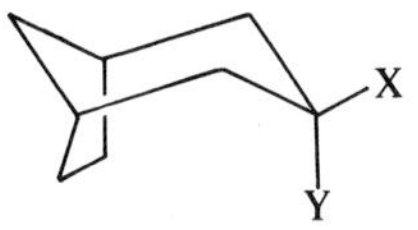

(**55**) a: X = Cl, Y = H
b: X = H, Y = Cl

(**54**). The 3α-chloride (**53**), as predicted, reacted by another mechanism, giving only substitution and elimination products at a rate comparable to the model (**55b**).

Neither direct nucleophilic participation by nitrogen nor frangomeric assistance is a possibility with the primary toluene-p-sulfonate (**56**), yet fragmentation occurs. The stepwise process [Eq. 29], with rate-limiting ionization directly to (**57**) which undergoes subsequent fragmentation, seems more probable to us than the mechanism involving the primary carbocation.[40]

Bordwell[45] has proposed an alternative mechanism to account for the rate increase called frangomeric assistance in some (or perhaps all) cases.*

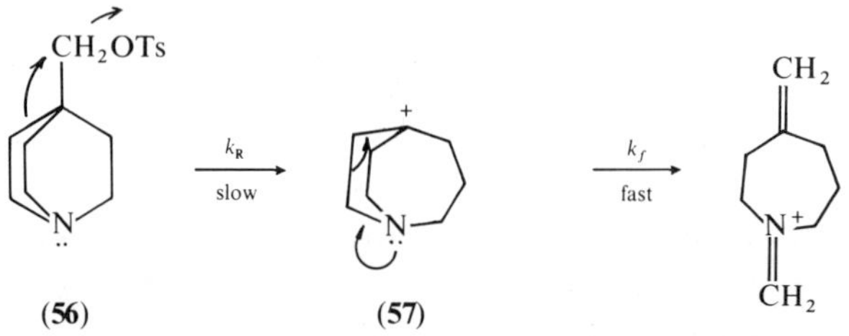

$$
(56) \xrightarrow[\text{slow}]{k_R} (57) \xrightarrow[\text{fast}]{k_f} \text{product} \quad (29)
$$

Since rate increases are not present in all examples where fragmentation occurs, e.g., (48), and since fragmentation is known to occur from carbocationic or carbocationic-like intermediates, e.g., (49), (51), or (57), a mechanism involving initial ion pairs is possible[45]; this mechanism is similar to that proposed by others,[46] but see Ref. 47. For example, 4-bromoquinuclidine (59) undergoes solvolysis in 80% ethanol at 40°C at a rate ∼ 10^5 times faster than 1-bromobicyclo[2.2.2]octane (58), yet (59) is some 10 times slower than t-butyl bromide. Grob[40] argues that the ionization of (59) is accelerated compared to bridgehead halides, which are normally inert. Hence Grob[40] concludes that ionization of (59) is concerted, giving (61)

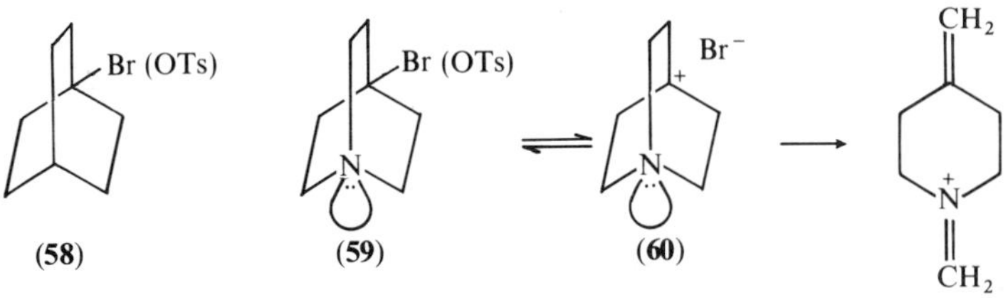

$$
(58) \quad (59) \rightleftharpoons (60) \rightarrow (61) \quad (30)
$$

directly. Bordwell[45] suggests that internal return (ion-pair return) is circumvented by fragmentation; hence a large rate increase is observed. Grob[40] noted that the ratio of the rate constants of the toluene-p-sulfonate and of the bromide (k_{OTs}/k_{Br}) is approximately 10^3 in (48) and 59). That the two substrates have similar k_{OTs}/k_{Br} rate ratios is taken to mean that bond breaking has advanced to the same degree in each example. A k_{OTs}/k_{Br} rate ratio of this magnitude generally has been assumed to mean that little nucleophilic assistance (neighboring group or solvent) is present.[48] Thus it

---

* We refer to this mechanism as the Bordwell–Shiner–Sneen mechanism for frangomeric assistance. Such a mechanism is also possible for anchimeric assistance (see p. 78).

seems to us that Bordwell's suggested mechanism merits strong consideration.

6.1.4. R_2N-5 and R_2N-6 Participation

From the rate data presented in the first part of this chapter it is clear that R_2N-5 and R_2N-6 anchimerically assisted reactions are even more favored than the reasonably facile R_2N-3 and R_2N-4 processes discussed above. It is not unexpected then that neighboring amino groups have been observed to intervene intramolecularly in a wide variety of reactions when five- or six-membered rings result. A few illustrative examples will be presented below in order to show the scope and limitations of these common reactions.

A large number of heterocyclic compounds of oxygen, nitrogen, and sulfur are prepared by condensation of di- or poly-functional compounds. Nitrogen-containing heterocycles by the thousands have been prepared by such reactions. The simplest such reaction is one leading to the preparation of a pyrrolidine or piperidine derivative. The synthesis of pyrrolidine may be accomplished by reaction of ammonia and 1,4-dichlorobutane in a stepwise process that may include an intramolecular displacement of chloride from the aminoalkyl chloride[1,2] [Eq. (31)] or of ammonia from the aminoalkyl amine hydrochloride[49] [Eq. (32)]; each process is known to occur.

(31)

(32)

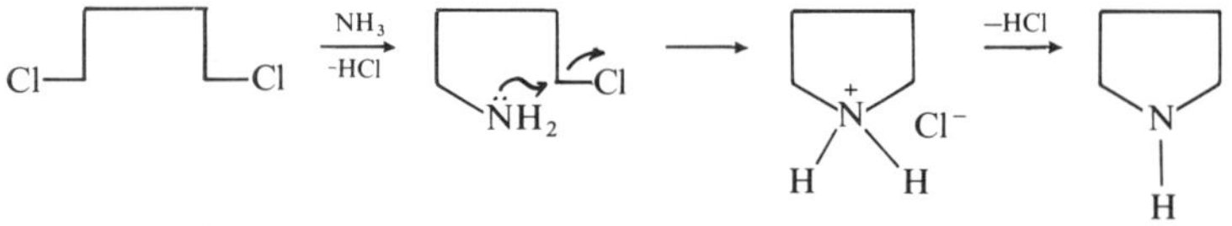

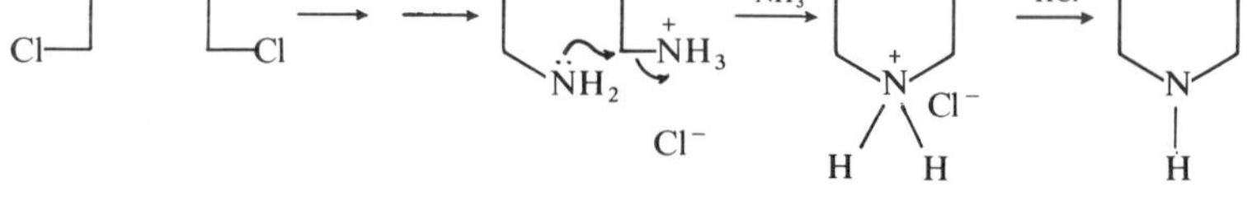

Undoubtedly a form of this intramolecular displacement reaction is involved in such syntheses as that of hexamethylenetetraamine (methenamine)[50] (**62**), which may be prepared simply by evaporating a mixture of formalin and concentrated ammonia solution, and of certain bicyclic amines, which can be prepared by the reductive cyclization of nitrodiesters such as the one shown in Eq. (34).[51,52]

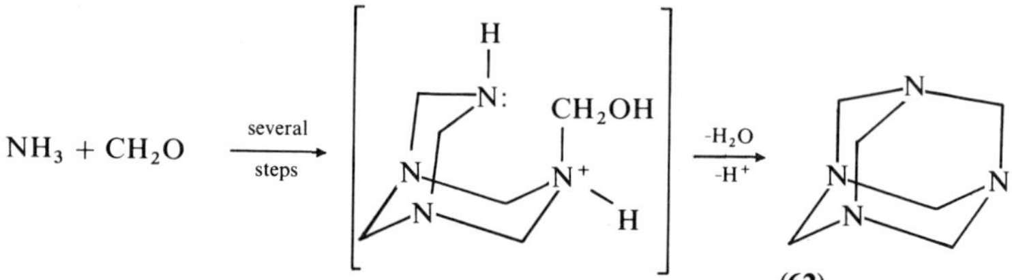

$$NH_3 + CH_2O \xrightarrow{\text{several steps}} [\text{(62)}] \xrightarrow[-H^+]{-H_2O} \quad (33)$$

$$\text{[nitrodiester with OMe groups]} \xrightarrow{H_2, PtO_2} \text{[bicyclic amine]} \quad (34)$$

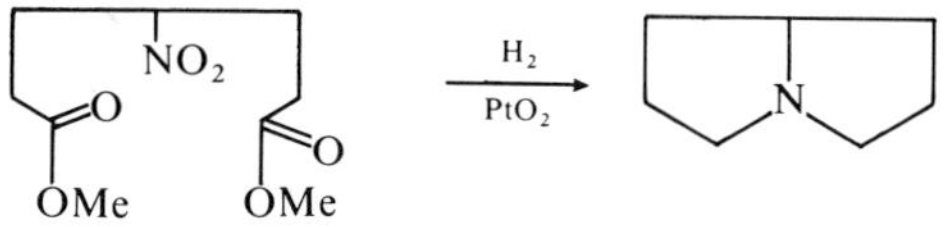

A limitation in the general synthesis has been described by Zwanenburg and Wynberg.[53] While a normal reaction is observed with the 1,4-dichloride (63) with ethylamine to form the fused pyrrolidine (65) apparently *via* the intermediate (64), which undergoes subsequent ring closure *via* N-5 participation, ring closure does not occur when groups are on the 2 and 5 positions of the thiophene ring. Thus (66) does not lead to (68), presumably as a result of a steric effect involving blocking of the leaving group by the ring chlorine, e.g., (67).[53]

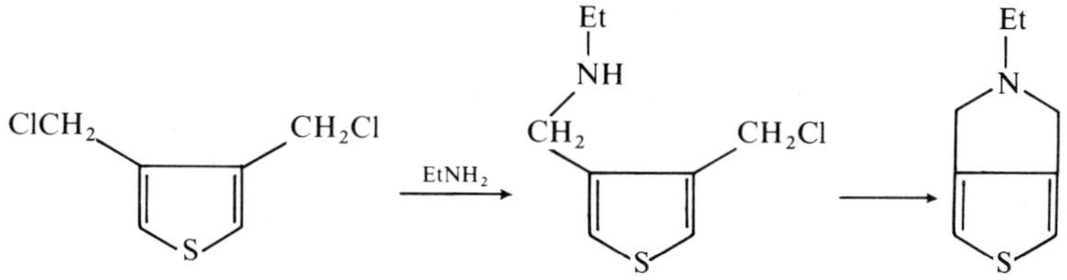

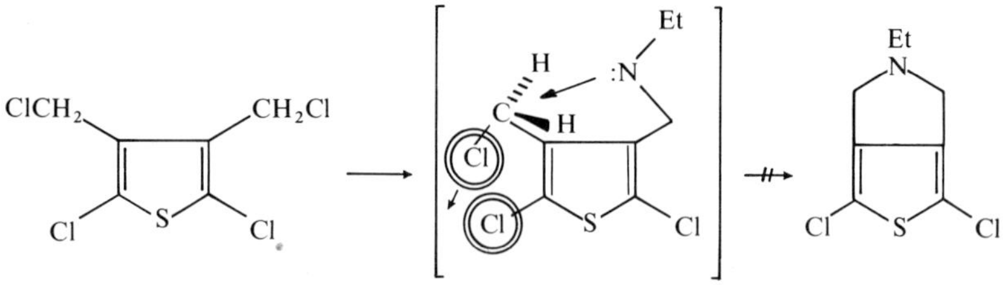

$$(63) \xrightarrow{EtNH_2} (64) \longrightarrow (65) \quad (35)$$

$$(66) \longrightarrow (67) \xrightarrow{-\!\!/\!\!\!\!\to} (68) \quad (36)$$

Participation by Nitrogen Groups

The activity of the amino group is also demonstrated by the facility with which it undergoes intramolecular Michael additions,[49,52] e.g., Eq. (37), and participates in hydrolysis reactions, e.g., Eqs. (38–40). In a more

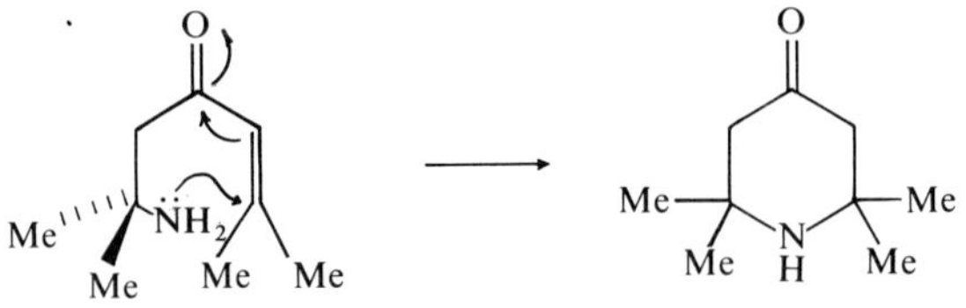

(37)

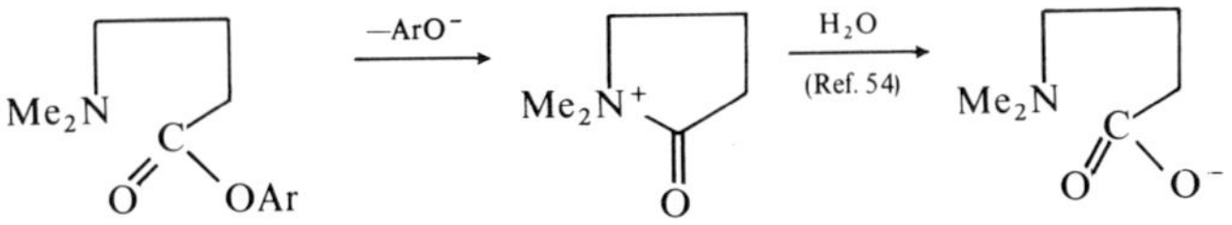

(38)

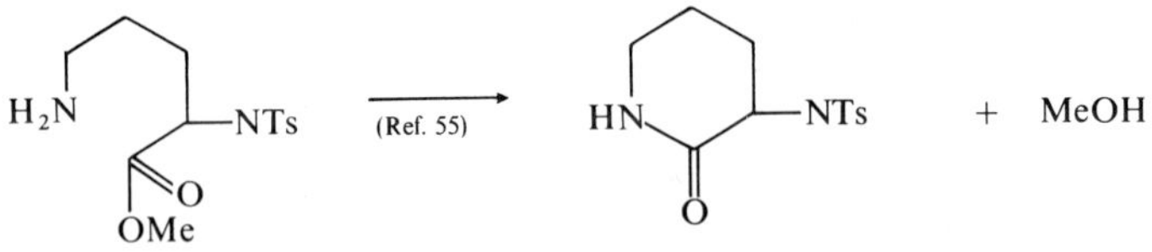

(39)

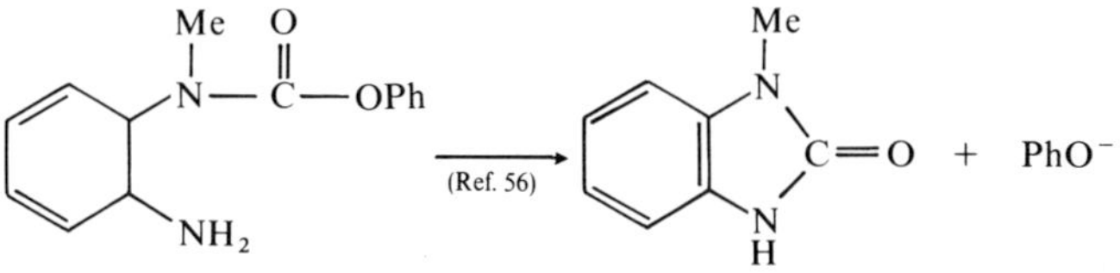

(40)

novel example, the tertiary amino group in ceveratrum esters participates, along with the axial hydroxyl group, in a rare example of intramolecular bifunctional general base–general acid catalysis of ester hydrolysis; i.e., see **(69)**[57]; neighboring group participation in hydrolysis reactions will be discussed more fully in Volume 2.

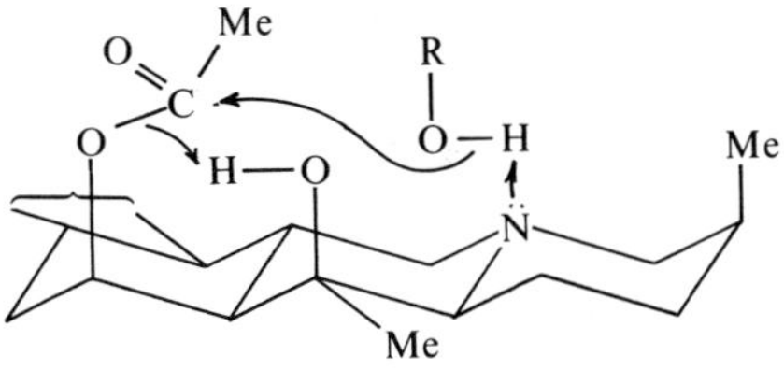

(69)

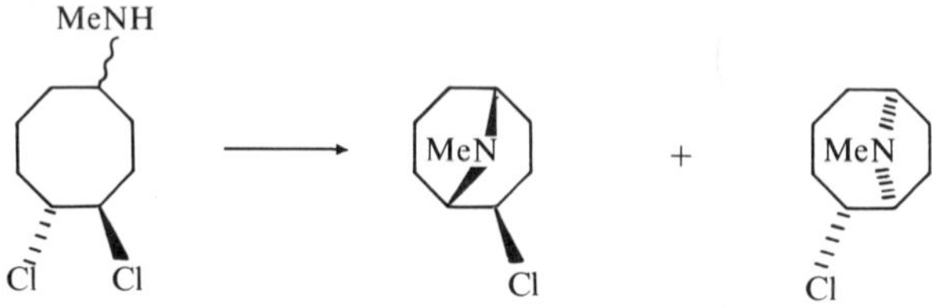

N-5 and N-6 participation may also lead to bicyclic heterocycles containing nitrogen. For example, 4,5-dichloro-N-methylcycloheptyl and octyl amines react with N-5 and N-6 participation to give products with inversion at the carbon undergoing displacement, e.g., Eq. (41).[58] In a similar type of reaction, heating of the amino alcohols (**70**), (**71**), or (**72**) with HI, HClO$_4$, or picric acid in ethanol led to ring closure, e.g., Eq. (42). However, under similar conditions, (**73**) led only to dehydration.[17,59]

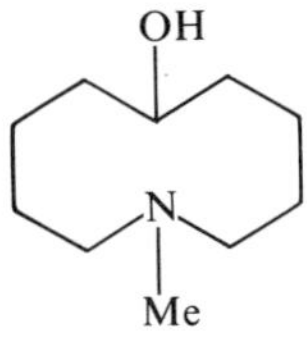

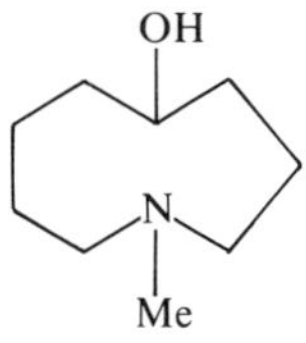

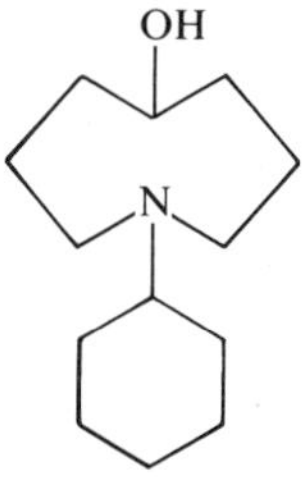

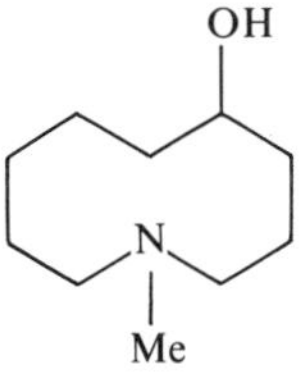

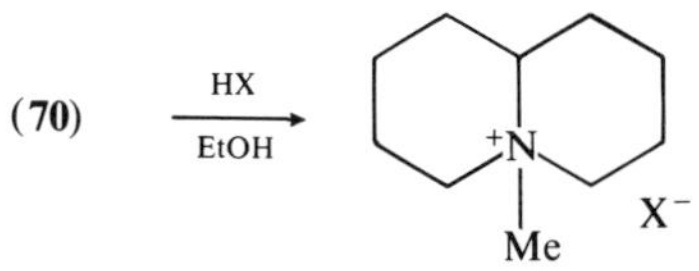

6.1.5. Miscellaneous Modes of Participation

Owing to the nucleophilicity of amines, the reaction of electrophiles, such as halogens, with olefinic amines may take a different course than that followed by alcohols or ethers, where concerted addition–cyclization normally occurs (see Chapter 4). For example, the bromination of 1-allylpyrazole (**74**) leads to the dibromide (**75**), which may be converted to diazapentalene (**76**) by N-5 cyclization and dehydrobromination.[60,61] In contrast, cyclization upon the addition of bromine and iodine does occur with 2-(3-buten-1-yl)pyridine (77) [Eq. (44)].[62] The rate constant for the addition of iodine to (77) at 20°C[63] is twice that for iodolactonization of $CH_2=CHCH_2CH_2CO_2^- Na^+$,[64] an olefin that also forms a five-membered ring. Staninets and Shilov[64,65] have determined that olefinic amines undergo concerted addition–cyclization only in the form of their free bases; hence cyclization can be prevented by halogenation in acidic medium.

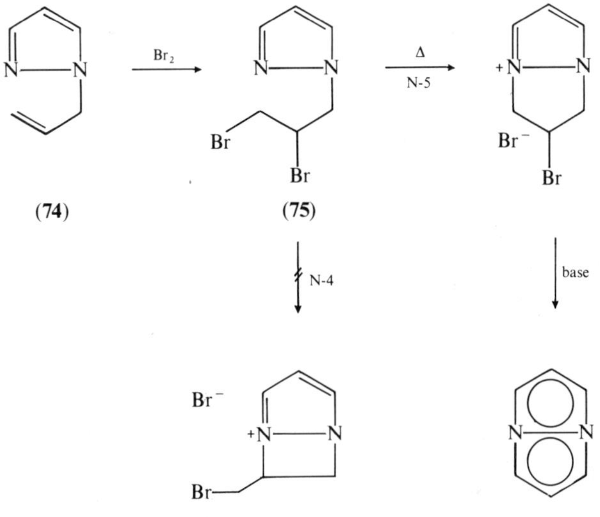

(43)

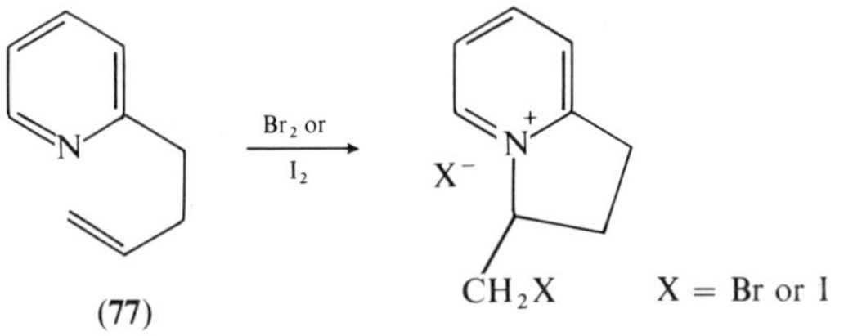

A novel type of anchimeric assistance involving nitrogen has been described by Miller et al.[66] The amine function in the alkaloid laudanosine (78) is said to result in anodic coupling at potentials 400–500 mV less positive than required for similar compounds without the amine functionality. This interaction is conceptualized either as an electrophilic attack on the aromatic ring by the initially formed aminium ion or homoconjugation between the aminium ion and the aromatic ring[66] (Scheme 5).

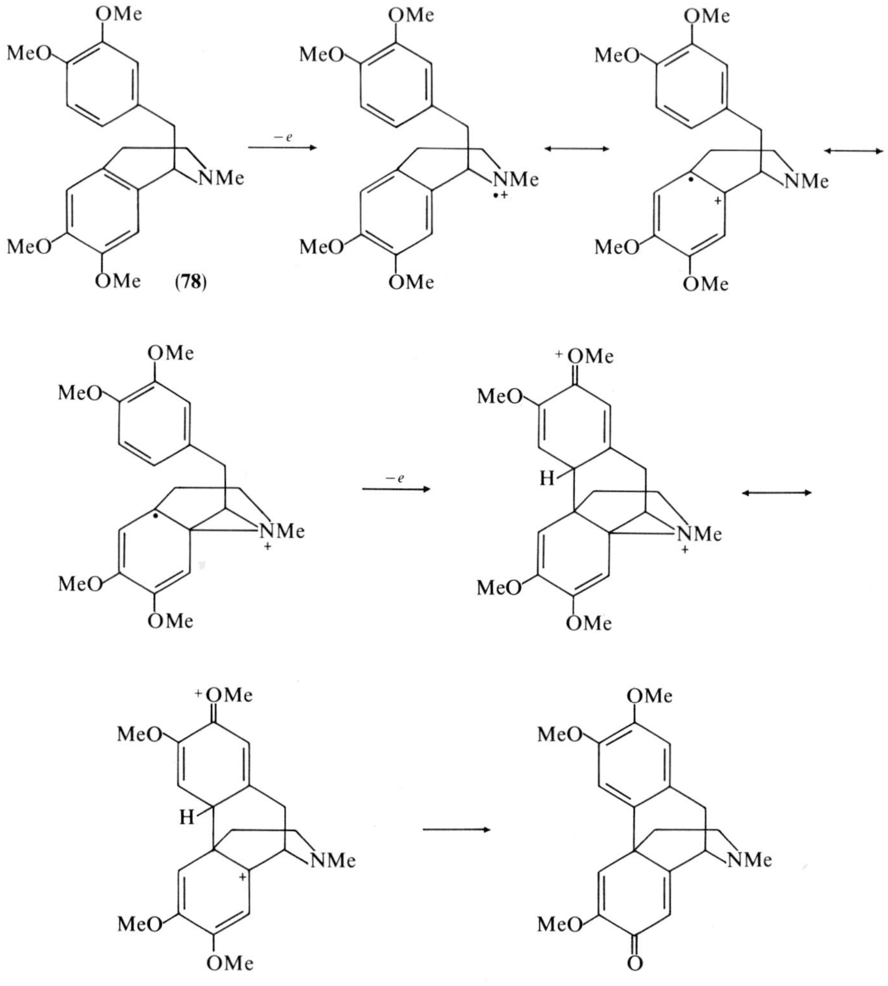

Scheme 5

Finally, we shall discuss the claim by Grob and his co-workers[67] that the dimethylamino group may, under certain circumstances, anchimerically assist displacement at tertiary carbon. In the previous section, the solvolysis of tertiary γ-N,N-dimethylaminoalkyl chlorides was discussed. Based primarily on the relative rate of solvolysis of the freely rotating tertiary amine (**48**) (Table 9) and its homomorph ($k/k_h = 0.52$), Grob[40] concluded that anchimeric assistance by nitrogen was absent. The next higher homolog of (**48**), i.e., (**79**), undergoes ethanolysis at a rate 0.89 times that of its homomorph to give a mixture of products [Eq. (45)]. Again only a slight rate-retarding effect of the neighboring amino group is evident. On

$$Me_2N{:}\overset{Me}{\underset{Me}{-}}Cl \quad \xrightarrow{80\% \text{ aq. EtOH}} \quad Me_2N\overset{(OH)}{\underset{MeMe}{-}}OEt$$
(79) (47%)

$$+ \quad Me_2N\overset{}{\underset{Me}{=}}Me \quad + \quad Me_2N\overset{}{\underset{}{=}}Me \quad + \quad Me_2\overset{+}{N}\overset{Me}{\underset{Me}{-}} \quad (45)$$

(37%) (16%)

replacement of the hydrogens on the β-carbon with methyl groups, as with the γ-N,N-dimethylamino compounds (Table 9), k/k_h values exceeding unity were measured (Table 10). Consistent with the rate indicated, the amount of pyrrolidinium salt increased to 77%. When a spiro-pentamethylene substituent [i.e., **(80)**] was used to restrict rotation (see Chapter 2, pp. 58–70) cyclization to **(81)** was the sole product with **(80a)**.[67] Grob et al.[67] considered two alternative mechanisms to account for the difference in the solvolytic behavior of **(80a)** and its freely rotating analogs. It was reasoned that the rate acceleration, as measured by k/k_h, could be explained by nucleophilic nitrogen participation in the transition state, i.e., **(82)**, leading to the pyrrolidinium ion as an intermediate.

Alternatively, the Bordwell–Shiner–Sneen ion-pair assistance mechanism[45-47,69] (see p. 248), i.e., Eq. (47), was put forth. In the latter case, rapid ionization to ion pairs occurs. Ionization of **(80b)** is assumed to be accompanied by rapid internal return (k_{-1} is fast). With **(80a)**, however,

Table 10. *Rates of Solvolysis of Tertiary δ-Aminoalkyl- and Homomorph Alkyl Chlorides in 80% Aqueous Ethanol at 30°C*[a]

Substrate	$k \times 10^5$ (sec^{-1})	k_{rel}	k/k_h
$Me_2NCH_2CH_2CH_2CMe_2Cl$[b]	2.07	0.89	0.89
$Me_2CHCH_2CH_2CH_2CMe_2Cl$[b]	2.32	1.0	
$Me_2NCH_2CMe_2CH_2CMe_2Cl_2Cl$[c]	372	160	7.20
$Me_2CHCH_2CMe_2CH_2CMe_2Cl$[c]	51.7	22.3	

[a]Rates courtesy of Professor C. Grob.
[b]Reference 68.
[c]Reference 67.

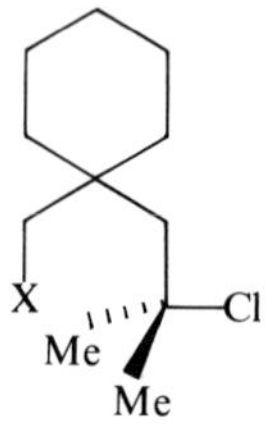

(80) a: X = NMe$_2$
b: X = CHMe$_2$

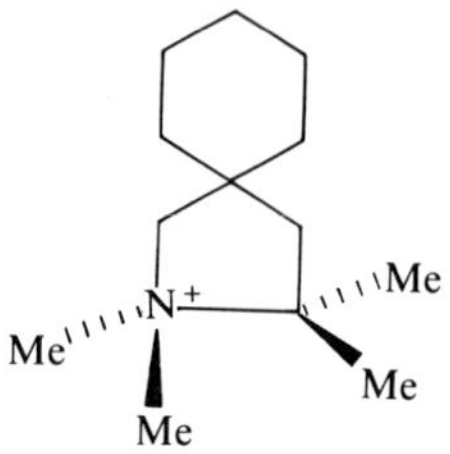

(81)

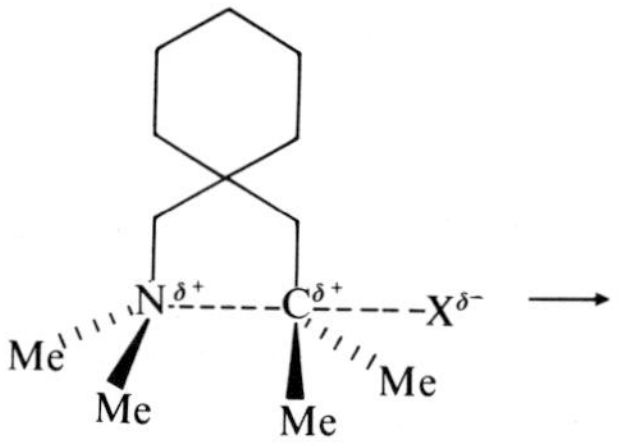

(82) → (81) (46)

(80a) 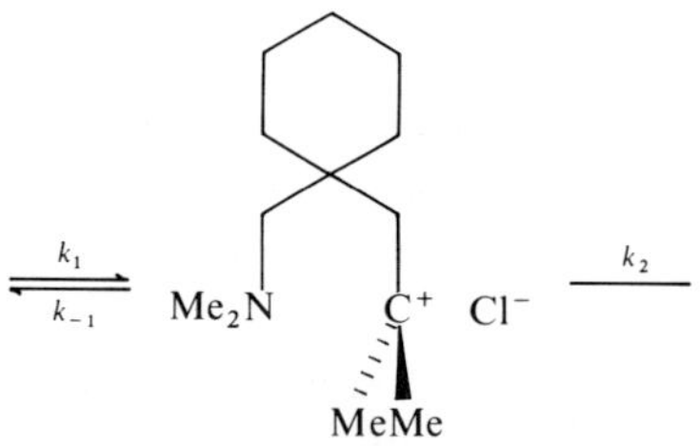 $\underset{k_{-1}}{\overset{k_1}{\rightleftharpoons}}$ [Me$_2$N ... C$^+$ MeMe] Cl$^-$ $\xrightarrow{k_2}$ (81) (47)

reaction to form (81) competes with internal return. Obviously, the restricted rotation favors both pathways relative to freely rotating analogs.

To reach a decision on the mechanism for participation by nitrogen in (80a), Grob et al.[67] prepared and solvolyzed the series of tertiary aryl-substituted p-nitrobenzoates (83) and (84) (Table 11). A plot of log k for

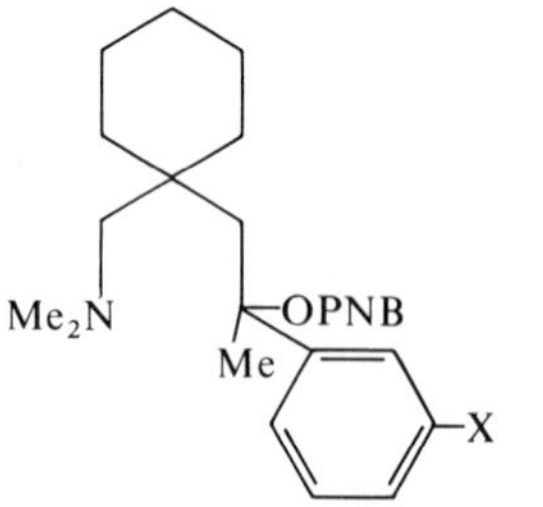

(83)

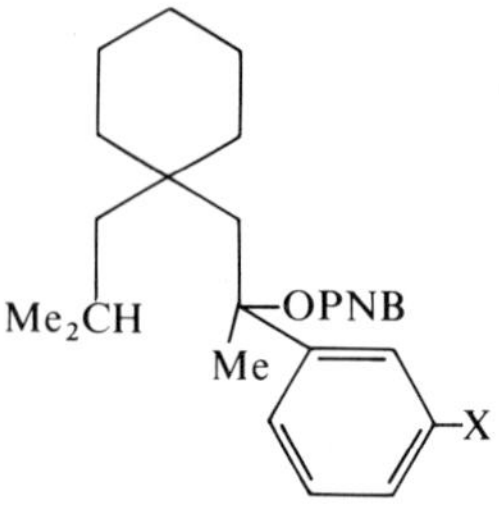

(84)

	X	k/k_h
a:	Me	25
b:	H	39
c:	OMe	43
d:	F	384

Table 11. Rates of Solvolysis of Tertiary δ-Aminoalkyl and Homomorph Alkyl p-Nitrobenzoates in 80% Aqueous Ethanol[a]

Substrate	X	Temp. (°C)	$k \times 10^4$ (sec^{-1})	$\Delta H^\ddagger$ (kcal mole^{-1})	$\Delta S^\ddagger$ (cal mole^{-1} °K^{-1})
(83a)	m-Me	30	49.6	20.69	−2.94
(83b)	H	30	40.6	20.38	−4.38
(83c)	m-OMe	30	25.5	21.21	−2.59
(83d)	m-F	30	9.05	21.97	−2.10
(84a)	m-Me	49.9	16.5	[b]	[b]
(84b)	H	49.9	8.34	23.85	−0.91
(84c)	m-OMe	49.9	5.08	[b]	[b]
(84d)	m-F	49.9	0.243	28.10	+5.07

[a] Data courtesy of Professor C. A. Grob.
[b] Rates were measured at only one temperature.

the model esters **(84a)**–**(84d)** against σ_m^+ constants yielded an experimental ρ^+ value of −4.33, which is in agreement with ionization of a normal tertiary substrate with substantial carbocationic character in the transition state, i.e., **(85)**. For comparison purposes, the ρ^+ value for solvolysis of the similar ester **(86)** in 80% aqueous acetone is −4.76 at 25°C.[70] The solvolysis of **(83a)**–**(83d)** produced cyclic product nearly quantitatively. The log k values for **(83a)**–**(83d)** gave a better correlation with σ_m (as opposed to σ_m^+) constants ($\rho = -1.91$), yet our own plot of the data indicates a modest curvature toward a more negative ρ value on the slow end (i.e., m-F).

         Me₂CH   C$^{\delta+}$----ŌPNB   Me—C—OPNB
                 Me   Me
                  Ar   Ar
                 (85)   (86)

From the magnitude of the ρ value there seems to be little doubt about the involvement of the amino group in the ionization of **(83)**. A plot of log k for the cyclization of the amine **(83b)** in ethanol–water mixture against Grunwald–Winstein Y values led to an m value 0.236. This value for sensitivity to a change in solvent ionizing power is far from that observed for limiting solvolyses[48] and more consistent with nucleophilic involvement.[48,71] Grob et al.[67] concluded that the ρ and m values for **(83a)**–**(83d)** were evidence for solvolysis via the traditional anchimerically assisted mechanism rather than by the Bordwell–Shiner–Sneen mechanism; that is, Eq. (46) rather than Eq. (47) better describes the process.

There is need for caution in accepting the above conclusion. The guidelines for the interpretation of data produced from the probes used by Grob et al. in arriving at their conclusion are essentially based on traditional S_N1 and S_N2 mechanisms and not on the existance of "hidden" ion-pair return.* If hidden ion-pair formation is as common as envisioned by some,[45-47,69] the interpretations of m and ρ values may be based, at least in part, on the reaction of ion pairs and not neutral substrates. The direction of the curvature of the log k versus σ_m plot for (**83a**)–(**83d**) (if the curvature is real) could be taken as evidence for the ion-pair mechanism. Hence all that can be said until the question of the general involvement of ion pairs is resolved is that nitrogen participation is demonstrated.

6.2. Nitrile Groups

The nitrogen atom in a nitrile is nucleophilic enough for such groups to participate as neighboring groups in carbonium ion reactions. Few examples, however, are known. The most probable reaction in which one might observe nitrile group participation is in an intramolecular Ritter reaction.[72] This has been observed in the conversion of the olefinic nitrile (**87**) to the lactam (**88**) in the presence of polyphosphoric acid.[73] The same effect is assumed to occur in the Beckmann cleavage of the oxime (**89**) to

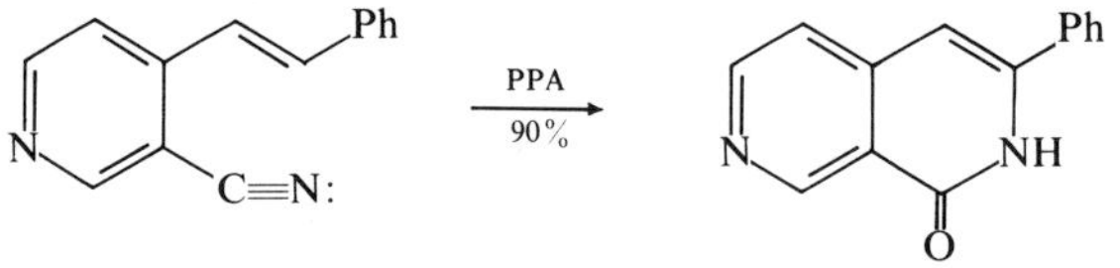

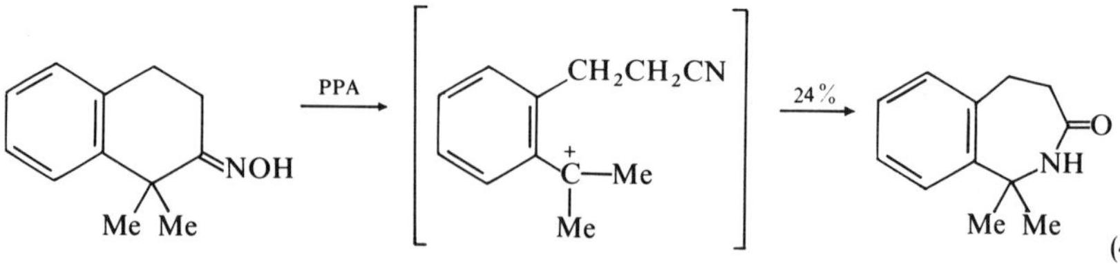

*The term *hidden* ion-pair return has been given to solvolytic examples where fast, reversible ionization to intimate ion pairs is not detectable.[46,47]

produce the lactam (**91**) via intermediate (**90**).[74] In certain cases, however, other processes are favored. Thus, reaction of (**92**) with polyphosphoric acid leads to the ketone (**93**) presumably by π participation as shown.[75]

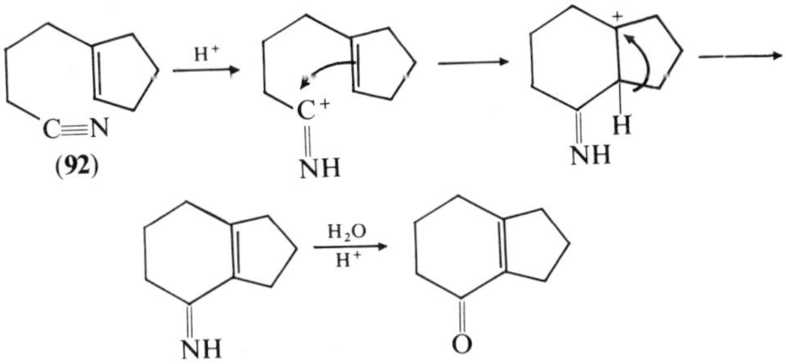

(50)

Participation by the cyano group in diazonium ion reactions has been reported.[80]

6.3. Hydrazone Groups

There are numerous condensation reactions involving a hydrazine derivative and a difunctional substrate that lead to pyrazoles or pyrazolines. These reactions have been reviewed elsewhere.[76] Typical of reactions involving participation by the hydrazone group are examples involving the ring opening of aziridines or epoxides. The stereochemistry of such

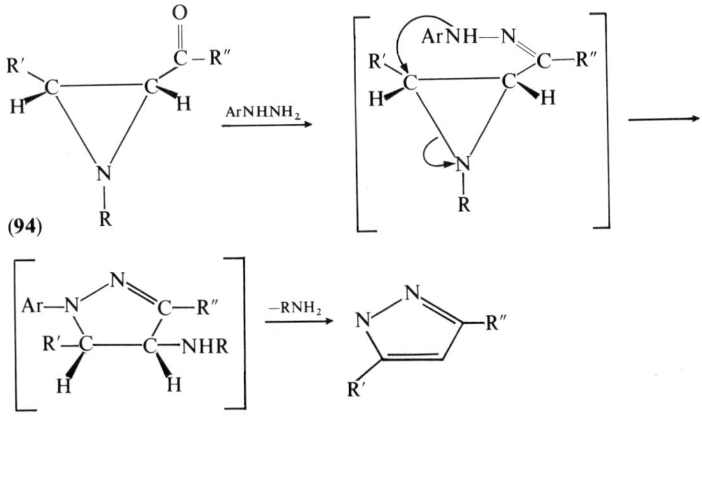

(51)

reactions have been studied by Cromwell and his co-workers and are postulated to occur as shown for the aziridine (94).[77]

A similar process is observed in the formation of pyrazolines from Mannich bases. The process apparently involves an elimination–cyclization mechanism[78] (Scheme 6) and not intramolecular displacement of the dimethylamino group in (95).

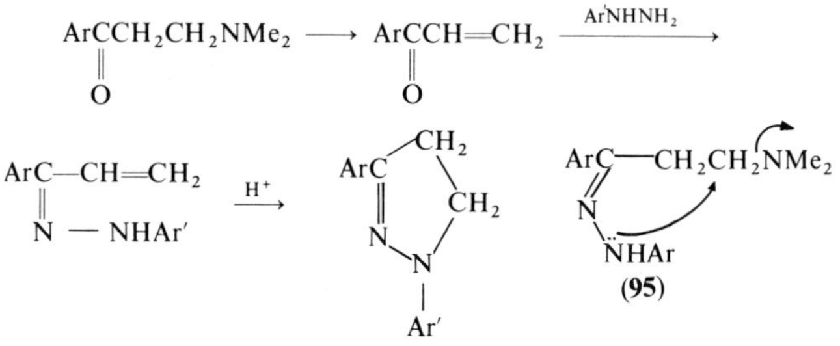

Scheme 6

References

1. H. Freundlich and H. Kroepelin, *Z. Phys. Chem.*, **122**, 39 (1926).
2. G. Salomon, *Helv. Chim. Acta*, **19**, 743 (1936).
3. R. Bird, A. C. Knipe, and C. J. M. Stirling, *J. Chem. Soc. Perkin Trans. 2*, 1215 (1973).
4. C. S. Dewey and R. A. Bafford, *J. Org. Chem.*, **32**, 3108 (1967).
5. E. L. Eliel, *Stereochemistry of Carbon Compounds*, McGraw-Hill Book Company, New York (1962), pp. 265–269.
6. A. Streitwieser, Jr., *Solvolytic Displacement Reactions*, McGraw-Hill Book Company, New York (1962).
7. P. D. Bartlett, S. D. Ross, and C. G. Swain, *J. Am. Chem. Soc.*, **69**, 2971 (1947).
8. E. M. Schultz and J. M. Sprague, *J. Am. Chem. Soc.*, **70**, 48 (1948); J. F. Kerwin, G. E. Ullyot, R. C. Fuson, and C. L. Zirkle, *ibid.*, **69**, 2961 (1947); E. M. Fry, *J. Org. Chem.*, **30**, 2058 (1965).
9. R. C. Fuson and C. L. Zirkle, *J. Am. Chem. Soc.*, **70**, 2760 (1948).
10. C. F. Hammer and S. R. Heller, *Chem. Commun.*, 919 (1966).
11. C. K. Ingold, *Structure and Mechanism in Organic Chemistry*, 2nd ed., Cornell University Press, Ithaca, N.Y. (1969).
12. J. M. Harris, *Prog. Phys. Org. Chem.*, **11**, 89 (1974).
13. E. R. Alexander, *Principles of Ionic Organic Reactions*, John Wiley & Sons, Inc., New York (1950), p. 99.
14. J. Hine, *Physical Organic Chemistry*, McGraw-Hill Book Company, New York (1956), pp. 121–122.
15. H. Gilman and A. P. Hewlett, *Recl. Trav. Chim. Pays-Bas*, **51**, 93 (1932); see also L. A. Paquette, R. W. Begland, and P. C. Storm, *J. Am. Chem. Soc.*, **90**, 6148 (1968).
16. D. S. Tarbell and J. R. Hazen, *J. Am. Chem. Soc.*, **91**, 7657 (1969).
17. N. J. Leonard, *Rec. Chem. Prog.*, **26**, 211 (1965).

18. N. J. Leonard and G. L. Shoemaker, *J. Am. Chem. Soc.* **71**, 1762 (1949).
19. V. R. Gaertner, *J. Org. Chem.*, **35**, 3952 (1970).
20. J. A. Deyrup and C. L. Moyer, *Tetrahedron Lett.*, 6179 (1968).
21. R. H. Higgins, F. M. Behlen, D. F. Eggli, J. H. Kreymborg, and N. H. Cromwell, *J. Org. Chem.*, **37**, 542 (1972).
22. R. H. Higgins and N. H. Cromwell, *J. Am. Chem. Soc.*, **95**, 120 (1973).
23. A. G. Hortmann and D. A. Robertson, *J. Am. Chem. Soc.*, **94**, 2758 (1972).
24. P. v. R. Schleyer, P. LePerchec, and D. J. Raber, *Tetrahedron Lett.*, 4389 (1969).
25. T. Okutani, A. Morimoto, and K. Masuda, *Third International Congress of Heterocyclic Chemistry, Sendia, Japan, Aug. 1971, Abst. No. D-23-6*.
26. J. A. Deyrup and S. C. Clough, *J. Org. Chem.*, **39**, 902 (1974).
27. J. M. Conia, *Angew. Chem. Int. Ed. Engl.*, **7**, 570 (1968).
28. S. Archer, T. R. Lewis, M. R. Bell, and J. W. Schulenberg, *J. Am. Chem. Soc.*, **83**, 2386 (1961).
29. J. W. Huffman, T. Kamiya, and C. B. S. Rao, *J. Org. Chem.*, **32**, 700 (1967); G. Buchi, D. L. Coffen, K. Kocsis, P. E. Sonnet, and F. E. Ziegler, *J. Am. Chem. Soc.*, **88**, 3099 (1966); these reactions are treated in Volume 2.
30. R. O. Hutchins and L. Rua, Jr., *J. Org. Chem.*, **40**, 2567 (1975).
31. H. O. Krabbenhoft, J. R. Wiseman, and C. B. Quinn, *J. Am. Chem. Soc.*, **96**, 258 (1974); P. G. Gassman, R. L. Cryberg, and K. Shudo, *ibid.*, **94**, 7600 (1972); W. P. Meyer and J. C. Martin, *ibid.*, **98**, 1231 (1976); T. A. Wnuk and P. Kovacic, *ibid.*, **97**, 5807 (1975); H. Böhme and K. Osmers, *Chem. Ber.* **105**, 2237 (1972).
32. F. L. Scott and R. N. Butler, *in: Carbonium Ions* (G. Olah and P. von R. Schleyer, eds.), Vol. IV, Chap. 30, John Wiley & Sons, Inc. (Interscience Division), New York (1973).
33. M. Mihalic, V. Šunjić, and F. Kajfež, *Tetrahedron Lett.*, 1011 (1975).
34. B. Capon, *Chem. Rev.*, **69**, 478 (1969).
35. R. D. Guthrie and D. Murphy, *J. Chem. Soc.*, 5288 (1963); see also J. Cleophax, S. D. Gero, and R. D. Guthrie, *Tetrahedron Lett.*, 567 (1967).
36. R. M. Allen and G. W. Kirby, *Chem. Commun.*, 1121 (1971).
37. K. Abe, Y. Nakamura, M. Onda, and S. Okuda, *Tetrahedron*, **27**, 4495 (1971).
38. J. C. Sheehan and J. W. Frankenfield, *J. Org. Chem.*, **27**, 628 (1962).
39. V. R. Gaertner, *J. Org. Chem.*, **32**, 2972 (1967); **33**, 523 (1968).
40. C. A. Grob, *Angew. Chem. Int. Ed. Engl.*, **8**, 535 (1969); see also C. A. Grob and P. W. Schiess, *ibid.*, **6**, 1, (1967), and Ref. 79.
41. C. A. Grob and F. A. Jenny, *Tetrahedron Lett.*, No. 23, 25 (1960).
42. C. A. Grob and W. Schwarz, *Helv. Chim. Acta*, **47**, 1870 (1964).
43. W. Klyne and V. Prelog, *Experientia*, **16**, 521 (1960).
44. A. T. Bottini, C. A. Grob, E. Schumacher, and J. Zergenyi, *Helv. Chim. Acta*, **49**, 2516 (1966); see also C. A. Grob, W. Kunz, and P. R. Marbet, *Tetrahedron Lett.*, 2613 (1975) for experiments dealing with regioselectivity of fragmentation.
45. F. G. Bordwell, *Acc. Chem. Res.*, **5**, 374 (1972).
46. R. A. Sneen, *Acc. Chem. Res.*, **6**, 46 (1973); R. A. Sneen and J. W. Larsen, *J. Am. Chem. Soc.*, **91**, 6031 (1969); V. J. Shiner, Jr., and W. Dowd, *ibid.*, **91**, 6528 (1969); V. J. Shiner, Jr., R. D. Fisher, and W. Dowd, *ibid.*, **91**, 7748 (1969).
47. T. W. Bentley, S. H. Liggero, M. A. Imhoff, and P. v. R. Schleyer, *J. Am. Chem. Soc.*, **96**, 1970 (1974).
48. J. L. Fry, C. J. Lancelot, L. K. M. Lam, J. M. Harris, R. C. Bingham, D. J. Raber, R. E. Hall, and P. v. R. Schleyer, *J. Am. Chem. Soc.*, **92**, 2538 (1970); but see J. Slutsky, R. C. Bingham, P. v. R. Schleyer, W. C. Dickason, and H. C. Brown, *ibid.*, **96**, 1969 (1974), relative to steric effects on the OTs/Br rate ratio.
49. R. O. C. Norman, *Principles of Organic Synthesis*, Methuen & Co. Ltd. London (1968), pp. 304–323.

50. E. M. Smolin and L. Rapoport, *s-Triazines and Derivatives*, John Wiley & Sons, Inc. (Interscience Division), New York (1959), pp. 545–596.
51. N. J. Leonard, L. R. Hruda, and F. W. Long, *J. Am. Chem. Soc.*, **69**, 690 (1947); N. J. Leonard, D. L. Felley, and E. D. Nicolaides, *ibid.*, **74**, 1700 (1952); and other papers in this series; also see the review in Ref. 17.
52. E. D. Bergmann, D. Ginsberg, and R. Pappo, *Org. React.*, **10**, 179 (1959).
53. D. J. Zwanenburg and H. Wynberg, *J. Org. Chem.*, **34**, 333, 340 (1969).
54. T. C. Bruice and S. J. Benkovic, *J. Am. Chem. Soc.*, **85**, 1 (1963).
55. E. F. Curragh and D. T. Elmore, *J. Chem. Soc.*, 2948 (1962).
56. T. H. Fife, J. E. C. Hutchins, and M. S. Wang, *J. Am. Chem. Soc.*, **97**, 5878 (1975).
57. S. M. Kupchan, S. P. Eriksen, and Y.-T. S. Liang, *J. Am. Chem. Soc.*, **88**, 347 (1966).
58. J. W. Bastable, J. D. Hobson, and W. D. Riddell, *J. Chem. Soc. Perkin Trans. 1*, 2205 (1972).
59. A. J. Sisti and D. L. Lohner, *J. Org. Chem.*, **32**, 2026 (1967).
60. S. Trofimenko, *J. Am. Chem. Soc.*, **87**, 4393 (1965).
61. T. W. G. Solomons and C. F. Voigt, *J. Am. Chem. Soc.*, **87**, 5256 (1965); **88**, 1992 (1966).
62. V. I. Staninets and E. A. Shilov, *Ukr. Khim. Zh.*, **31**, 1286 (1965); *Chem. Abs.*, **64**, 12626 (1966).
63. V. I. Staninets and E. A. Shilov, *Ukr. Khim. Zh.*, **34**, 1132 (1968); *Chem. Abs.*, **70**, 67318 (1970).
64. E. N. Rengevich, V. I. Staninets, and E. A. Shilov, *Dokl. Akad. Nauk SSSR*, **146**, 111 (1962); *Chem. Abs.*, **58**, 3285 (1963).
65. V. I. Staninets and E. A. Shilov, *Russ. Chem. Rev.*, **40**, 272 (1971) (a review of addition–cyclization reactions).
66. L. L. Miller, F. R. Stermitz, J. Y. Becker, and V. Ramachandran, *J. Am. Chem. Soc.*, **97**, 2922 (1975).
67. C. A. Grob, K. Seckinger, S. W. Tam, and R. Traber, *Tetrahedron Lett.*, 3051 (1973).
68. D. Currell, C. A. Grob, and S. W. Tam, *Helv. Chim. Acta*, **50**, 349 (1967).
69. F. G. Bordwell and T. G. Mecca, *J. Am. Chem. Soc.*, **94**, 2119 (1972); **94**, 5829 (1972); **97**, 123 (1975); **97**, 127 (1975); F. G. Bordwell, P. F. Wiley, and T. G. Mecca, *ibid.*, **97**, 132 (1975); F. G. Bordwell and G. A. Pagani, *ibid.*, **97**, 118 (1975).
70. E. N. Peters and H. C. Brown, *J. Am. Chem. Soc.*, **95**, 2397 (1973).
71. P. v. R. Schleyer, W. F. Sliwinski, G. W. Van Dine, U. Schöllkopf, J. Paust, and K. Fellenberger, *J. Am. Chem. Soc.*, **94**, 125 (1972).
72. L. I. Krimen and D. J. Cota, *Org. React.*, **17**, 213 (1969).
73. J. M. Bobbitt and R. E. Doolittle, *J. Org. Chem.*, **29**, 2298 (1964).
74. R. T. Conley and R. J. Lange, *J. Org. Chem.*, **28**, 210 (1963).
75. R. T. Conley and B. E. Nowak, *J. Org. Chem.*, **26**, 692 (1961); also see T. Sasaki, S. Eguchi, and M. Sugimoto, *Bull. Chem. Soc. Jpn*, **44**, 1382 (1971).
76. L. C. Behr, R. Fusco, and C. H. Jarbol, in: *Heterocyclic Compounds—Pyrazoles, Pyrazolines, Pyrazolidines, Indazoles and Condensed Rings* (R. H. Wiley, ed.), Vol. 22, John Wiley & Sons, Inc. (Interscience Division), New York (1967).
77. N. H. Cromwell, N. G. Barker, R. A. Wankel, P. J. Vanderhorst, F. W. Olson, and J. H. Anglin, *J. Am. Chem. Soc.*, **73**, 1044 (1951); N. H. Cromwell and R. J. Mohrbacher, *ibid.*, **75**, 6252 (1953).
78. F. L. Scott, S. A. Houlihan, and D. F. Fenton, *Tetrahedron Lett.*, 1991 (1970).
79. W. J. le Noble, H. Guggisberg, T. Asano, L. Cho, and C. A. Grob, *J. Am. Chem. Soc.*, **98**, 920 (1976), have reported activation volume studies on some substrates which undergo competitive fragmentation, solvolysis, elimination, and N-4 participation.
80. J. R. Beck and J. A. Yahner, *J. Org. Chem.*, **41**, 1733 (1976).

Author Index

Abe, K., 261 (37)
Abikar, K., 223 (11)
Adams, T., 193 (134)
Adelstein, G., 191 (53)
Agtarap, A., 192 (74)
Akagi, M., 225 (64)
Alexander, E.R., 260 (13)
Alfred, E., 74 (96)
Allen, L.C., 71 (11, 16)
Allen, R.M., 261 (36)
Allinger, N.L., 75 (153), 119 (33)
Allred, E.L., 18 (66), 120 (71), 190 (1, 2, 3, 7), 191 (30)
Anderson, C.B., 121 (82)
Arganbright, R.P., 224 (13, 39)
Armstrong, R., 17 (26), 224 (19)
Arnaud, P., 72 (69)
Arnold, R.T., 18 (61)
Asano, T., 262 (79)
Ashe, A.J., 17 (23), 72 (83), 119 (31)
Asperger, S., 120 (66)
Ausloos, P., 71 (20)

Bader, R.F.W., 192 (97)
Bafford, R.A., 260 (4)
Baird, R., 18 (67), 74 (114), 120 (80)
Baker, J.F., 194 (145)
Ballinger, P., 192 (96)
Barker, N.G., 262 (77)

Barker, R., 193 (108, 109, 110, 111)
Barnett, W.E., 75 (136)
Bartell, L.S., 72 (69), 119 (33)
Bartels, R., 17 (37)
Bartlett, P.D., 16 (8), 17 (23, 39), 18 (55, 61), 118 (31), 121 (86), 191 (54), 223 (1), 224 (18), 260 (7)
Barton, W.H., 18 (50), 192 (77)
Basalay, R.J., 224 (33)
Basolo, F., 117 (1)
Bastable, J.W., 262 (58)
Bastian, B.N., 190 (10)
Battiste, M.A., 119 (40, 48)
Beauchamp, J.L., 121 (96)
Beck, J.R., 262 (80)
Becker, J.Y., 262 (66)
Beckman, E.D., 18 (59), 262 (6)
Bedenbaugh, A.O., 193 (123)
Bedennec, G., 192 (90)
Begland, R.W., 191 (43, 45, 48), 260 (15)
Behlen, F.M., 261 (21)
Behr, L.C., 262 (76)
Bell, M.R., 261 (28)
Bell, R.P., 75 (140), 119 (50), 120 (62)
Belloli, R., 118 (21)
Bellucci, G., 193 (119)

Bel'skii, V.E., 225 (60)
Bender, M.L., 16 (3, 11), 117 (1), 118 (13), 121 (98), 192 (83)
Benkovie, S.J., 16 (11), 74 (121), 262 (54)
Bennett, G.M., 18 (51), 74 (104), 223 (3)
Bennett, W., 17 (26), 224 (19)
Bentley, T.W., 120 (58, 75), 261 (47)
Bentrude, W.G., 118 (15), 225 (55)
Berchtold, G.A., 225 (58)
Berke, T.D., 71 (8)
Bernhard, R., 17 (23), 118 (31)
Berry, W.A., 18 (51)
Berson, J., 16 (18)
Bert, G., 193 (119)
Berwin, H.J., 17 (22), 72 (58, 62)
Bethell, D., 121 (88)
Beverly, G.M., 223 (10)
Bienvenue-Goetz, E., 118 (3), 193 (120)
Bingham, R.C., 72 (65), 118 (30), 120 (75, 76), 261 (48)
Birch, A.J., 191 (31)
Bird, R., 74 (102, 106), 223 (6), 260 (3)
Bischaf, P.K., 71 (31)
Blandamer, M.J., 190 (17)

*The number in parentheses is the reference number on the page cited.

Blicke, F.F., 18 (62), 192 (66)
Blickenstaff, R.T., 194 (145)
Block, E., 223 (4), 224 (17)
Block, J.H., 194 (136)
Blumbergs, P., 191 (62)
Bly, R.K., 193 (123)
Bly, R.S., 120 (74), 193 (123)
Bobbitt, J.M., 262 (73)
Bodor, N., 119 (33)
Bodot, H., 192 (90, 94)
Boerma, J.A., 224 (31)
Bogar, A.Q., 190 (12)
Bogdanowicz, M.J., 224 (24)
Böhme, H., 18 (53), 74 (124), 223 (2), 261 (31)
Bolton, R., 118 (4)
Bonazza, B.R., 121 (92)
Borbrànski, B., 121 (83)
Borchardt, R.T., 75 (146), 194 (137)
Bŏrcić, S., 72 (55), 119 (51), 120 (55, 67)
Borden, W.T., 194 (149)
Bordwell, F.G., 74 (105), 120 (68), 223 (4), 261 (45, 69)
Borner, P., 192 (70)
Boschan, R., 18 (60)
Bottini, A. T., 261 (44)
Boyd, R.H., 119 (33)
Bradbury, W.C., 75 (149)
Bradshaw, R.W., 192 (81)
Bramen, W.T., 74 (105), 223 (4)
Braun, A.M., 16 (8)
Breslow, R., 16 (6), 224 (32)
Brice, C., 18 (64)
Brimacombe, J.S., 190 (11), 191 (39)
Brindell, G.D., 224 (39)
Brookhart, M., 118 (6)
Brown, A., 194 (136)
Brown, F., 16 (17)
Brown, H.C., 16 (19), 17 (23), 70 (1), 72 (44, 61), 118 (11, 29, 31), 119 (40, 41, 42, 43,

Brown, H.C. (cont'd) 44, 45, 46), 120 (77), 193 (126), 224 (39), 261 (48), 262 (70)
Brown, K.S., 71 (24)
Brown, R.S., 17 (22), 72 (58, 63)
Bruice, T.C., 16 (11), 73 (87, 89), 74 (121), 75 (148, 149), 121 (83), 194 (136), 262 (54)
Buchanan, J.G., 75 (135), 190 (12), 193 (99)
Buchi, G., 261 (29)
Buckingham, D.A., 193 (114), 194 (146)
Buckles, R.E., 16 (1), 17 (33), 18 (48)
Buddenbaum, W.E., 120 (54)
Buchler, E., 194 (154)
Bullock, C., 191 (64)
Buss, V., 71 (11, 16)
Butler, A.R., 121 (89)
Butler, P.E., 223 (10), 224 (50)
Butler, R.N., 261 (32)
Byrd, J.E., 193 (128)

Cadogan, J.I.G., 194 (152)
Caldin, E.F., 18 (56)
Cannie, J., 194 (142)
Caple, R., 224 (14)
Capman, M.L., 191 (37)
Capon, B., 16 (7), 18 (73), 72 (65), 73 (87, 89), 74 (98, 119, 125, 139), 75 (139), 121 (89), (89), 190 (8), 191 (37) 192 (79), 193 (136), 194 (138, 147), 261 (34)
Capozzi, G., 224 (22)
Cargill, R.L., 192 (73)
Carroll, B.L., 72 (69)
Carroll, J.T., 121 (99), 194 (141)
Casey, C., 118 (20), 192 (74)
Cerrini, S., 121 (83)
Chadwick, A.F., 18 (957)
Challis, J.A., 194 (152)

Chaney, M.O., 224 (36)
Chanley, J.D., 18 (63)
Chapman, N.B., 72 (65)
Chapuis, J., 193 (122)
Chau, M.M., 225 (57)
Chen, P.H., 190 (28)
Cherest, M., 193 (111)
Ching, E.A., 190 (11)
Ching, O.A., 191 (39)
Chloupek, F.J., 16 (19), 121 (98)
Chodkiewicz, W., 191 (32)
Choi, L., 262 (79)
Cholod, M.S., 75 (137)
Chow, Y.-L., 121 (98)
Cinquini, M., 74 (109)
Clark, D.T., 71 (13)
Clayton, R.B., 193 (105)
Clevenger, J.V., 17 (20)
Clinton, N.A., 17 (22), 72 (58, 63)
Closs, G.L., 72 (69)
Closson, W.D., 191 (41), 192 (75)
Clough, S.C., 261 (26)
Coffen, D.L., 261 (29)
Coffey, F., 74 (103)
Cohen, B., 17 (42)
Cohen, J.F., 74 (133)
Cohen, L.A., 75 (145, 146), 194 (137)
Coke, J.L., 121 (102)
Collins, C.J., 71 (35, 55)
Coloma, S., 74 (109)
Conley, R.T., 262 (74, 75)
Core, S.K., 190 (27)
Corey, E.J., 193 (107), 224 (17)
Corny, M., 193 (100)
Corse, J., 18 (69), 70 (1), 72 (53), 190 (23), 224 (45)
Cota, D.J., 194 (139), 262 (72)
Coulson, C.A., 71 (39), 73 (74)
Counsell, R.E., 191 (53)
Coward, J.K., 194 (156)
Coxon, J.M., 193 (101)
Cram, D.J., 17 (23, 25), 118 (18)

Author Index

Crawford, R.J., 225 (53)
Creary, X., 75 (144)
Creighton, A.M., 225 (65)
Cristol, S.J., 224 (13, 14, 39)
Cromartie, T.H., 192 (82)
Cromwell, N.H., 72 (69), 261 (21, 77)
Crooks, J.E., 120 (62)
Cross, F.J., 193 (13)
Crothers, D.M., 73 (89)
Crouzet, P., 192 (90), 225 (62)
Crowdle, J.H., 193 (104)
Cruickshank, P.A., 75 (154)
Curragh, E.F., 74 (122), 262 (55)
Currell, D., 262 (68)

Dafforn, G.A., 73 (89), 120 (65), 194 (136)
Dagnani, M.J., 120 (67)
Dahm, R.H., 16 (7)
Danishefsky, S., 74 (130, 132), 194 (159)
Dannenberg, J., 71 (8)
Darieli, R., 74 (109)
Dauksas, V., 190 (29)
Davis, C.E., 193 (114)
Davis, R.E., 71 (6)
Deaken, D.M., 223 (11)
DeGroot, A., 224 (31)
de la Mare, P.B.D., 118 (4)
Delisi, C., 73 (89)
DeMare, G.R., 72 (69)
Dembinskiene, J., 190 (29)
de Meijere, A., 72 (69)
De Member, J.R., 71 (25)
de Moura Campos, M., 18 (61)
Denny, D.B., 192 (72)
de Salas, E., 16 (16)
DeTar, D.F., 74 (93), 119 (33), 193 (135)
deWaard, E.R., 224 (44)
Dewar, M.J.S., 71 (9, 31, 39), 72 (50), 119 (33)
Dewey, C.S., 260 (4)
Deyrup, C.L., 119 (40)
Deyrup, J.A., 261 (20, 26)
Diaz, A.F., 71 (23), 118 (6), 120 (72), 121 (101)

Dickason, W.C., 120 (77), 261 (48)
Dickerson, J.P., 191 (62)
Diegman, G.A., 224 (47)
Dines, M.B., 224 (17, 31)
Dixon, D.A., 71 (18)
Donaldson, M.M., 118 (29)
Drago, R.S., 117 (1)
Drayer, D., 224 (38)
Drinnan, C.V.A., 224 (30)
Dubois, J.E., 118 (3), 193 (120)
Dunkin, I.R., 191 (51, 52), 193 (115)
Dunn, A.R., 224 (28)
Durham, L.J., 190 (21)
Dyas, C., 73 (86)
Dyckes, D., 121 (81)
Dynak, J., 74 (130, 132)

Eastlick, D.T., 194 (152)
Eastman, R.H., 72 (69)
Eaton, D.F., 72 (64)
Eberson, L., 75 (151), 121 (81)
Ebner, C.E., 16 (8)
Edison, D.H., 120 (66)
Edwards, L.J., 18 (62)
Eggli, D.F., 261 (21)
Eguchi, S., 193 (130), 262 (75)
Eliason, R., 120 (55, 56)
Eliel, E.L., 118 (8), 194 (155), 260 (6)
Elmore, D.T., 74 (122), 262 (55)
Emmel, R.T., 71 (26)
Emmons, W.D., 223 (7)
Engelman, C., 73 (86)
Engler, E.M., 119 (33)
Enkssen, S.P., 262 (57)
Epstein, W.W., 192 (72)
Eriksen, S.P., 194 (145)
Eschenmoser, A., 74 (131)
Evans, E., 71 (24)
Evans, R.A., 190 (20)
Evans, W.P., 18 (44)
Ewald, A.H., 192 (87)

Factor, A., 193 (124)
Fahey, R.C., 120 (61)

Fairweather, R., 16 (6)
Fanta, P.E., 16 (5)
Farazmand, I., 74 (98), 192 (79)
Farooq, S., 74 (131)
Favini, G., 119 (33)
Fedeli, W., 121 (83)
Feliu-Otero, L.A., 224 (47, 48, 49)
Felkin, F., 193 (111)
Fellenberger, K., 120 (79), 262 (71)
Felley, D.L., 262 (51)
Fentiman, A.F., 70 (1), 119 (36, 37)
Fenton, D.F., 74 (110), 262 (78)
Fiato, R.A., 119 (48)
Fife, T.H., 262 (56)
Fisher, H., 71 (28)
Fisher, R.D., 120 (64), 261 (3, 46)
Fisher, T.H., 225 (56)
Fisher, W.F., 118 (7)
Fishman, M., 75 (154)
Fitzgerald, P.H., 225 (54)
Fluendy, M.A.D., 75 (140)
Flynn, E.J., 74 (110)
Fodor, G., 121 (83)
Foote, C.S., 72 (68), 118 (25)
Foster, D.M., 193 (114)
Foster, J.P., 120 (67)
Fountaine, J.E., 121 (106)
Fourner, J.L., 194 (143)
Frank, D.L., 121 (104)
Frankenfield, J.W., 261 (38)
Franzus, B., 194 (148)
Frateu, F., 71 (10)
Freundlich, H., 17 (37), 74 (10)
Friedman, M., 194 (145)
Friedrich, E.C., 121 (82)
Frundlich, H., 260 (1)
Frush, H.L., 17 (36)
Fry, A., 192 (86)
Fry, E.M., 260 (8)
Fry, J.L., 71 (36), 118 (30), 119 (33), 120 (75), 261 (48)
Fusco, R., 262 (76)
Fuson, R.C., 260 (8, 9)

Gaertner, V.R., 261 (19, 39)
Gagnaire, D., 191 (41)
Galli, C., 74 (128)
Gandour, R.D., 75 (138)
Ganem, B., 194 (157)
Garner, H.K., 17 (33, 34)
Garratt, S., 224 (32)
Gassman, P.G., 70 (1), 119 (35, 36, 37), 191 (58, 59, 60), 261 (3)
Gates, M., 121 (104)
Gaylord, N.G., 193 (104)
Geanangel, R.A., 71 (26)
Gero, S.D., 261 (35)
Giacin, J., 192 (72)
Gibson, M.S., 192 (81)
Gilman, H., (34, 43), 260 (15)
Gindler, E.M., 18 (63)
Ginsberg, D., 262 (6)
Girot, I., 192 (85)
Gleim, R.D., 194 (158)
Glick, R., 18 (60), 74 (96), 120 (71), 190 (1)
Glinski, R.P., 191 (62)
Goering, H.L., 17 (20), 224 (12)
Gokel, G.W., 191 (35)
Gold, V., 71 (4)
Golinkin, H.S., 190 (17)
Goodman, A.L., 72 (69)
Goodman, L., 18 (60), 191 (37), 193 (117)
Goodson, T., 224 (36)
Gordon, A.J., 120 (60)
Gortler, L.B., 121 (86)
Gould, C.W., 17 (34)
Gould, E.S., 117 (1)
Gould, V., 121 (88)
Graeve, R., 120 (60)
Graff, M.A., 72 (69)
Grammaccioni, P., 119 (33)
Gratz, R.F., 224 (29)
Gravitz, N., 121 (83)
Gray, D., 192 (75)
Gray, G.R., 191 (38)
Green, G.E., 192 (76)
Green, M.J., 224 (25)
Gregoriou, G.A., 120 (63)
Greiner, R.W., 18 (50), 74 (97), 192 (77)

Grob, C.A., 16 (8), 18 (72), 118 (14), 262 (79)
Grossman, N.R., 120 (69)
Gruetzmacher, R.R., 118 (12)
Grunwald, E., 16 (4), 17 (33, 34), 18 (58, 67, 71), 70 (1), 72 (53), 117 (1), 120 (75), 121 (85), 192 (89), 224 (45)
Guggisberg, H., 262 (79)
Guillom-Dron, D., 191 (32)
Guillory, J.P., 72 (69)
Guthrie, R.D., 261 (35)
Gutovski, G.E., 191 (62)
Guyberg, R.L., 261 (31)
Gylys, J.A., 193 (121)

Haddon, V.R., 121 (104)
Haga, M., 225 (64)
Hagen, E.L., 71 (32, 34)
Halevi, E.A., 119 (50)
Hall, R.E., 120 (64, 75), 261 (48)
Halpern, J., 193 (127, 128)
Halpern, Y., 71 (25)
Hamaide, N., 190 (14)
Hammer, C.F., 260 (10)
Hammett, L.P., 117 (1)
Hammond, G.S., 75 (142)
Hanessian, S., 191 (62)
Hansen, C., 16 (4), 17 (33)
Hanstein, W., 17 (22), 72 (58, 62)
Hardwick, T.J., 18 (43)
Harihara, P.C., 71 (17, 30)
Harness, J.A., 224 (35)
Harper, J.J., 120 (75), 118 (19, 29)
Harris, C.L., 118 (7)
Harris, D.O., 73 (89), 194 (136)
Harris, J., 17 (42)
Harris, J.M., 18 (22), 70 (2), 118 (24, 30), 120 (64, 73, 75), 121 (103), 190 (25), 260 (12), 261 (48)
Hartman, F.C., 191 (38), 193 (108)
Hartnagel, J., 18 (52)
Hartshorn, M.P., 193 (101)
Haseltine, R., 121 (90)

Hatch, E., 74 (130)
Hauakami, J.H., 193 (126)
Hay, A.W., 18 (43)
Haywood-Farmer, J., 119 (38, 40)
Hazen, J.R., 190 (9, 13), 191 (44, 57), 260 (16)
Heap, N., 192 (76)
Heard, D.D., 193 (111)
Heathcoat, F., 18 (51), 74 (104), 223 (3)
Heck, R., 18 (65, 66, 67, 68), 74 (96, 115, 116), 120 (71), 121 (104), 190 (1, 23), 224 (45)
Hehre, W.J., 71 (11, 15), 72 (46, 47, 52)
Heilbronner, E., 119 (49)
Heimbach-Johász, S., 18 (62), 192 (67)
Heine, H.W., 18 (50, 59), 74 (97, 100), 192 (77, 78, 95)
Heitz, R.M., 224 (39)
Heller, S.R., 260 (10)
Helmkamp, G.K., 223 (11)
Henbest, H.B., 193 (102, 105, 106)
Henderson, R.B., 17 (32), 190 (18)
Hendrickson, W., 190 (19)
Henrichs, P.M., 190 (6)
Henrick, Q., 74 (118)
Henry, P.M., 193 (129)
Hepner, R.R., 118 (3)
Herk, R., 74 (96)
Hess, Jr. B.A., 17 (23), 73 (83), 119 (31)
Hewlett, A.P., 191 (43), 260 (15)
Hiberty, P.C., 72 (47)
Higgins, R.H., 261 (21, 22)
Himmler, W.A., 193 (104)
Hine, J., 75 (137), 117 (1), 260 (14)
Hirsch, J.A., 193 (131)
Hoare, D.G., 193 (133)
Hobson, J.D., 262 (58)
Hock, A.L., 18 (51)
Hoffman, R., 71 (5), 72 (51, 52), 73 (71), 119 (47)

Hogeveen, H., 74 (109)
Hogg, D.R., 223 (10)
Hojo, M., 224 (34)
Holiday, E.R., 18 (54), 223 (1)
Honig, M.L., 120 (60, 67)
Hooley, S.R., 193 (121)
Horn, D.E., 120 (75)
Hortmann, A.G., 261 (23)
Horzi, Z., 224 (46)
Hosokawa, T., 194 (154)
Houanee, J.W., 194 (153)
Hough, L., 191 (64)
Houlihan, S.A., 262 (78)
Howdon, M.E.H., 72 (41)
Howe, K.L., 224 (12)
Hruda, L.R., 262 (51)
Hudson, B.G., 193 (110, 111)
Hudson, G.V., 72 (69)
Hudson, R.F., 194 (150)
Huffman, J.W., 261 (29)
Hughes, E.D., 16 (13, 17)
Hughes, N.A., 191 (61, 63), 224 (35)
Huisman, H.O., 224 (44)
Humski, K., 120 (64)
Hutchins, J.E.C., 262 (56)
Hutchins, R.O., 194 (155), 261 (30)
Hutchinson, E.G., 191 (31)

Ichi, T., 224 (34)
Ikegami, S., 119 (40), 193 (126)
Illuminati, G., 74 (128, 129)
Imhoff, M.A., 120 (58), 261 (47)
Ingold, C.K., 75 (147), 119 (31), 260 (11), 16 (10, 13, 17), 17 (31)
Ingraham, L.L., 16 (2), 17 (32), 18 (71), 190 (5, 22, 24)
Ingrosso, G., 193 (119)
Inoi, T., 224 (46)
Ireland, R.E., 224 (26)
Irie, T., 118 (6), 119 (39)
Isbell, H.S., 17 (34)
Isvasyuk, N.V., 225 (61)
Itani, H., 224 (15)
Ito, M., 72 (69)

Jackman, L.M., 121 (104)
Jackson, T.E., 192 (73)
Jacob, E.J., 119 (33)
Jacobus, J., 193 (109)
Jagow, R.H., 120 (61)
Janoschek, R., 71 (10)
Jarbel, C.H., 262 (76)
Jencks, W.P., 16 (11), 73 (87, 90, 91), 117 (1), 119 (50), 121 (84), 192 (83, 85), 194 (136)
Jenny, E.F., 18 (70)
Jenny, F.A., 16 (8), 261 (41)
Jensen, F.R., 17 (22), 72 (55)
Jensen, J.H., 75 (137)
Jerrow, J.L., 192 (75)
Jeuell, C.L., 72 (44, 56)
Jewett, J.G., 71 (12), 72 (54)
Jindal, S.P., 118 (12)
Johnson, P.Y., 225 (58, 59, 60)
Johnson, W.S., 16 (12)
Jones, A.J., 121 (90)
Jones, M.G., 121 (102)
Jones, N.D., 224 (35, 36)
Jullien, J., 192 (94)

Kabakoff, D.S., 72 (43)
Kaiser, E.M., 191 (34)
Kajfez, F., 261 (33)
Kamata, K., 191 (36)
Kamen, M.D., 225 (63)
Kamiya, T., 261 (29)
Kaplan, L., 224 (32, 37)
Karabatos, C.D., 71 (36), 72 (69)
Karle, I.L., 75 (150)
Karle, J.M., 75 (150)
Karplus, M., 119 (33)
Kato, H., 71 (7)
Katritzky, A.R., 117 (1)
Kavarnos, S.J., 191 (41)
Kawikami, J.H., 224 (39)
Keana, J., 16 (6)
Keeley, D., 224 (24)
Keene, F.Q., 194 (146)
Kelly, D.P., 72 (44, 56)
Kelly, R.E., 118 (21)

Kelly, W.J., 118 (12)
Kerwin, J.F., 260 (8)
Kiarmann, A., 190 (14, 15)
Kimball, G.E., 16 (14)
King, J.F., 223 (11)
King, R.A., 75 (137)
Kingsbury, W.D., 225 (66)
Kinsman, D.V., 191 (42), 224 (21)
Kirby, A.J., 75 (152)
Kiriyama, T., 193 (130)
Kispert, L.D., 73 (86)
Kleinfelter, D.C., 190 (28)
Klinger, H.B., 72 (69)
Klootsterziel, H., 118 (22)
Klopman, G., 72 (49)
Klyne, W., 261 (43)
Knipe, A.C., 74 (112, 113), 120 (68), 223 (5), 260 (3)
Kocsis, K., 261 (29)
Kohen, F., 191 (53)
Kohnastam, G., 74 (99), 192 (88)
Kohrman, R.E., 225 (60)
Kollman, H., 71 (14, 28, 29)
Komeno, T., 224 (15)
Kondo, Y., 194 (143)
Konopka, W.P., 194 (151)
Koshland, D.E., 73 (88), 193 (133, 136)
Kosower, E.M., 72 (69), 117 (1)
Kovacic, P., 193 (132)
Koza, E., 225 (60)
Krabbenloft, H.O., 261 (31)
Kramer, G.M., 121 (91)
Krapcho, A.P., 120 (75)
Kraus, G., 191 (36)
Krestounikoff, H., 17 (37)
Kreymborg, J.H., 261 (21)
Kriman, L.I., 262 (72)
Krimer, M.Z., 223 (11)
Kroepelin, H., 17 (37), 74 (101), 260 (1)
Kuhn, D.A., 192 (84)
Kunz, W., 261 (44)
Kupchan, S.M., 194 (145), 262 (57)
Kutzelnigg, W., 71 (19)
Kwart, H., 224 (38, 39)
Kwiatkouski, G.T., 191 (41)

La Follette, D., 224 (32)
Lallemand, J.Y., 74 (134)
Lalor, F.J., 194 (154)
Lam, L.K.M., 120 (75), 261 (48)
Lamaty, G., 120 (54)
Lamb, J.T., 119 (35)
Lambert, J.B., 118 (9)
Lancaster, P.W., 75 (152)
Lancelot, C.J., 17 (23), 118 (18, 19), 120 (75), 121 (103), 261 (48)
Lane, J.F., 18 (59)
Lankin, D.C., 224 (36)
La Perchee, P., 261 (24)
Lapparte, S., 18 (66)
Large, D.G., 190 (12)
Large, R.J., 262 (74)
Larsen, J.W., 192 (71), 261 (46)
Laszlo, P., 72 (58)
Latham, W.A., 71 (11, 15, 17)
Laurent-Dieuzeide, E., 192 (90), 225 (62)
Leach, S.J., 18 (62)
Lee, C.C., 17 (30), 71 (22, 37)
Leffek, K.T., 120 (67)
Leffler, J.E., 117 (1)
Lemieux, R.U., 18 (64), 192 (65)
le Noble, W.J., 262 (79)
Leonard, N.J., 260 (17), 261 (18, 51)
Letourneau, F., 121 (83)
Lewis, A.J., 193 (101)
Lewis, E.S., 75 (141)
Lewis, S.N., 223 (7)
Lewis, T.R., 261 (28)
Lhomme, J., 70 (1)
Liang, C.L., 72 (56)
Liang, G., 121 (90)
Liang, Y.-T.-S., 194 (145)
Lichtenstein, H.J., 18 (49), 193 (98)
Lieske, C.N., 194 (153)
Liggero, S.H., 120 (58, 75), 261 (47)
Lilley, D.M.J., 71 (13)

Lindegren, C.R., 16 (2), 17 (32), 190 (22, 24), 193 (106)
Lindley, H., 18 (62)
Lindsay, K.L., 18 (61)
Liotta, C.L., 118 (7)
Lipscomb, W.N., 71 (18)
Liu, J.H., 193 (132)
Liu, K.-T., 119 (40, 41), 193 (126), 224 (39)
Livingston, C.M., 225 (52)
Llewellyn, J.A., 120 (67)
Lo, D.H., 119 (33)
Lochner, D.L., 262 (59)
Lok, R., 194 (156)
Long, F.A., 192 (96)
Long, F.W., 262 (5)
Lotspeich, F.U., 190 (26, 27)
Loukas, S.L., 120 (63)
Lucas, H.J., 16 (15), 17 (33, 34), 118 (3)
Lucente, G., 121 (83)
Lumb, J.T., 70 (1), 119 (35)
Lustgarten, R.K., 70 (1)
Lüttke, W., 72 (69)
Lüttringhaus, A., 74 (27)

Maccagnani, G., 74 (109)
MacConaill, R.J., 194 (154)
MacMillan, J.G., 191 (59, 60)
Malloy, T.P., 121 (105)
Manava, R.M., 191 (33)
Mandava, N., 121 (83)
Mandolini, L., 74 (128, 129)
Marbel, P.R., 261 (44)
Marieni, C., 119 (33)
Marita, H., 191 (42)
Marshall, H., 16 (2, 32), 190 (24)
Marshall, J.L., 191 (59, 60)
Martel, R., 193 (121)
Martin, J.C., 118 (15), 191 (54), 224 (18, 33), 225 (55, 56, 57), 261 (31)
Masci, B., 74 (129)
Mason, T.J., 17 (22)
Mastrorilli, E., 193 (119)
Masuda, K., 261 (25)

Mateer, R.A., 120 (74)
Mateescu, G.D., 121 (90)
Matsen, F.A., 72 (69)
Matusak, C.A., 74 (117)
Maynard, J.A., 194 (152)
Mazur, R.H., 17 (27)
Mazza, F., 121 (83)
McCabe, C.L., 18 (46)
McCabe, P.H., 225 (52)
McCasland, G.E., 190 (21)
McCrary, T.J., 118 (9)
McDonald, R.S., 75 (152)
McDowell, S., 74 (119)
McEwen, W.E., 121 (106)
McFarlane, F.E., 121 (102)
McGowan, J.C., 118 (22)
McKema, J.C., 75 (136)
McKillop, T.F.W., 71 (39)
McManus, S.P., 16 (5), 117 (1), 118 (24), 119 (32), 121 (95, 97, 99), 192 (71), 194 (141)
McMaster, B.N., 71 (3)
McMillan, I., 225 (67)
Mebane, R., 194 (155)
Mecca, T.G., 262 (69)
Meehan, G.V., 224 (27)
Meerwein, H., 190 (16), 192 (70)
Meisels, G.G., 71 (26)
Melander, L., 119 (50)
Melby, E.G., 121 (93)
Melloni, G., 224 (22)
Meresak, W.A., 193 (131)
Merritt, R.F., 121 (100)
Meyer, W.P., 261 (31)
Meyers, A.I., 191 (36)
Mihelich, E.D., 191 (36)
Miller, A.D., 18 (50), 74 (97), 192 (77)
Miller, K., 71 (28)
Miller, L.L., 262 (66)
Miller, M.A., 119 (33)
Miller, R.K., 224 (39)
Miller, W.B.T., 120 (62)
Millward, B.B., 193 (106)
Milstein, S., 75 (145), 194 (137)
Mislow, K., 120 (60)
Mitchell, F.W., 17 (33)

Mitra, R.B., 193 (107)
Mizukmi, Y., 74 (120)
Mo, Y.K., 121 (93)
Modena, G., 224 (22, 23)
Mohler, H., 18 (52)
Mohrbacker, P.J., 262 (77)
Mollere, P.D., 119 (47)
Monkoule, I., 193 (121)
Monkovic, I., 193 (122)
Montanari, F., 74 (109)
Moriarty, R.M., 194 (134)
Morimoto, A., 261 (25)
Morita, H., 224 (20)
Morse, B.K., 18 (69), 70 (1), 72 (53)
Morti, S.A., 72 (69)
Morton, J.W., 191 (34)
Mosses, A.N., 18 (51), 223 (3)
Motell, E.L., 119 (52)
Mourning, M.C., 121 (102)
Mousseron, M., 192 (94)
Moyer, C.L., 261 (20)
Mueller, W.H., 223 (8, 10), 224 (17, 31, 50, 51)
Murr, B.L., 120 (54)
Murthy, A.S.N., 71 (16)
Music, J.F., 72 (69)
Myhre, P.C., 71 (24)

Nakamura, Y., 261 (37)
Nakatsuji, H., 71 (7)
Namenworth, J., 72 (43)
Nattaghe, A., 192 (90)
Naumann, M.O., 190 (21)
Neumann, W., 17 (37)
Nevell, T.P., 16 (16)
Newman, M.S., 117 (1)
Nickerson, R.G., 193 (103)
Nicolaides, E.D., 262 (51)
Nikoletie, M., 120 (67)
Nilsson, H., 18 (45)
Nordlander, J.E., 118 (9, 12)
Norman, R.O.C., 261 (49)
Norton, C., 118 (5)
Novak, E.R., 190 (4)
Nowak, B.E., 262 (75)
Noyce, D.S., 190 (10), 192 (68)
Nunes, M.A., 194 (136)

Oae, S., 191 (42), 224 (20)
Ogston, A.G., 18 (54)
Ogstrom, A.G , 223 (1)
Okuda, S., 261 (37)
Okutani, T., 261 (25)
Olah, G.A., 71 (25, 33), 72 (44, 56, 57), 73 (83, 84), 121 (87, 90, 93), 261 (32)
Olson, W.F., 262 (77)
Onanga, N., 74 (134)
Ong, J.H., 120 (59)
Osmers, K., 261 (31)
Ottley, D.J., 192 (87)
Owen, L.N., 192 (94), 225 (65)
Owsley, D.C., 223 (11)

Pacak, J., 193 (100)
Pacsu, E., 18 (57)
Page, M.I., 18 (74), 73 (90), 194 (136)
Pandit, U.K., 75 (148)
Pappo, R., 262 (52)
Paquette, L.A., 191 (43, 45, 46, 47, 48, 49, 50, 51, 52), 193 (115), 224 (27), 260 (15)
Parcell, A., 74 (118)
Parks, A.T., 72 (69)
Partrick, C.R., 73 (89)
Pascone, J.M., 70 (1)
Pasto, D.J., 74 (107)
Paust, J., 120 (79), 262 (71)
Payne, G.B., 75 (143)
Pearson, R.G., 117 (1)
Peet, J.H., 194 (144)
Peet, N.P., 192 (73)
Penty, M., 74 (99), 192 (88)
Pepe, H.O., 193 (104)
Perkins, M.J., 121 (89)
Perron, Y.G., 193 (121)
Perst, H., 192 (71)
Peters, E.N., 70 (1), 72 (61), 119 (42, 43, 44, 45, 46), 262 (70)
Peters, J.W., 194 (151)
Peters, R.A., 223 (1)
Peterson, P.E., 74 (103), 118 (20, 21), 120, (75), 190 (6), 192 (74)

Petrovich, J.P., 121 (81)
Petterson, R.C., 224 (36)
Pews, R.G., 223 (11)
Pfeiffer, G.V., 71 (12)
Philpot, J. St. L., 18 (54), 223 (1)
Philpotts, A.R., 18 (49), 193 (98)
Piccolini, R.J., 72 (40)
Pierre, J.L., 72 (69)
Pilcher, G., 74 (95)
Pincock, R.E., 119 (40)
Pittman, C.U., 16 (5), 73 (86), 119 (32), 121 (32), 192 (72)
Plumbergs, P., 194 (153)
Pocker, Y., 193 (112, 113)
Pond, D.M., 192 (73)
Porret, D., 18 (47)
Port, G.N.J., 73 (89), 194 (136)
Povarenkina, S.V., 225 (61)
Prelog, V., 18 (62), 192 (67), 261 (43)
Preuss, H., 71 (10)
Putz, G.J., 118 (9)

Raasch, M.S., 224 (40)
Raber, D.J., 120 (75), 261 (24, 48)
Radom, L., 71 (30), 72 (52, 59, 60)
Raftery, W.V., 74 (119)
Ramachandran, V, 262 (66)
Ranganayakulu, K., 17 (20)
Rao, C.B.S., 261 (29)
Rapoport, L., 262 (50)
Ravindranathan, M., 70 (1), 119 (42, 44, 46)
Rebbert, R.E., 71 (20)
Rees, C.W., 121 (89)
Regulski, T.W., 223 (9)
Reich, I.L., 71 (23), 120 (72)
Rengevich, E.N., 224 (43), 262 (64)
Rettig, M.F., 223 (11)
Reuben, J., 194 (136)
Richards, M.B., 17 (37)
Richards, W.G., 194 (136)

Richardson, A.C., 191 (64)
Richardson, W.H., 194 (151)
Richey, H.G.Jr., 17 (23), 73 (84), 191 (42), 224 (21)
Riddell, W.D., 262 (58)
Riemenschneider, J.L., 121 (90)
Ritchie, C.D., 119 (49)
Ritter, J.J., 194 (140)
Roberts, D.D., 190 (19), 192 (91, 93)
Roberts, J.D., 16 (14), 17 (26, 27, 29, 30), 71 (21), 72 (41), 73 (79, 80), 224 (19)
Robertson, D.A., 261 (23)
Robertson, R.E., 120 (59, 67), 190 (17)
Robertson, W.W., 72 (69)
Robinson, A.B., 225 (63)
Robson, R., 224 (35)
Rockett, B.W., 194 (144)
Rogers, G.A., 121 (83)
Romeo, A., 121 (83)
Ronald, B.C., 193 (113)
Rosenfeld, J., 71 (34)
Ross, S.D., 17 (39, 40), 260 (7)
Rua, S. Jr., 261 (30)

Sable H.Z., 75 (135), 193 (99)
Saeed, M.A., 224 (25)
Sager, W.F., 119 (49)
Salem, L., 72 (52)
Saleman, R.G., 194 (158)
Salomon, G., 17 (37, 38), 260 (2)
Sankey, G.H., 16 (7)
Sargent, G.D., 17 (22, 23), 72 (67), 119 (31)
Sargeson, A.M., 193 (114), 194 (146)
Sasaki, T., 193 (130), 262 (75)
Sasse, H.J., 192 (70)
Satterthwait, A., 192 (85)
Saunders, Jr. W.H., 120 (66)
Schadt, F.L., 120 (75), 121 (103)
Schaefer, J.P., 120 (67)

Scheppele, S.E., 119 (50)
Schiess, P.W., 261 (40)
Schleyer, P.v.R., 17 (23), 71 (9, 11, 30), 72 (52, 59, 60, 66), 73 (78, 81, 83, 84), 118 (18, 19, 28, 30), 119 (33), 120 (58, 64, 75, 76, 79), 121 (87, 103), 261 (21, 32, 47), 262 (48, 71)
Schlosberg, R.H., 71 (25)
Schmid, G.H., 225 (54)
Schöllkopf, U., 120 (79) 262 (71)
Schowen, R.L., 75 (138), 192 (84)
Schreiber, K.C., 18 (69), 72 (53)
Schrodt, H., 192 (70)
Schulenberg, W., 261 (28)
Schultz, E.N., 260 (8)
Schumacker, E., 261 (44)
Schwartz, J.C.P., 117 (1)
Schwartz, W., 261 (42)
Scott, F.L., 194 (154)
Scott, M.K., 191 (49, 50)
Searles, Jr. S., 193 (103)
Seckinger, K., 18 (72), 118 (14)
Seckinger, R., 262 (67)
Sell, K., 18 (72), 74 (124), 223 (2)
Sendijarvie, V., 120 (64)
Sequeira, R.M., 224 (14)
Serue, M.P., 74 (107)
Sharp, J.J., 225 (63)
Shatavsky, M., 118 (5)
Shepard, J.C., 72 (69)
Sherman, P.D., 74 (108)
Shermergorn, I.M., 225 (61)
Shiau, W.-I., 121 (106)
Shilov, E.A., 117 (2), 193 (116), 224 (41, 42, 43), 262 (62, 63, 64, 65)
Shin, J.H., 73 (85)
Shiner, Jr., V.J., 72 (54), 119 (50), 120 (54, 64), 261 (46)
Shoemaker, G.I., 261 (18)
Shoppe, C.W., 17 (24)
Shulz, D.N., 121 (106)

Sicher, J., 193 (111)
Sieck, L.W., 71 (20)
Siegfried, W., 192 (78)
Silver, M.S., 118 (13)
Simonetta, M., 71 (38), 119 (33)
Simpson, W.J., 193 (121)
Sipos, F., 193 (111)
Sipp, K.A., 118 (21)
Sirokmann, F., 191 (62)
Sixma, F.L.J., 117 (1)
Sladowska, M., 121 (83)
Slater, J.C., 71 (39)
Slighton, E.L., 118 (7)
Slivinski, W.F., 120 (79), 262 (71)
Slocum, D.W., 191 (34)
Slutsky, J., 120 (77)
Slutsky, R.C., 261 (48)
Smart, B.E., 17 (22), 72 (55)
Smit, W.A., 223 (11)
Smith, C.R., 75 (152)
Smith, H.A., 224 (26)
Smith, J.F., 16 (17)
Smith, J.H., 223 (10)
Smith, L., 18 (45)
Smith, L.O., 224 (14)
Smith, M.C., 16 (7)
Smith, M.R., 192 (92)
Smith, P.N., 192 (94)
Smolin, E.N., 262 (50)
Sneen, R.A., 261 (46)
Snyder, E.I., 72 (45), 194 (148)
Sobotka, H., 18 (63)
Sohn, W.H., 75 (136)
Solomons, T.W.G., 262 (61)
Sonnet, P.E., 261 (29)
Sonntag, A.C., 192 (72)
Sorenson, T.S., 17 (20), 121 (90)
Speakman, P.R.H., 191 (61)
Spille, J., 192 (70)
Spitzer, W.A., 224 (36)
Sprague, J.N., 260 (8)
Spurlock, L.A., 191 (55)
Spurlock, S.N., 223 (11)
Stamnets, V., 193 (116)
Stanck, J., 193 (100)
Staninets, V.I., 117 (2), 224 (41, 42, 43), 262 (62, 63, 64, 65)

Author Index

Stermitz, F.R., 262 (66)
Stevens, C.L., 191 (62)
Stevens, J.E., 18 (46)
Stirling, C.J.M., 74 (112, 113, 123), 120 (78), 223 (5, 6), 260 (3)
Stocken, L.A., 223 (1)
Stodey, R.H., 121 (96)
Stoodley, R.J., 224 (28), 225 (67)
Stork, G., 74 (133)
Storm, D.R., 73 (88), 193 (133), 191 (43, 45, 46, 47, 51), 260 (15)
Streicher, H.J., 73 (85)
Streitwieser, Jr. A., 16 (9), 118 (17), 119 (50), 120 (61, 65), 260 (6)
Sugimoto, M., 262 (75)
Sugimoto, T., 224 (16)
Sunko, D.E., 72 (55), 119 (51), 120 (55, 67)
Sustmann, R., 71 (9)
Suzuki, S., 120 (61)
Swain, C.G., 17 (39), 18 (55), 192 (82, 84, 97), 223 (1), 260 (7)
Symant, H.H., 191 (33)

Tabushi, I., 224 (16)
Taft, R.W., 118 (16)
Takagi, O., 194 (156)
Takeuchi, K., 119 (44)
Tam, S.W., 18 (72), 118 (14), 262 (67, 68)
Tamaru, Y., 224 (16, 34)
Tanida, H., 118 (6), 119 (39)
Tao, E.V.P., 118 (20)
Tarbell, D.S., 18 (61), 190 (4, 13), 191 (44), 260 (16)
Tarle, M., 120 (57)
Taub, D., 193 (118)
Taylor, K.G., 191 (62)
Tejima, S., 225 (64)
Tenud, L., 74 (131)
Thacker, D., 194 (138)
Thaler, W.A., 223 (10)
Thomar, B.R., 190 (10)
Thompson, A.L., 18 (43)

Thompson, G., 118 (20), 192 (74)
Thompson, H.B., 119 (33)
Tichy, M., 193 (111)
Tidwell, T.T., 193 (125)
Tienan, T.O., 71 (20)
Tillmanns, F.J., 194 (140)
Tinker, H.B., 193 (127)
Tipson, R.S., 17 (35)
Tomie, M., 120 (55)
Tonellato, U., 224 (22, 23)
Tao, E.V.P., 192 (74)
Traber, R., 18 (72), 118 (14), 262 (67)
Trahanovsky, W.S., 72 (48)
Traylor, T.G., 17 (22), 72 (58, 62, 63, 64), 193 (124, 125)
Traynham, J.G., 192 (93)
Tribble, M.T., 119 (33)
Trifan, D., 17 (21), 118 (10)
Tritle, G.L., 119 (40)
Trofimenko, S., 262 (60)
Trost, B.M., 224 (24)
Tse, K.K., 120 (74)
Tsuji, A., 74 (120)
Tsuji, J., 224 (46)
Tsuji, T., 118 (6), 119 (39), 224 (15)
Turner, E.G., 18 (51)
Twigg, G.H., 18 (49), 193 (98)

Uda, H., 193 (10)
Ugi, I.K., 191 (35)
Ulliyot, G.E., 260 (8)

Valazquez, R.A., 224 (36)
Valkovich, P.B., 191 (35)
van Artsdalen, E.R., 17 (42)
Vanderhorst, P.J., 262 (77)
Van Dine, G.W., 120 (79), 262 (71)
van Tamelen, E.E., 16 (12)
Veazey, R.L., 120 (74)
Velkou, M.R., 120 (63)
Vitullo, V.P., 120 (69, 70)
Vloom, W.J., 224 (44)
Vogel, P., 71 (34)
Voight, C.F., 262 (61)

Volz, H., 73 (85)
von Felton, W.C., 121 (104)
Vorachek, J.H., 71 (26)
Vorob'eva, E.A., 223 (11)

Wachs, R.H., 192 (80)
Wahl, Jr. G.H., 120 (60)
Walker, E., 223 (1)
Waller, F.J., 120 (75)
Wang, M.S., 262 (56)
Wankel, R.A., 262 (77)
Ward, H.R., 74 (108)
Ward, J.P., 191 (56)
Ware, D.W., 121 (95)
Warner, J.C., 18 (46)
Warshel, A., 119 (33)
Watson, H.B., 16 (16)
Webster, B.C., 71 (39)
Weeks, D.P., 75 (144)
Weingarten, H.I., 192 (68)
Weissberger, A., 117 (1)
Welinder, H., 75 (151)
Welvart, Z., 72 (58)
Wendish, D., 72 (69)
Wendler, N.L., 193 (118)
Westerman, P.W., 121 (93)
Westheimer, F.H., 119 (34), 120 (53)
White, A.M., 71 (33)
Whitehouse, R.D., 224 (25)
Whiten, J.L., 71 (27)
Whitham, G.H., 192 (76)
Wiberg, K.B., 17 (23), 73 (83, 92), 117 (1), 119 (31, 50, 52), 192 (69)
Wieting, R.D., 121 (96)
Wilder, Jr., P., 224 (29, 30)
Wiley, P.F., 262 (69)
Wilgis, F.P., 120 (70)
Wilkinds, C.L., 223 (9)
Willer, R.L., 194 (155)
Williams, D.L.H., 118 (3, 23), 193 (120)
Williams, J.E., 71 (11, 16)
Wilson, C.L., 16 (16)
Wilson, H., 75 (141)
Wilt, J.W., 121 (105)
Winkler, C.A., 18 (43)
Winstein, S., 16 (1, 2, 4, 15), 17 (21, 32, 33, 34), 18 (48, 58, 60, 65, 66,

Winstein, S. (cont'd)
67, 68, 69, 70, 71), 70
(1, 23), 71 (38), 72 (40,
42, 53), 74 (96, 114,
115, 116), 118 (3, 5, 6,
10), 120 (71, 72, 76,
80), 121 (82, 85, 101,
104), 190 (1, 2, 3, 5, 7,
18, 22, 23, 24), 191 (30),
192 (89), 193 (106), 224
(45)
Winstom, L.O., 18 (46)
Winter, R.E.K., 120 (60)
Wise, L.D., 224 (27)
Wise, W.S., 18 (49)
Wiseman, J.R., 261 (3)

Witkop, B., 194 (143)
Witsiepe, W.K., 193 (103)
Wnuk, T.A., 261 (31)
Wohl, R.A., 194 (142)
Wohlgemuth, K., 74 (127)
Wolfenden, J.H., 18 (56)
Wong, H., 193 (122)
Wong, N., 121 (90)
Woo, C., 225 (53)
Woodward, R.B., 72 (68), 118 (5)
Wright, Jr., W.B., 18 (62), 192 (66)
Wylde, J., 192 (90), 225 (62)
Wynberg, H., 117 (1), 224 (31), 262 (53)

Yahner, J.A., 262 (80)
Yamana, T., 74 (120)
Yonezawa, T., 71 (7)
Yoshida, Z., 224 (16, 34)

Zalkow, V., 75 (153)
Zeller, J., 70 (1)
Zergenyi, J., 261 (44)
Ziegler, F.E., 261 (29)
Ziegler, K., 74 (126, 127)
Zienty, M.F., 18 (62)
Zirkle, C.L., 260 (8, 9)
Zurawski, B., 71 (19)
Zurn, L., 74 (118)
Zwanenburg, D.J., 262 (53)

Subject Index

Ab initio calculations, 20-22, 24-25, 27-29, 33
Acetoxonium ions (*See* Bridged ions, from ester group participation)
Acidity function, 174
Addition–cyclization reactions (*See* Nucleophilic participation in addition reactions)
Alkaloids
  anodic coupling of laudanosine, 253-254
  fragmentation of tropone derivatives, 247
  hydrolysis of ceveratrum esters, 251
  solvolysis of codeinone and neopinone derivatives, 242-243
Alkanes, cyclization of (*See* Cycloalkanes)
Ambident neighboring groups, 123, 194
Amines, effect of alkyl group substitution on the formation of cyclic, 61, 63, 65
Anchimeric assistance
  compared to resonance, 146
  competition with resonance, 141, 152, 153
  definition of, 3, 6-9
  effect of leaving group on measured, 51-57, 147-148
  effect of substrate structure on (*See* Ring closure rates)
  field effects on, 146, 148, 152
  –I effects on, 82-85, 108, 115, 136-137, 147-148, 150, 152, 155, 255
  in tertiary substrates, 15, 254, 258
  involving σ participation, 8, 81, 111, 152-153, 202-205
  –M effects on, 137-144
  measurement of, 13-16

Anchimeric assistance (*cont'd*)
  model compound selection, 8, 79-89
  quantitative expression of, 15-16
  See also k_Δ Process, Solvent assistance, and Driving force for participation
Anhydrides, the effect of alkyl substitution on the rate of formation of, 59-62, 70
Anodic coupling, 253-254
2-Arylethyl cations, 27-28, 89, 98-101, 107-108
ASMO calculations, 28
Attenuation factor, ϵ
  definition of, 84
  use in assigning a k_Δ mechanism, 84-85, 160-162
Azacarbocations, 240
Azido group, as amine precursors, 242
Aziridines, synthesis of, 229-231, 242

Beckmann cleavage, 259
Bell-clapper rearrangement, 208
Bicyclobutonium ion (*See* Cyclopropylmethyl cation)
Bordwell–Shiner–Sneen mechanism, 78, 248-249, 255-258
Bridged ions
  detection by isotopic scrambling, 23, 26, 108, 113, 132-133, 135, 151, 202-205
  detection by NMR, 26-27, 42, 110-112, 206, 209, 239
  detection by stereochemical studies, 5, 11-12, 79, 90-91, 111-112, 126-129, 151-152, 197-198, 233, 236-240
  factors affecting the ring-opening of, 127-136, 194, 196-197, 232-235

273

Bridged ions (cont'd)
   from amide group participation, 4
   from amino group participation, 11, 227-262
   from aryl participation, 11-12, 27-28, 89, 98-101, 107-108, 114-117
   from carbon participation, 6-9, 111, 202
   from carboxylate group participation, 5
   from ester group participation, 4, 11-12
   from ether and hydroxy group participation, 12, 123-194
   from halogen group participation, 5, 27-28, 85, 111-112, 160-161
   from sulfur group participation, 195-225
   in equilibrium with open carbocations, 5, 23-24, 112, 155, 206-208
   MO calculations of, 20-30
   non-bonded interactions on going to, 133, 183
   rate-limiting destruction of, 127-202
   spectrophotometric detection, 110, 223
   See also Phenonium ions, Cyclopropylmethyl cation, and Norbornyl cation
σ-Bridged ion, 6-10
Brønsted plot, 163, 165-167
tert-Butyl chloride, ethanolysis of, 8

$C_2H_5^+$ species, 20-24
$C_3H_7^+$ species
   heats of formation of, 26
   MO calculations of, 24-26
   NMR studies of, 26
   relative energies of, 25
$C_4H_7^+$ cation, 27-29, 39-43
Camphene hydrochloride
   rearrangement of, 6
   solvolysis of, 8
Carbohydrates (See Nucleophilic participation)
Catalysis
   acidic, 3, 4
   basic, 3, 4
   intermolecular, 3
   nucleophilic, 3
   See also Intramolecular, General acid, and general base catalysis
Charge transfer complexes, 35-37
Chlorine isotope effects, 165-166
2-Chloroethyl cation, 27-29
CNDO calculations, 20-22, 24-25, 28-29
CNDO/2 calculations, 27

Common ion rate depression, 127
Conformational requirements for participation (See Nucleophilic participation)
Counterion effects, 127
Cycloalkanes, effect of alkyl substitution on formation of, 58-59, 70
Cyclobutanone derivatives, synthesis of, 203
Cyclopropylmethyl cation
   calculations on the structure of the, 27-29
   chemical studies on, 39-43
   chemical studies on derivatives of the, 39-43, 149, 153, 202, 203

Deamination (See Nucleophilic participation)
Degenerate rearrangements, 6-9, 23, 26-30, 42, 112, 197-198, 206-208
σ-Delocalized ion (See Synartetic ion, σ-Bridged ion)
α-Deuterium isotope effects, 97-100, 102, 103, 152
β-Deuterium isotope effects, 97-100, 246
γ-Deuterium isotope effects, 98-100
Dialkoxycarbocations, formation by rearrangement of α-carbonyl oxonium ions, 159
Diazapentalene, 252-253
Dicarboxylic acids
   cyclization of, 59-61
   cyclization of derivatives of, 59-61
   rotamer distribution, 68
Dioxolan-2-ylium ions (See Bridged ions, from ester group participation)
Downhill process (See k_R Process)
Driving force for participation, 15, 123, 133, 154, 160, 202, 205, 209, 221, 240
   See also Ring closure rates

Edge participation, 91
α-Effect, 187-188
Effective molarity, 15-16
EHT calculations, 20, 22, 24, 27-28
Electron spectroscopy (ESCA), 110-111
Electrophilic participation
   by nickel, 4
   in the hydrolysis of nitriles, 4
Electrostatic stabilization, 5, 101-102, 109
Elimination reactions, as a rate-limiting competing process under solvolytic conditions, 81, 104, 167, 170, 189, 244-246
   See also Nucleophilic participation

Subject Index

Entropy effects on ring closure rates, 43-50
Enzyme mechanisms, 5, 183
Epoxide synthesis, 61, 66, 168, 177-178
Ethers, effect of alkyl group substitution on the formation of cyclic, 61-65
Ethyl cation, structure of the, 20-24

Field effects, 80
2-Fluoroethyl cation, 27-29
Fragmentation
 activation volume in, 262
 Bordwell–Shiner–Sneen mechanism, 247-249
 of β-amino mixed anhydrides, 243
 of γ-aminopropyl derivatives, 243-249
 of 4-bromobicyclo[2.2.2]octane, 248
 of 3-β-chlorotropone,
 of the ethylene ketal of 7-keto-*endo*-2-norbornyl toluene-*p*-sulfonate, 156
 of ionized 3-hydroxypropyl derivatives, 170-172
 of quinuclidine derivatives, 247-249
 of 4-toluene-*p*-sulfonyloxymethyl-1-azabicyclo[2.2.2]octane, 248
 stepwise, 247-249
 stereochemistry of, 247
 stereoelectronic requirements of, 247
Franck–Condon principle, 35
Frangomeric assistance, 246-248
Free radical reactions
 participation by the iodo group in, 218-219
 participation by π groups in, 218-219
 participation by sulfur groups in, 210, 214, 218-220
 steric acceleration in, 218-219
Foote–Schleyer method, 86-87, 150

G-n symbolism, 12
Gabriel synthesis, 163
General acid catalysis, 101, 113, 186, 188, 251
General base catalysis, 101, 163-166, 186, 188, 251
Glutaric anhydride derivatives, effect of alkyl group substitution on the formation of, 60-62
Grob fragmentation, 244
Grunwald–Winstein plot, 102
 See also m Values

Halonium ions
 equilibria with carbocations, 5, 112
 formed by electrophilic addition, 57, 78, 111
 rearrangement of, 112
 ring opening of, 129
 spectroscopic studies of, 111-112
 See also Bridged ions, from halogen group participation
Hammett treatment, 84, 90-94, 114-117, 199, 256-257
Hexamethylenetetraamine, 249-250
Hidden return, 258
 See also Ion pairs
Hoffman elimination, 181-182, 189
Homoallylic participation, 10, 79, 91-94, 98, 134, 147, 153
Homoaromatic systems, 27
Homoconjugated systems, 27, 253
Hückel calculations, 20, 26-27, 30
Hydride transfer, 95-96, 153, 211
Hydrogen abstraction, 85-86
Hydrogen bonding, 186
 See also General acid catalysis
Hydrolysis
 of 2-chloroethanol, 164-165, 167
 of 4-chloro-1-butanol, 163-167
 of chloromethylmercaptophosphonates, 220
 of dimethyl acetals of glucose and galactose, 183
 of haloalkylamines, 227-229
 of halohydrins, 163-167
 of imino chlorides, 209-210
 of laudanosine, 251-252
 of 3-methyluridine, 185
 of mustard gas, 195, 231
 of penicillin intermediates, 209-210
 of phosphate esters, 188-189
 of phthalamic acid, 113
 of phthalimidium cation, 109
Hyperconjugation
 as a factor in β-deuterium isotope effects, 97
 as an effect separate from bridging, 3, 31-43
 by C–C groups, 31-43
 by C–H groups, 31-32
 effect of multiple groups, 31-32
 effect of strain, 33-43
 geometrical requirements for, 31-43

I-Strain, 168
INDO calculations, 20, 22, 24, 29
Internal return (*See* Ion pairs)
Intimate ion pairs, 127-128, 131, 246
Intramolecular catalysis
  by carboxyl groups, 82, 101, 113
  by hydroxyl groups, 82, 101, 187
  definition of, 3
  in the hydrolysis of phthalamic acid, 113
  pH dependence in, 101
Iodolactonization, 57, 77-78, 253
Ion pairs
  as contributors to anchimeric assistance, 78, 248-249, 255-258
  devaluation of titrimetric rate constants because of, 152
  effect of leaving group on the return of, 134-135
  effect on k_Δ/k_s rate ratio, 14
  effect on rates and products, 81, 127-128, 138, 144, 202
  formation from free ions, 127
  in fragmentation processes, 247-249
  in the solvolysis of arylthioalkyl halides, 196-198
  in the solvolysis of methoxy-substituted alkyl esters, 126-134, 138-139
  isomerization *via,* 81, 127, 157, 197-198, 232-235
  return from, 14, 81. 101-102, 115-116, 126-131, 133-134, 138-139, 152, 155, 157, 159, 196, 198
  return of, 134-135
Isobornyl chloride, solvolysis of, 6-7
Isotope scrambling (*See* Bridged ions)
Isoxazoles and isoxazolines, synthesis of, 189

k_c Process, 7, 13-16, 19, 86, 89, 94, 105, 114, 160, 246
k_Δ Process, 7, 13-16, 19, 84, 85-89, 94, 98-107, 114-117, 127-129, 132, 133, 136, 152-154, 156, 160-161, 178, 244-246
k_e Process, 170, 244-246
k_f Process, 156, 244-246
k_R Process, 8, 13-16, 114, 152-153, 157
k_s Process, 14-16, 19, 81, 84, 86, 87, 98-107, 114-116, 127-129, 132-133, 136, 138, 153, 178, 244-246
Kinetic control in product formation, 79, 108, 117, 128-131, 152, 232-235

Kinetic isotope effects, 95-103, 132

Lactones
  effect of alkyl group substitution on formation of, 61, 63-66, 68, 69
  formation of upon halogenation, 77-78
  from halide displacement in α-halocarboxylate salts, 5, 109
Laticyclic participation, 176-177
Leveling effect
  of aryl groups, 90-94
  of perfluoro substitution on oxygen, sulfur, and amino groups, 215
Limiting solvolysis (*See* k_c Process)
Lysozyme, 221

m Values 102, 105, 106, 257-258
Mannich base, 189, 260
Markovnikov process, 198
Mechanistic criteria in solvolysis reactions, 106-108
σ_I^* Method, 87-89
α-Methyl/hydrogen ratios, 87-89
Michael addition, 251
Microscopic reversibility, principle of, 233
Migratory aptitudes, 174-176
MINDO/2 calculations, 20, 25, 29
MINDO/3 calculations, 20, 25, 30

Neighboring group participation
  definition of, 3
  in relation to biogenesis, 5
  in relation to enzyme reaction mechanisms, 5
  See also Nucleophilic participation, Electrophilic participation
NMR spectroscopic observation of intermediates, 110-112
  See also Bridged ions
NNDO calculations, 20, 22, 25, 29
Nonclassical ions, 10, 31
Norbornyl cation
  calculations on the structure of the, 30
  derivatives of the, 6-8, 33-34, 150-153, 156
  from solvolysis of esters and halides, 8-9
  spectroscopic studies of the stable ion, 111
Nortricyclyl esters, 41-42
Nucleophilic participation
  as a function of electron demand, 15, 39, 89-94, 147

Subject Index

Nucleophilic participation (*cont'd*)
  by a *cis* neighboring group, 169
  by amide groups, 4, 240
  by amino groups, 61, 63, 176, 179, 211
  by aryl groups, 82, 83, 89-94, 98-101, 104, 107-108, 114-115
  by aryloxy and benzyloxy groups, 143-145
  by carbanionic carbon, 54-57
  by carbonyl oxygen in haloalkyl esters, 159
  by C–C π-bonds, 79, 90-94, 98, 177-178
  by C–C σ-bonds, 8, 39, 81, 90-94, 111, 152-153, 202-206
  by ester groups, 124
  by halogeno groups, 5, 27-28, 85, 111-112, 160-161
  by hydrazone groups, 259-260
  by hydrogen, 81, 104, 167, 220
  by ionized and un-ionized hydroxyl groups, 61, 63-69, 168, 172, 173, 177, 182, 185, 243
  by ionized or un-ionized thiol groups, 220-223
  by multiple neighboring groups, 206-209
  by n, π, and σ groups, 10, 13, 123-124
  by oxygen groups in cyclic ethers, 145-154
  by oxygen in acetals and ketals, 154-158
  by remote cyclopropyl groups, 91
  by solvent (*See k_S Process*)
  by the acetoxy group, 4, 85, 108-109
  by the acetoxyl alkyl oxygen, 158
  by the carboxylate group, 5, 59-62, 77-78, 101, 113
  by the cyano or nitrile group, 85, 258-259
  by the epoxy group, 145-146, 161-162
  by the hydroperoxide group, 187-188
  by the hydroxyl group, 161-187
  by the methoxy group, 80-81, 94, 97, 103, 125-143, 154-161
  by the methyl group, 104
  by the oxime group, 188-190
  by the trifluoroacetoxy group, 85
  by thioacetal groups, 209
  by thioamide groups, 124
  by thioether groups, 195-220
  competition between amide and hydroxyl groups, 179
  competition between ether groups and a double bond, 134-147

Nucleophilic participation (*cont'd*)
  competition between homoallylic participation and oxygen participation, 134-147
  competition between hydroxyl and amine groups, 176, 179, 243
  competition between ionized alcohol and thiol groups, 222
  competition between the methoxy group and aryl groups, 137-141, 212
  competition between sulfur and carbon groups, 199-200, 203
  competition between thioether groups and aryl groups,
  competition between uncatalyzed and general base catalyzed in, 165
  competition with elimination, 81, 104, 167, 170, 189, 220, 244-246
  competition with fragmentation, 244-249
  competition with resonance, 141, 152-153, 206-207
  consequences of ion pairs (*See* Ion pairs)
  consequences of ring closure, 11-12, 146, 232-235
  dimerization involving, 231-232
  historical survey, 10
  in addition reactions, 4, 5, 10, 77-78, 84, 85, 113-114, 160-162, 177-181, 210-216, 252-253
  in alkylation reactions, 143
  in azabicyclic derivatives, 240-241
  in carbene reactions, 181
  in cyclopropane cleavage, 148-149
  in deamination of a diazoketone, 144
  in elimination reactions, 181-182, 216-218
  in fragmentation reactions (*See* Fragmentation)
  in Grignard reactions, 143
  in halogen addition reactions, 5, 77-78, 111, 113-114, 177-179, 210-211, 214-215, 252-253
  in hydrolysis reactions, 10, 101-102, 113, 124, 180-189, 195-210, 220, 251
  in metallation reactions, 143
  in oxabicyclic derivatives, 150-153, 200-202, 206
  in oxymercuration reactions, 160-162, 179
  in oxythallation reactions, 180
  in peptide synthesis, 221

Nucleophilic participation (cont'd)
  in perester decompositions, 218-219
  in photolytic reactions, 219-220
  in polymerization reactions, 221
  in quarternization of phosphines and arsines, 94
  in reduction reactions, 124, 143, 187, 253-254
  in ring closure reactions to form carboxylic acid derivatives, 59-70, 77-78, 113, 182-186
  in ring-opening reactions, 54-57, 61, 66
  in solvolysis reactions, 4, 79-94, 101-109, 125-159, 163-177, 181-190, 195-210, 220-225, 227-252, 254-260
  in sulfenyl chloride additions, 198-199, 210-211, 215-216
  in tertiary substrates, 15, 254-258
  in the addition of sulfenyl halides to dienes, 216
  in the addition to heteroatom-substituted norbornenes, 213-215
  in the Beckmann cleavage, 259
  in the deamination of aminoalcohols, 174-176
  in the pinacol rearrangement, 174-176
  in the preparation of pyrrolidine and piperidine derivatives, 249-252
  in the presence of α-aryl groups, 89-94
  in the reactions of carbohydrate derivatives, 11, 133, 135, 143-144, 156-158, 168-169, 183, 209, 222, 242
  in the reactions of phosphate esters, 10
  in the rearrangement of alkoxyacyl halides, 158-159
  in the reductive cyclization of nitrodiesters, 250
  in the Ritter reaction, 184, 258
  in thiabicyclic derivatives, 152, 199-206
  stepwise reactions involving, 4-9, 77-78, 102, 113-114, 129-130, 137, 141, 145, 147-153, 157, 160-181, 183-184, 213, 254-260
  stereochemistry of, 5, 11-12, 79, 90-91, 98, 128-129, 135, 151-152, 158, 167-168, 174, 177, 189, 194, 199-201, 203-206, 216, 233, 236-241
  stereoelectronic requirements of, 7, 19, 54-57, 79, 80, 98, 123-124, 150, 152-153, 158, 167, 174, 177, 193, 194, 199, 204-205, 241

Orbital steering, 183
Organometallic compounds
  electrophilic participation by nickel in, 4
  nucleophilic assistance in the attack on α-methoxyketones by, 143
  nucleophilic assistance in the formation of, 143
  nucleophilic coordination of stannane derivatives by the methoxy group, 176
  nucleophilic participation by cobalt-bound nitrogen, 176
  nucleophilic participation by cobalt-bound oxygen, 176
  nucleophilic participation by the methoxy group in quarternization of phosphines and arsines, 94
Oxide migrations, 61, 66, 169-170

Penicillin derivatives, 209-210, 223
Peptide synthesis, 221
pH Dependence in intramolecular catalysis, 101
Phenonium ions, 11-12, 27-28, 89, 98-101, 107-108
Photolysis
  of alkylthio tosylhydrazones, 220
  of α-dione sulfides, 220
  of γ-keto sulfides, 219-220
Pinacol rearrangement, 174-176
Piperidine derivatives, synthesis of, 249-252
PNDDO$^+$ calculations, 30
Polarimetric method, 152
Primary isotope effect, 95-96, 165-166, 246
1-Propyl cation, structure of the, 24-26, 39
Protein synthesis, 221
Proton transfer, 58, 95-96
Protonated cyclopropanes, 24-26, 149-207
Pyrazoles and pyrazolines, synthesis of, 259-260
Pyrrolidine derivatives, synthesis of, 249-250

Racemization (See Bridged ions)
Raman spectroscopy, 110-111
Rate-limiting elimination (See Elimination reactions)
Rate ratios
  comparison of cis/trans, 80-81, 167, 199-200

Subject Index

Rate ratios (*cont'd*)
comparison of *exo/endo,* 81, 150-153, 200-202
comparison of α-Me/H, 87-89
comparison of *ortho/para,* 81, 101, 216
comparison of R-Br/R-Cl, 106-107
comparison of *syn/anti,* 39, 79, 90-94, 188-189, 199
comparison of R-OTs/R-Br, 105-106, 248
Rearrangements (*See* Ion Pairs)
Ring closure rates
comparison with ring−opening, 131
the effect of alkyl group substitution on, 58-70, 126, 135, 229-231, 255
the effect of chain rigidity on, 54, 182, 183
the effect of conjugative stability on, 51, 54, 197
the effect of electronic factors on, 50-58, 125-127, 172-173, 187-188
the effect of enthalpy changes on, 43-50, 182, 183, 197, 228
the effect of entropy on, 43-50, 54, 182, 197, 235
the effect of ion pair return on (*See* Ion pairs)
the effect of ring size on, 44-54
the effect of rotamer population on, 68-70
the effect of steric factors on, 43-70, 125-127, 144-145, 174, 182-183, 205, 246, 250
the effect of the leaving group on, 51-57
the effect of the neighboring group on, 51-54, 145
the effect of transition state strain on, 50, 69, 155, 182-183, 235-238, 246
Ring opening of onium ions, 127-136, 194, 196-197
Ritter reaction, 184, 258
Rotamer population, 68-70

SCF calculations, 20, 22
Secondary isotope effect, 95, 97-101
Solid-phase synthesis, 221
Solvent assistance
competition with anchimeric assistance, 14-15, 114-115, 127-129, 138, 153, 160, 163-167, 170, 178
in the solvolysis of 2-adamantyl derivatives, 105

Solvent assistance (*cont'd*)
in the solvolysis of methoxy-substituted alkyl derivatives, 127-129
in the solvolysis of 2-norbornyl derivatives, 105
in the solvolysis of β-phenylethyl derivatives, 83-84, 114-115
in the solvolysis of 2-propyl derivatives, 105
Solvent effects
on electrostatic stabilization, 101-102
on halonium ion-carbocation equilibria, 112
on intermediate stability, 110-112
on ion pairing, 101-102
on the magnitude of observed k_Δ/k_S values, 19, 81, 83-85, 88, 89, 95, 98-107, 114-115, 127-129, 135, 137, 154, 163, 168, 170, 178, 210
on the ring opening of bridged intermediates, 110-112, 127-129, 134
Solvent isotope effect, 95, 163, 165, 168
Solvent-separated ion pairs, 102, 127-128, 131
Solvolysis
of 2-acetoxycyclohexyl derivatives, 4, 12
of acyclic acetals, 154-156
of 1-adamantyl derivatives, 41
of 2-adamantyl derivatives, 40-105
of β-aminoalkyl derivatives, 243-249, 254-258
of γ-aminoalkyl derivatives, 227-243
of β-9-anthrylethyl derivatives, 108
of 3-aryl-2-butyl derivatives, 10, 84, 98, 101
of 4-aryl-1-butyl derivatives, 12, 116
of 1-aryl-1-cyclopentyl esters, 92-93
of 1-aryl-1-cyclopropylethyl esters, 91- 94
of 2-aryl-3-methyl-2-butyl esters, 91-94
of *syn*-7-aryl-*anti*-7-norbornenyl-*p*-nitrobenzoates, 90, 92, 93
of 7-aryl-7-norbornyl-*p*-nitrobenzoates, 90-93
of arylthioalkyl halides, 196-197
of azabicyclic derivatives, 249-251
of azetidine derivatives, 235-239
of aziridine derivatives, 235-239
of benzonorbornenyl derivatives, 91-93
of benzyl derivatives, 97
of benzyloxy-substituted alkyl derivatives, 143-144
of bicyclic acetals and ketals, 155-157
of 2-chlorocyclohexanethiols, 220-221

Solvolysis (*cont'd*)
  of 2-chloromethylthiirane, 202
  of cholesteryl derivatives, 10, 98
  of codeinone and neopinone derivatives, 242-243
  of cyclic acetals, 158
  of cyclic and bicyclic β-thioalkyl derivatives, 199-201
  of cyclic ether derivatives, 145-148
  of cyclic thioethers, 206-207
  of cyclobutyl derivatives, 235-239
  of cyclopropylmethyl derivatives, 40-43, 146, 202-203, 235-239
  of 8,9-dehydroadamantyl derivatives, 40, 105
  of epichlorohydrin, 145-146, 202
  of halohydrins in aqueous media, 163-169
  of 3-halo-1-propanol derivatives, 171-172
  of β-*p*-hydroxyphenylethyl bromide, 107
  of hydroxy-substituted cycloalkyl derivatives, 167-168
  of hydroxy-substituted steroidal acetates, 186
  of methoxy-substituted alkyl derivatives, 97, 125-137
  of methoxy-substituted cycloalkyl derivatives, 132-133, 135-136
  of 2-norbornenyl esters, 91-94
  of *anti*-7-norbornenyl esters, 79, 98
  of 2-norbornyl derivatives, 6-9, 33-34, 81, 91-94, 150-151, 200-202, 205
  of oxabicyclic derivatives, 152, 199-206
  of 1-phenyl-2-propyl derivatives, 83
  of secondary acyclic alkyl derivatives, 87
  of tertiary cycloalkyl derivatives, 87-89
  of thiabicyclic derivatives, 152, 199-206
  of γ-thiobenzyl derivatives, 206-209
  of 5-*O*-toluene-*p*-sulfonyl arabinose diethyl thioacetal, 209
  of 2,2,2-triphenylethyl derivatives, 89
  of 8-vinyl-8-bicyclo[3.2.1]octyl derivatives, 39
Special salt effect, 127, 131, 133-134, 138-139
Spectroscopic methods, 110-112
Stereochemistry of Participation (*See* Nucleophilic participation)
Stereoelectronic requirements for participation (*See* Nucleophilic participation)
Steric acceleration
  in cyclization reactions, 183
  in ionization, 87, 89, 141, 236-237
  in perester decompositions, 82, 218

Steric retardation of ionization, 9, 81, 87, 91, 126-127
Steroids
  intramolecular general acid–general base catalysis in the hydrolysis of esters of, 186
  nucleophilic participation in the solvolysis of, 10, 98, 147, 171, 200, 203
Strain energy
  of cyclic amines, 50
  of cyclic ethers, 50
  of cyclic hydrocarbons, 50
Succinic anhydride derivatives, effect of alkyl group substitution on formation of, 59-61
  from perester decompositions, 219
  in solvolytic reactions, 198-199
  in the addition of sulfenyl halides to alkenes, 198-199
Synartesis, definition of, 6-9
Synartetic acceleration, 8
Synartetic ions
  calculated gas phase stability, 20-30
  definition of, 6-9

Taft equation, 83
Taft treatment, 83, 85, 87, 103-104, 180
Temperature coefficient of activation, 167
Terpenes
  neighboring group effects in biogenesis, 5
  neighboring group effects in the reactions of, 6-8, 86, 87
Tetrahedral intermediate, 60, 69, 109
Thermodynamic control in product formation, 79, 108, 117, 128-131, 232-235
Thietanes, synthesis of, 222
Titrimetric rate constant (*See* Ion pairs)
Transannular participation, 146-150, 153, 160-161, 206-207, 216, 234, 252
Tricyclobutonium ion (*See* Cyclopropylmethyl cation)

β, γ-Unsaturated ketone rearrangements, effect of β-ether or amine group on the acid catalyzed reaction, 159

Vertical stabilization, 34-43
Volume of activation, 167, 262

Wagner–Meerwein rearrangement, 199-200

Y values, 102, 257